T0192258

Biological Processing of Solid Waste

Edited by
Sunil Kumar
Zengqiang Zhang
Mukesh Kumar Awasthi
Ronghua Li

CRC Press
Taylor & Francis Group
Boca Raton London New York

CRC Press is an imprint of the
Taylor & Francis Group, an **Informa** business

CRC Press
Taylor & Francis Group
6000 Broken Sound Parkway NW, Suite 300
Boca Raton, FL 33487-2742

First issued in paperback 2021

© 2019 by Taylor & Francis Group, LLC
CRC Press is an imprint of Taylor & Francis Group, an Informa business

No claim to original U.S. Government works

ISBN-13: 978-1-03-209375-8 (pbk)
ISBN-13: 978-1-138-10642-0 (hbk)

Contents

Preface

It has been estimated by the World Bank that at least 1.3 billion tonnes (Gt) of solid waste is produced by various anthropogenic activities around the world each year, and this amount is expected to rise to 2.2 Gt by 2025. There is an urgent need to recover value from solid waste rather than commit it to landfills or other disposal methods. Even greater quantities of solid waste are generated by manufacturing, construction, and mining industries, and global concerns over the decline of the planet's finite natural resources have seen a change of emphasis toward resource recovery through reuse, recovery, and recycling of all waste.

Biological Processing of Solid Waste brings together a group of experts involved in various aspects of waste management to give a wide-ranging overview of the problems that need to be addressed in achieving greater efficiencies in resource recovery from waste and the potential solutions to these problems. Cutting-edge disposal fees, recycling subsidies, and taxes on landfill need a sustainable solution that solve the problem of waste management. The book begins with an interesting chapter by Professor Zhang and Dr. Wang of the Northwest A&F University, China, who summarize the evolution of waste characterization and biological solid waste disposal studies around the world. Although Elena Cristina Rada popularized the practice of excavating waste and its bioprocessing in his book *Waste Management and Valorization: Alternative Technologies*, the process of waste characterization and bioprocessing has been evolving for over 100 years as governments and academics wish to understand what value might be obtained from solid waste. Chapter 1 of *Biological Processing of Solid Waste* reviews the range of advanced concepts of biological solid waste disposal and the international market for bioprocessing of solid waste that provide significant economic and environmental options for converting waste materials into useful and high-value materials, fuels, and chemicals.

Chapters 2 and 3 focus on the three Rs policy for different solid waste management and resource recovery from landfill solid waste. Chapter 3 details precisely the types of landfill, land mining options, and the modern energy recovery from the renewable landfill.

Chapter 4 provides an overview of global organic waste and then focuses on the composting and environmental benefits of resource recovery as sources of important resource waste especially. This chapter looks at the fundamental processes that drive composting and the key features needed for the delivery of a quality product. The biological degradation process is analyzed in detail, with particular attention to factors controlling the reaction rate. The organic fraction of solid waste is shown to be capable of yielding organic fertilizer and chemicals and a wide range of critical elements.

The related biodegradation processes that occur in the anaerobic digestion of municipal solid waste are discussed in Chapter 5. The emphasis is given on China and India, but the discussion is relevant to all developing and developed countries where attention is shifting from waste disposal to modern anaerobic digestion and energy recovery from solid waste.

The existing waste literature has considered the problems associated with illegal disposal and improper bioprocessing of solid waste when waste disposers are the only agents in greenhouse gas emissions through biological processing of solid waste. Chapter 6 emphasizes greenhouse gas emissions through biological processing of solid waste, the second source of asymmetric information that occurs when only waste generators and bioprocessing have information on the contents of waste. The authors estimate greenhouse gas emissions from various biogenic treatments and the problems that arise when bioprocessing of solid wastes are present in the industry, and propose an optimal policy scheme that will rectify those problems. This optimal policy involves mitigation of greenhouse gas emissions during the waste disposal and management.

Chapter 7 focuses on the evolution of mineral nutrients in solid waste during bio-drying and implications of their subsequent transfer during combustion. The kinetic model demonstrates solid waste bio-drying and resource recovery.

Chapter 8 evaluates the recent trend of bioprocessing antibiotic residues and resistant genes in solid waste in light of the difference between antibiotic residues and source-generated solid waste.

Chapter 9 summarizes the modern technology development of bioprocessing of mining solid waste and resource recovery. Chapter 10 presents emerging technologies to remediate organic pollutants and emphasizes the health hazards of organic pollutants, approaches to evolution of catabolic pathways and microbial adaptation to organic pollutants, metabolic engineering and biocatalytic applications, chemical approaches, biological approaches, involvement of microorganisms and genetically engineered microorganisms, and advantages and disadvantages of bioremediation.

Chapters 11 and 12 describe the commonly used and emerging cost-effective additives for heavy metal immobilization during bioprocessing of solid waste and recent development in the treatment of petroleum hydrocarbon and oily sludge from the petroleum industry.

We believe that *Biological Processing of Solid Waste* will be interesting and useful to all concerned about the ever-growing volume of bioprocessing of waste generated by both developed and developing countries, and anyone wishing to know what can be done to minimize waste and to recover value from it. In particular, the book will be useful to policy makers, teachers, and students in science and engineering programs and resource management courses, and those engaged in the municipal solid waste, mining, and agricultural industries.

Dr. Sunil Kumar
Senior Scientist and Head
Technology Development Centre
CSIR-NEERI, Nagpur- 440 020
Maharashtra, India

Prof. Zengqiang Zhang
Professor
College of Natural Resources and Environment
Northwest A&F University
Yangling, Shaanxi Province 712100, PR China

Dr. Mukesh Kumar Awasthi
College of Natural Resources and Environment
Northwest A&F University
Yangling, Shaanxi Province 712100, PR China
Department of Biotechnology
AKS University
Satna, India

Dr. Ronghua Li
College of Natural Resources and Environment
Northwest A&F University
Yangling, Shaanxi Province 712100, PR China

Editors

Dr. Sunil Kumar is a Senior Scientist in Solid and Hazardous Waste Management Division of CSIR-NEERI, Nagpur. Dr. Kumar has extensive experience in the field of solid and hazardous waste management. His major research interests include municipal solid waste management, hazardous waste management, and environmental impact assessment. He is a member of the Curriculum Development Committee for M. Tech program in Environmental Engineering at Punjab State Technical University, Bhatinda, India, and M. Sc. in Environmental Science and M. Tech in Environmental Engineering program at National Institute of Technology, Rourkela, India. He is honorary director of the Institute of Chartered Waste Managers, Jaipur, India. He has carried out 20 important national and internationally sponsored projects. He has authored more than 80 publications, including three books and eight book chapters, with h index of 24, i-10 index of 54 and 3606 citations (Google scholar) He transferred several technologies to industries and has done industrial consultancy for various projects in Indian/international industries. He is the editor, associate editor, or editorial board member of many international journals, including *Bioresource Technology Journal of Hazardous, Toxic and Radioactive Waste, Env Chemistry Letter, Environmental Monitoring and Assessmen* , and *Environmental Monitoring and Assessment.*

Dr. Zengqiang Zhang is a professor at College of Natural Resources and Environment, Northwest A&F University, Yangling, Shaanxi Province, People's Republic of China. Professor Zhang is one of the leading personalities in the field of environmental science and technology. His major research interests are in the areas of environmental microbiology, composting, bioremediation, phytoremediation of heavy metal contaminated soil, wastewater treatment, and nutrient recovery and mitigation of greenhouse gas emissions during the organic waste composting. He has written two books, 12 technical reports, six book chapters, 220 original and review papers, and 50 research communications at international and national conferences, with h index of 80 and > 500 citations (Google scholar). He has transferred several composting technologies to industry and has done industrial consultancy for about a dozen projects for national and international industries. He has completed 22 national, 15 province, and more than 30 industrial consultancy projects. Professor Zhang is the recipient of many national and international awards and fellowships, which include Fellow of Science and Technology Progress of Shaanxi EPA (2007), Science and Technology.

Dr. Mukesh Kumar Awasthi is an assistant professor in the Department of Biotechnology, AKS University, Satna, India, a constituent private university working under the umbrella of Madhya Pradesh Private Regulatory Commission, Bhopal, India. Dr. Awasthi has also been a postdoctoral fellow at College of Natural Resources and Environment, Northwest A&F University, Yangling, Shaanxi Province, People's Republic of China since 2015. His major research interests includes a wide range of microbiology, environmental science, biotechnology, and bioengineering topics, including solid waste management and organic waste composting, water and wastewater treatment, hazardous and municipal waste management, environmental impact assessment, and environmental audit and climate change. He has authored more than 48 publications, including one book and three book chapters. He has transferred technologies to industry and served as principal or coprincipal investigator of two projects in Indian and international industries. Dr. Awasthi is the recipient of many national and international awards, including Young Life Foundation Award 2016–2017 at Northwest A&F University, best paper awards for poster presentations in the 99th Indian Science Congress Association (2012) and Asia Pacific Conference on Solid Waste Management, Hong Kong and Postdoctoral Meet (2016). He is also an editorial board member of the *Journal of Basic and Applied Mycology* and the *International Journal of Environmental Sciences*.

Dr. Ronghua Li is an associate professor in the College of Natural Resources and Environment, Northwest A&F University, Yangling, Shaanxi Province, People's Republic of China. Dr. Li has extensive experience in the field of solid and hazardous waste management, biomass management, and nutrient recovery. His major research interests include a wide range of control methods for heavy metals mobility in soil. Recent research efforts have also focused on heavy metals passivation during animal manure composting by natural minerals additions. He is also working on competitive adsorption/desorption behavior between engineering nanoparticles and heavy metal ions in soil-water systems. Other research areas include beneficial use of wastes, biochar properties and functions, and metal adsorption by nanoparticles. Dr. Li has published more than 70 papers and three patents. He is a reviewer of many journals, including *Bioresource Technology*, *Waste Management*, *Journal of Applied Polymer Science*, *Journal of Colloid Interface*, *Chemosphere*, *International Journal of Biological Macromolecules*, *Environmental Technology*, and *Journal of Nanotechnology*. He transferred several technologies to industry and has done industrial consultancy for various projects in China and international industries.

Contributors

Amjad Ali
College of Natural Resources
  and Environment
Northwest A&F University
Yangling, People's Republic of China

Shashi Arya
Technology Development Centre
CSIR, National Environmental
  Engineering Research Institute
Nagpur, India

Mukesh Kumar Awasthi
College of Natural Resources
  and Environment
Northwest A&F University
Yangling, People's Republic of China

Sanjeev Kumar Awasthi
College of Natural Resources
  and Environment
Northwest A&F University
Yangling, People's Republic of China

Digambar Chavan
Technology Development Centre
CSIR, National Environmental
  Engineering Research Institute
  (CSIR-NEERI)
Nagpur, India

Hongyu Chen
College of Natural Resources
  and Environment
Northwest A&F University
Yangling, People's Republic of China

Jishuang Chen
Key Laboratory of Basic
  Pharmacology and Joint
  International Research
  Laboratory of Ethnomedicine of
  Ministry of Education
Bioresource Institute for Healthy
  Utilization
Zunyi Medical University
Zunyi, People's Republic of China

Yumin Duan
College of Natural Resources
  and Environment
Northwest A&F University
Yangling, People's Republic of China

Pratibha Gautam
Department of Environmental
  Science and Technology
Shroff S.R. Rotary Institute of
  Chemical Technology
Ankleshwar, India

Di Guo
College of Natural Resources
    and Environment
Northwest A&F University
Yangling, People's Republic of China

Pradip S. Jadhao
EISD, CSIR, National Environmental
    Engineering research institute
    (CSIR-NEERI)
Nagpur, India

Archana Jain
Sri Satya Sai University of
    Technology and Medical Sciences
Sehore, India

Qi Jia
Key Laboratory of Basic
    Pharmacology and Joint
    International Research
    Laboratory of Ethnomedicine of
    Ministry of Education
Bioresource Institute for Healthy
    Utilization
Zunyi Medical University
Zunyi, People's Republic of China

Ankur Khare
EISD, CSIR, National Environmental
    Engineering Research Institute
    (CSIR-NEERI)
Nagpur, India

Sunil Kumar
Technology Development Centre
CSIR, National Environmental
    Engineering Research Institute
    (CSIR-NEERI)
Nagpur, India

Kanchan Kumari
Environmental Impact
    and Sustainability Division
CSIR, National Environmental
    Engineering Research Institute
    (CSIR-NEERI)
Nagpur, India

Jiao Li
College of Natural Resources
    and Environment
Northwest A&F University
Yangling, People's Republic of China

Ronghua Li
Northwest A&F University
Xianyang, People's Republic of China

Tao Liu
College of Natural Resources
    and Environment
Northwest A&F University
Yangling, People's Republic of China

Saket Mishra
Madhya Pradesh Pollution Control
    Board
Bhopal, India

Rucha V. Moharir
Technology Development Centre
CSIR, National Environmental
    Engineering Research Institute
Nagpur, India

Xiuna Ren
College of Natural Resources
    and Environment
Northwest A&F University
Yangling, People's Republic of China

Rena
Technology Development Centre
CSIR, National Environmental
    Engineering Research Institute
    (CSIR-NEERI)
Nagpur, India

Surendra Sarsaiya
Key Laboratory of Basic
    Pharmacology and Joint
    International Research
    Laboratory of Ethnomedicine of
    Ministry of Education
Bioresource Institute for Healthy
    Utilization
Zunyi Medical University
Zunyi, China

and

Sri Satya Sai University of
    Technology and Medical Sciences
Sehore, India

Siratun Montaha S. Shaikh
EISD, CSIR, National Environmental
    Engineering research institute
    (CSIR-NEERI)
Nagpur, India

Jingshan Shi
Key Laboratory of Basic
    Pharmacology and Joint
    International Research
    Laboratory of Ethnomedicine of
    Ministry of Education
Bioresource Institute for Healthy
    Utilization
Zunyi Medical University
Zunyi, People's Republic of China

Fuxing Shu
Key Laboratory of Basic
    Pharmacology and Joint
    International Research
    Laboratory of Ethnomedicine of
    Ministry of Education
Bioresource Institute for Healthy
    Utilization
Zunyi Medical University
Zunyi, People's Republic of China

Ranjan Singh
Department of Botany
    and Microbiology
St. Aloysius College (Autonomous)
Jabalpur, India

Ankaram Snehalata
Department of Zoology
Vasantrao Naik Mahavidyalaya
Aurangabad, India

Zilin Song
College of Natural Resources
    and Environment
Northwest A&F University
Yangling, People's Republic of China

Fazli Wahid
Department of Agriculture
University of Swabi
Swabi, Pakistan

Meijing Wang
College of Natural Resources
    and Environment
Northwest A&F University
Yangling, People's Republic of China

Quan Wang
College of Natural Resources
    and Environment
Northwest A&F University
Yangling, People's Republic of China

Ran Xiao
College of Natural Resources
    and Environment
Northwest A&F University
Yangling, People's Republic of China

Gaihe Yang
College of Agronomy
Northwest A&F University
Yangling, People's Republic of China

Zengqiang Zhang
College of Natural Resources
    and Environment
Northwest A&F University
Yangling, People's Republic of China

Junchao Zhao
College of Natural Resources
    and Environment
Northwest A&F University
Yangling, People's Republic of China

1

Concept of Biological Solid Waste Treatment

Quan Wang, Xiuna Ren, Junchao Zhao, Hongyu Chen, Meijing Wang, and Zengqiang Zhang

CONTENTS

1.1 Introduction

With high-speed development in society, the economy, and agriculture, huge quantities of biological solid waste (i.e., organic fractions of municipal solid waste, agriculture waste, animal manure, sewage sludge, etc.) are produced every year (Campuzano and Gonzalez-Martinez 2016; Feng et al. 2017). Biological solid waste contains a lot of organic matter and nutrients, which could serve as a potential soil fertilizer, but it also contains pathogenic microorganism, heavy metals, and other hazardous components (Bernal et al. 2009).

Unsuitable management of this kind of solid waste causes a series of environmental problem such as greenhouse gas emissions, surface water and ground water contamination, and heavy metals pollution of the soil, and thus a threat to humans (Hong et al. 2017). Effectively dealing with and recycling organic solid waste has become a significant issue all over the world.

In recent decades, various technologies such as landfilling, incineration, composting, anaerobic digestion, and pyrolyzation had been proposed and installed to dispose of organic waste (Münster et al. 2015; Liu et al. 2017). Among all treatments, landfilling and incineration are the two major treatments in many countries. On the other hand, the government and the community of scientific experts also formulated some polices and criteria to support and accelerate management of biological solid waste (Yap and Nixon 2015; Jovanovic et al. 2016). However, the environmental effects of solid waste treatment has also become an important considered factor. Developing and perfecting the technology and criteria are essential to satisfying the current demand for organic waste management and environmental protection. This chapter will introduce the concept of biological solid waste, and evaluate the technical, social, economic, and environmental feasibility for management of different solid wastes.

1.2 Solid Waste Hierarchy

In general, solid waste refers to the solid and semisolid waste arising from production, construction, and other human activities. It includes several types, for example, (1) waste that loses the original value of use; (2) waste that does not lose the value of use but is abandoned; (3) toxic or harmful items, in gas, liquid, or solid form, that is placed in containers; and (4) substances related to the solid waste management (Sharholy et al. 2008; Kolekar et al. 2016).

The quantity of solid waste has increased rapidly in recent years; it is estimated that the amount of solid waste will increase from 2.78 billion tons to 4.16 billion tons by 2025 (Melikoglu et al. 2013). Solid waste is generated from different sources, which means that the components of solid waste are complex (AbdAlqader and Hamad 2012; Irwan et al. 2012). Solid waste can be sorted into various types. For example, on the basis of composition, solid waste can be divided into organic waste and inorganic waste; in terms of the form, it can be classified as solid waste, semisolid waste, liquid, and gaseous waste. In the light of the pollution characteristics, it can be divided into hazardous waste and general waste. On the basis of the classification of waste sources, solid waste can be divided into four categories, including municipal solid waste, industrial solid waste, agricultural solid waste, and hazardous waste, and this classification method is widely used in China.

The management rule of solid waste is various because of the different toxic of solid waste. The hazardous waste must be managed by expert company which is allowed by government. Besides, municipal solid waste is supposed to transport in a fixed place. Furthermore, the transport of solid waste needs to prevent dust and water. It is not complicated now in China, which need to better in the future.

1.3 Criteria for Evaluation of Solid Waste Treatment Orientation and Processing

Solid waste management is a complex process that involves multiple stakeholders such as government, industries, citizens, and experts (De Feo and De Gisi 2010). In accordance with the different compositions and properties of the solid waste, various methods (landfilling, anaerobic digestion, composting, waste-to-energy technology, etc.) could be carried out to dispose of solid waste (Hong et al. 2017). However, choosing a waste management technique is usually with location, capacity of the treatment plant, and technology. Furthermore, the suitable or optimal solid waste treatment option should fulfill the different criteria related to the economic costs and benefits, environmental impacts (e.g., human health risks, resource depletion, and environmental pollutions), and social factors (Jovanovic et al. 2016).

The stakeholders always tend to select the method that maximize benefits and minimize costs. Due to the range and diversity of the waste treatment options and the changes in people's attitudes, a single criterion for evaluation of solid waste treatment orientation and processing cannot satisfy the decision-makers' objectives (Yap and Nixon 2015). Today, the stakeholder will not only assess the economic benefits but also consider the environmental impacts of waste treatment. In order to design and implement the optimal solid waste treatment and satisfy all stakeholders, multiple criteria, including environmental, economic, and social criteria are proposed for helping the stakeholders decide in relation to human health and environmental and economic development (Soltani et al. 2015).

1.4 Technical and Social Feasibility for Solid Waste Treatment

Huge quantities of solid waste are produced every day, and the quantity of municipal solid waste in China has reached 178.6 million tons. Disposal and recycling of solid waste has become the major issue all over the world (Sonle and Louati 2016). Various kinds of technology such as landfill, anaerobic digestion, composting, thermal plasma technology, and incineration have been carried to manage the solid waste (Yang et al. 2013; Wu et al. 2014; Campuzano and Gonzalez-Martinez 2016).

Among all solid waste treatments, landfills are the most widely used method for their simple processing, large handling capacity, and low investment. And approximately 90% and 54% of waste are landfilled in China and United States, respectively (Feng et al. 2017). However, landfills occupy plenty of land resources and also cause air, soil, and water pollution. The negative effect of landfill sites for ecology and public health has attracted people's attention (Wang et al. 2015). Leachate from landfill sites contains many kinds of pollutants, such as heavy metals, organic contaminants, pathogens, and hazardous chemicals, which threaten the ecology of soil and aqueous systems (Han et al. 2016). Moreover, the toxic gases and greenhouse gases generated from landfill sites also aggravate the greenhouse effect and harm residents' health. Hence, using landfills for solid waste has been limited gradually. Incinerating municipal solid waste is increasingly popular in China and is used on a large scale in many developed countries such as the United States (16%), Japan (75%), and Germany (40%–50%) (Ma et al. 2016). The incineration method could significantly reduce the solid mass and volume of the original waste, and recover the energy, but the investment of the incineration plant is high. And because of the different characters of the solid waste, the calorific value cannot meet the actual demand. On the other hand, the toxic and hazardous gas (mercury, polychlorinated dibenzo-p-dioxins and polychlorinated dibenzofurans (PCCD/Fs)) emissions from the incinerator also pollutes the ambient air and ecological system (Li et al. 2017). More reliable and practical technology should be carried out to improve the incineration of solid waste, and the exhaust gas also should be purified before emission. Among all methods, composting is one of the most preferred and profitable technologies for disposing solid waste; it can transform the organic fraction of solid waste into the sanitary and nutrient-rich final product called compost (Li et al. 2012).

With the increase in the amount of organics in the solid waste, the nutrients and the quality of the compost has been gradually increased. However, the composition of the solid waste is complex; it can contain plastic, glass, batteries, construction waste, food waste, and some other organic and inorganic waste. This character of the solid waste influences the quality of the compost and increases the difficulty of the operation. In consequence, solid waste should be collected and sorted before composting (Ma et al. 2016).

1.5 Cost Feasibility and Perfectibility of Biological Solid Waste Treatment

With encouragement from economic and societal factors, the technologies utilized for disposing of organic waste were also improved in the last ten years. These technologies focus on not only managing the solid waste but also recycling the solid waste and obtaining the energy (Münster et al. 2015). The treated capacity of landfill is high and the investment is low, but it has many disadvantages, such as landscape deterioration, dust, and leachate production, and the contaminating gas emissions have attracted people's attentions (Fernández-González et al. 2017). In order to reduce the environmental impact of a landfill site, a sanitary landfill method was put forward. Additional equipment is set up to deal with the leachates and the gases, which will obviously increase the cost of the sanitary landfill technology. While the gas collection from the sanitary landfill site contains a lot of methane, the utilization of this gas to produce the energy (electricity and heat) will slightly reduce the cost of running the landfill system. Although the sanitary landfill technology could deal with a lot of solid waste, it also occupies plenty of land resources (Ma et al. 2016). Incinerating the biological solid waste could obviously decrease the volume and mass of the solid waste and also generate heat. However, the toxic emissions (dust, dioxin, furan, mercury, arsenic, and lead) from the incineration plant will harm people' health and pollute the environment (Li et al. 2017). In order to reduce the environmental pollution and promote energetic efficiency, many technologies such as flue gas cleaning system and wet and dry cooling system are used. Although these technologies increase the initial investment, operation, and maintenance costs of an incineration plant, they offer more environmental and economic benefits and thus raise the financial performance (Zhao et al. 2016).

Anaerobic digestion and composting are the major biological solid waste treatment methods all over the world. They could transform organic waste to bioenergy and compost. In comparison with the landfill and incineration method, anaerobic digestion and composting present better environmental and economic benefits and can also generate highly valued by-products such as methane and compost (Li et al. 2012; Carlsson et al. 2015).

Anaerobic digestion of biological solid waste can break down organic matter and produce the energy source for electricity generation or heat. The biogas residue and biogas slurry can be used as soil amendment or fertilizer (Jain et al. 2015). Many researchers have improved the production efficiency of biogas and reduced the costs of anaerobic plant via adjusting the operating parameters, optimizing reactor design, and pretreating organic waste (Long et al. 2012; Carlsson et al. 2015; Campuzano and Gonzalez-Martinez 2016). The utilization of composting technology to dispose of biological solid waste has increased significantly in last decades due to its simple operation, low investment, high value-added and low running cost, and reduced environmental impact. Economic analysis of composting systems normally include capital, labor, energy, maintenance costs, biomass, construction costs, and the income from the compost sales (Lim et al. 2016). Compared to other waste management techniques, investment in the composting system is lower, and selling the final product may bring greater benefit. The payback period of setting up the composting plant is shorter (Cukjati et al. 2012). During the last 10 years, many effective methods, such as using the various types of bulking agents (Bernal et al. 2009; Zhang and Sun 2016), increasing the air flow rate (Chowdhury et al. 2014), and incorporating chemical and mineral additives (Li et al. 2012; Awasthi et al. 2016a; Jiang et al. 2016), have been used to improve the quality of compost and decrease the composting period. With the composting time decreasing and the quality of the final product increasing, the economic benefits of a composting plant are significantly improved.

1.6 Consistency with Macroscopically Centralized and Decentralized Plan

Macroscopically centralized and decentralized solid waste treatments are the major waste management patterns. The application of macroscopically centralized waste treatment could dispose of a large quantity of solid waste, enhance solid waste management, and reduce the costs of facilities and staffing (Gutierrez et al. 2017). The centralized solid waste treatment increases the investment of the collection, separation, and the environmental pressure of the centralized solid waste treatment site. Compared to the centralized waste treatment plan, the decentralized solid waste treatment plan could reduce the transport cost, environmental pollution, traffic, noise, and the amounts of solid waste stored in temporary deposits. Meanwhile, the decentralized solid waste treatment could better consider the local condition and consequently increase the efficiency of solid waste management and reduce the energy and environmental rates (Righi et al. 2013). However, the decentralized solid waste treatment increases the cost of the facilities, construction, and labor, and the handling capacity of decentralized solid waste treatment is lower than it is in the solid waste centralized treating. Although the centralized and decentralized solid waste treatments have advantages and disadvantages, the main target is consistency in dealing and utilizing the solid waste. The economic, social, and environmental benefits of the centralized and decentralized solid waste treatment plants need to be considered carefully after setup. In addition, the local practical situation, the composition and character of solid waste, as well as the technology are also very important factors (Chen and Liu 2017).

1.7 Previous Experience Affecting Biological Processing of Solid Waste

Organic waste, such as household waste, food waste, municipal sludge, human or animal manure, agriculture, light industry, and food processing industry wastes, has an important role in solid waste. Organic solid waste has

TABLE 1.1

Optimized Range of Physicochemical Parameters

Item	Range
Temperature	45°C–60°C
Moisture	40%–60%
Aeration rate	5%–15%
Organic matter	20%–80%
C/N	25:1–35:1
Particle size	12–60 cm

high organic content and perishable deterioration. The emission of organic waste without treatment results in high concentrations of leachate and odor, and groundwater, surface water, and air pollution. Furthermore, some organic solid wastes (such as manure) also carry a large number of pathogens that spread disease (Awasthi et al. 2014; Sukholthaman and Sharp 2016).

The characteristics of organic solid waste are particularly suitable for composting. Composting transforms organic waste into safe and stable substances (Chowdhury et al. 2014; Manu et al. 2017). The final product of compost can serve as agricultural fertilizer, which has a positive impact on the crop production (Guo et al. 2012). Therefore, composting is one of the most effective ways to achieve organic waste stabilization with harmless and resource-based treatment (Jiang et al. 2016).

The physiochemical parameters of composting are crucial factors in the performance of composting. In order to improve the efficiency and decrease the secondary environmental pollution of composting, different kinds of approaches were proposed to regulate the physical and chemical parameters such as temperature, moisture, pH, carbon/nitrogen ratio, particle size, and others (Li et al. 2013; Juarez et al. 2015). The appropriate ranges for these parameters are listed in Table 1.1.

1.8 Economic and Technological Development for Biological Processing of Solid Waste

With the development of industry and agriculture and the improvement in human living standards, a large volume of solid waste has been generated in the world, which has resulted in a series of environmental problems requiring suitable treatment methods (Morita and Sasaki 2012; Zhang and Sun 2016). Biological processing of solid waste has the advantages of simple operation, low cost, and high efficiency. It has been widely used in solid waste treatment and played an important role. Today, composting and anaerobic fermentation have been accepted as the economically feasible technologies for recycling solid waste (Wang et al. 2016; Zhen et al. 2017).

Composting is a microbiological treatment process. Biodegradable organic matter is converted into the stable humic substance or decomposed into carbon dioxide by microorganisms during composting process (Bernal et al. 2009). Compost is an ecological fertilizer that can improve soil structure, enhance soil fertility, and promote the growth of plants (Wang et al. 2017).

In 1925, British agronomist Albert Howard invented the anaerobic fermentation, that was the earliest scientific exploration of composting technology. Subsequently, Bangalore established Bangalore-Composting method, which was devoted to promoting the aerobic composting process. Since then, large scale research on organic solid waste composting technology was established around the whole word. At present, organic solid waste composting has been widely applied in Holland and France; India and China also attach great importance to composting and have achieved rapid expansion (Yuan et al. 2016). Composting technology could effectively treat the organic solid waste and produce a nutrient-rich and sanitary soil fertilizer or conditioner (Awasthi et al. 2016b; Wang et al. 2017). Meanwhile, biological waste can also be utilized effectively. However, there are still some drawbacks that will inhibit the development of composting, such as high heavy metals mobility, greenhouse gas emissions, nitrogen loss, and antibiotic resistance gens residue (Awasthi et al. 2016b; Wang et al. 2016). The addition of different mineral additives, such as medical stone, biochar, bentonite, and zeolite, is one of the most useful methods for mitigating these negative effects and has been widely studied and proposed in recent years (Li et al. 2012, 2016; Awasthi et al. 2016b; Wang et al. 2016).

Anaerobic fermentation is an environmentally friendly process in which organic matter is decomposed by anaerobic microorganisms in specific anaerobic conditions, and part of the carbon material is transformed to CH_4 and CO_2 (Appels et al. 2008). The application of anaerobic fermentation of organic waste has a long history. In 1896, an anaerobic digestion tank was built in a small city in England to deal with the sewage sludge; the biogas was used as a fuel for lighting.

In comparison with other treatments of solid waste management, anaerobic fermentation offers signification advantages, including:

1. Biogas produced can be used for heat and electricity generation.
2. The process of digestion without oxygen saves the consumption of equipment and power.
3. Anaerobic fermentation is suitable for the treatment of heavy polluted water and organic waste, and the digestate can be used as agricultural fertilizer, feed, or composting materials.

4. Anaerobic fermentation is a lower-cost and simply technology system to generate energy for rural areas.

5. Anaerobic fermentation reduces the volume and weight of waste and mitigates the emission of odors.

However, application of anaerobic organic solid waste fermentation was often limited by its length of time and low digestion efficiency. Many previous researchers have demonstrated the advantage of pretreatment of organic waste (such as sludge), which can accelerate the anaerobic digestion process. Different kinds of pretreating methods have been studied to improve the biodegradability of organic waste. The combination of alkaline pretreatment and mechanical method such as high-press homogenization proved to be effective in increasing the efficiency of anaerobic sludge digestion and accelerating the production of biogas (Fang et al. 2014).

1.9 Collection, Separation, and Natural Characteristic of Solid Waste for Biological Processing

The main collection method of garbage includes fixed container collection method and mobile container collection method. Most use garbage trucks to load the garbage at each container, then the container is emptied and put back in place. This collection makes the composition of solid waste very complex (Kolekar et al. 2016). Especially in China, most of the garbage is without classification. Under these circumstances, pretreatment is particularly important. Pretreatment processes include crushing; sorting; screening; mixing; and adjusting nutrient, moisture, and other physical properties. Pretreatment has two main roles:

1. It can remove what cannot be composted. When municipal solid waste is used as compost material, rubbish often contain coarse particles and noncompostable substances, such as stones, plastics, and metal. The presence of these substances can affect the normal operation of the waste disposal machinery, increase the volume of the fermentation bin for composting, and affect the quality of the compost product. Therefore, before the composting, the raw materials should be sorted and removed.

2. It adjusts the nutritional content of raw materials and physical properties.

Due to the properties of solid waste, it is usually treated with biological methods, such as composting and anaerobic digestion. Anaerobic digestion transforms organic solid waste into CH_4, CO_2, and H_2O as well as the cellular requirements.

And it mainly includes three stages. During the first stage of hydrolysis, complex insoluble polymers are transformed into simple dissolving monomers through various enzymes. In the second stage, the end production of the first progress (like volatile fatty acids (VFAs), alcohol, amino acid, saccharides) transforms into acetic acid and hydrogen through hydrogen-producing acetogenic bacteria. After that, the first two stages converts the products into CH_4 through methanogens. Anaerobic fermentation can convert organic waste into bioenergy and is a good way to achieve organic material stabilization and render it harmless. Therefore, anaerobic fermentation not only has the function of pollutant treatment but also provides renewable energy production.

Composting is an aerobic microbial fermentation process. The growth of microorganisms need adequate and balanced nutrients (the general composting raw material content of 20%~80%) and moisture (general composting raw material moisture content of 40%–60%) (Thambirajah et al. 1995; Pietronave et al. 2004). The size of raw materials (composting material particles average suitable size of 12~60 cm), porosity, and other physical properties also have certain requirements (Gao et al. 2010). Solid waste composition is complex, and the nature of solid waste is different from batch to batch, so generally it cannot meet these requirements. Thus, the need for pretreatment of the organic matter content, moisture content, carbon and nitrogen ratio, pH, porosity, and other factors to meet the requirements of biological fermentation (efficient composting process and high-quality composting products) is important.

1.10 International Market for Bio-Processing of Solid Waste

Aerobic composting and anaerobic fermentation are common biological treatments of solid waste. Composting technology, a biological process, provides a harmless and value-added end product (Bernal et al. 2009). Compost is rich in nitrogen (N), potassium (K) and phosphorus (P), and widely used as an alternative to inorganic fertilizer (Doan et al. 2013). Compared with other treatments, the cost of organic waste composting is cheaper; a study about the products of a compost plant is shown in Table 1.2 (Dimitris and Robert 2004). For instance, the utilization of compost can be divided into three groups in Italy: 34% of high-class compost is sold as humus in garden centers; about 62% is regarded as an alternative for chemical fertilizers; and the rest, 4%, is used for the purpose of soil remediation (Rigamonti et al. 2010). Different composting plants have different market requirements, so the kinds of goods produced and their costs are also different. In the state of South Dakota in the United States, the rapid development of urban composting production sites now means that, the price of 1 cm size crushed material composting is $30/ton while the 0.5 cm size of the product is $25/ton, the transportation cost of compost is around 3,000–3,500 tons per year, with an annual revenue of about $0.1 million (Emerson 2005).

TABLE 1.2

Spent of Capital (C), Operate (O), Maintenance (M) of Compost Facilities

Project/Equipment	Cost ($)
Paving	180,000/hm²
Grading	12,500/hm²
Fencing	23/m
Land acquisition	3,100/hm²
Office cost	430/m²
Front and loader	C: 150,000; M: 1,000/y
Screens	C: 100,000; M: 500/y
Tub grinder	C: 180,000; M: 0.74/h

Compared with aerobic composting, anaerobic fermentation of solid waste could produce the biogases which can be used to generate the heat or energy. Relevant data show that organic solid waste can produce about 130 m³ methane per ton. Anaerobic treatment could improve the social and economic benefits and promote sustainable development of the environment. Biogas could be used for heating and burning power generation, and can be processed into industrial raw materials (Balat et al. 2008; Balat 2011). The digested sludge can be dewatered and rendered harmless, and then can be made into organic fertilizer or as auxiliary fuel in coal-fired power plants and cement factories.

Anaerobic fermentation could effectively dispose the solid waste and produce the high-valued product, which has been widely used for treating the sewage sludge in developed countries such as Europe and the United States. A lot of research has made great breakthroughs for anaerobic fermentation in technology, medicine, and so on. The European Union uses many sludge digestion facilities; 69% of sewage treatment facilities were equipped with sludge digestion and biogas plants. Because of soaring global energy prices, climate changes, and other factors, many countries are paying more attention to anaerobic digestion and biomass and energy utilization technologies.

1.11 Political Considerations for Maneuverability of Project

With rapid economic development, people's consumption levels have increased steadily, and the production amounts of solid waste have also increased year by year. At the same time, the requirements of recycling and harmless solid waste are still high; thus, governments also began to actively promote energy-saving emission reduction and promote full use of resources. Biological treatment of solid waste has seen rapid development because of its simple operation, low cost, and high economic value (Liu et al. 2017). In China, the

Ministry of Agriculture introduced Carry out the Organic Fertilizer of Fruit and Vegetable Tea Instead of Fertilizer Action Plan in 2017. This program advocated organic fertilizer to replace inorganic fertilizer, improved the utilization of organic fertilizers, and thus promoted the development of biological treatment of organic waste. Biological treatment of solid waste meets the requirements of the current era. Biological treatment can reduce the cost of waste disposal and secondary pollution, and the by-products can continue to be used as fertilizer, fuel, building materials, etc. It can achieve energy-saving emission reduction, enhance resource utilization, reduce carbon emissions, and mitigate climate change. Biological treatment of solid waste will receive the strong support of man governments in the near future.

1.12 Conclusion

Rapidly increasing quantities of solid waste has become a significant issue all over the world. Disposing of and recycling biological solid waste are significant processes for developing the sustainable waste management and green ecology. With the recent societal and economic developments, the compositions and characteristics of solid waste have become more complex, which increases the costs and difficulty of solid waste management. The techniques of solid waste treatment such as landfilling, incineration, composting, and anaerobic digestion have been improved to increase the efficiency and increase the benefits. Other methods have also been put forward to deal with solid waste in the last decade. Apart from the investment and economic benefit, the environmental and social implications also become crucial factors for selecting the technologies of solid waste treatment. The choice of solid waste treatment should take into account the composition and characteristics of the solid waste as well as the local environmental and economic conditions. A single criterion or policy for disposing of solid waste cannot satisfy all stakeholders, including governments, industries, citizens, and experts. The many criteria for solid waste management should also include economic, social, and environmental factors.

Acknowledgments

The authors are grateful for the financial support from a post-doctoral scholarship from Northwest A&F University (No. 154433), and The National Key Research and Development Program of China (2016YFD0800606). We are also thanks to our all laboratory colleagues and research staff members for their constructive advice and help.

References

AbdAlqader, A. and Hamad, J. 2012. Municipal solid waste composition determination supporting the integrated solid waste management in Gaza strip. *International Journal Environmental Science & Development* 3:172–176.

Appels, L., Baeyens, J., Degrève, J., and Dewil, R. 2008. Principles and potential of the anaerobic digestion of waste-activated sludge. *Progress in Energy & Combustion Science Combust* 34:755–781.

Awasthi, M.K., Pandey, A.K., Khan, J., Bundela, P.S., Wong, J.W.C., and Selvam, A. 2014. Evaluation of thermophilic fungal consortium for organic municipal solid waste composting. *Bioresoure Technology* 168:214–221.

Awasthi, M.K., Wang, Q., Huang, H., Ren, X., Lahori, A.H., Mahar, A., Ali, A., Shen, F., Li, R., and Zhang, Z. 2016a. Influence of zeolite and lime as additives on greenhouse gas emissions and maturity evolution during sewage sludge composting. *Bioresoure Technology* 216:172–181.

Awasthi, M.K., Wang, Q., Ren, X., Zhao, J., Huang, H., Awasthi, S.K., Lahori, A.H., Li, R., Zhou, L., and Zhang, Z. 2016b. Role of biochar amendment in mitigation of nitrogen loss and greenhouse gas emission during sewage sludge composting. *Bioresource Technology* 219:270–280.

Balat, M. 2011. Production of bioethanol from lignocellulosic materials via the biochemical pathway: A review. *Energy Conversion & Management* 52:858–875.

Balat, M., Balat, H., and Oz, C. 2008. Progress in bioethanol processing. *Progress in Energy & Combustion Science Combust* 34:551–573.

Bernal, M.P., Alburquerque, J.A., and Moral, R. 2009. Composting of animal manures and chemical criteria for compost maturity assessment: A review. *Bioresoure Technology* 100:5444–5453.

Campuzano, R., and Gonzalez-Martinez, S. 2016. Characteristics of the organic fraction of municipal solid waste and methane production: A review. *Waste Management* 54:3–12.

Carlsson, M., Holmstrom, D., Bohn, I., Bisaillon, M., Morgan-Sagastume, F., and Lagerkvist, A. 2015. Impact of physical pre-treatment of source-sorted organic fraction of municipal solid waste on greenhouse-gas emissions and the economy in a Swedish anaerobic digestion system. *Waste Management* 38:117–125.

Chen, Q., and Liu, T. 2017. Biogas system in rural China: Upgrading from decentralized to centralized? *Renewable & Sustainable Energy Reviews* 78:933–944.

Chowdhury, M.A., De, N.A., and Jensen, L.S. 2014. Potential of aeration flow rate and bio-char addition to reduce greenhouse gas and ammonia emissions during manure composting. *Chemosphere* 97:16–25.

Cukjati, N., Zupancic, G.D., Ros, M., and Grilc, V. 2012. Composting of anaerobic sludge: An economically feasible element of a sustainable sewage sludge management. *Journal of Environmental Management* 106:48–55.

De Feo, G. and De Gisi, S. 2010. Using an innovative criteria weighting tool for stakeholders involvement to rank MSW facility sites with the AHP. *Waste Management* 30:2370–2382.

Dimitris, P.K., and Rober, K.H. 2004. Life-cycle inventory of municipal solid waste and yard waste windrow composting in the United States. *Journal of Environment Engineering* 130:1390–1400.

Doan, T.T., Ngo, P.T., Rumpel, C., Nguyen, B.V., and Jouquet, P. 2013. Interactions between compost, vermicompost and earthworms influence plant growth and yield: A one-year greenhouse experiment. *Scientia Horticulturae* 160:148–154.

Emerson, D. 2005. Latest trends in yard trimmings composting. *Biocycle* 46:22.

Fang, W., Zhang, P., Zhang, G., Jin, S., Li, D., Zhang, M., and Xu, X.Z. 2014. Effect of alkaline addition on anaerobic sludge digestion with combined pretreatment of alkaline and high-pressure homogenization. *Bioresource Technology* 168:167–172.

Feng, S.J., Gao, K.W., Chen, Y.X., Li, Y., Zhang, L.M., and Chen, H.X. 2017. Geotechnical properties of municipal solid waste at Laogang Landfill, China. *Waste Management* 63:354–365.

Fernández-González, J.M., Grindlay, A.L., Serranobernardo, F., Rodríguezrojas, M.I., and Zamorano, M. 2017. Economic and environmental review of waste-to-energy systems for municipal solid waste management in medium and small municipalities. *Waste Management* 67:360–374.

Gao, M.C., Li, B., Yu, A., Liang, F.Y., Yang, L.J., and Sun, Y.X. 2010. The effect of aeration rate on forced-aeration composting of chicken manure and sawdust. *Bioresoure Technology* 101:1899–1903.

Guo, R., Li, G., Jiang, T., Schuchardt, F., Chen, T., Zhao, Y., and Shen, Y. 2012. Effect of aeration rate, C/N ratio and moisture content on the stability and maturity of compost. *Bioresoure Technology* 112:171–178.

Gutierrez, M.C., Serrano, A., Siles, J.A., Chica, A.F., and Martin, M.A. 2017. Centralized management of sewage sludge and agro-industrial waste through co-composting. *Journal Environmental Management* 196:387–393.

Han, Z., Ma, H., Shi, G., He, L., Wei, L., and Shi, Q. 2016. A review of groundwater contamination near municipal solid waste landfill sites in China. *Science Total Environment* 569–570:1255–1264.

Hong, J., Chen, Y., Wang, M., Ye, L., Qi, C., Yuan, H., Zheng, T., and Li, X. 2017. Intensification of municipal solid waste disposal in China. *Renewable & Sustainable Energy Reviews* 69:168–176.

Irwan, D., Basri, N.E.A., and Watanabe, K. 2012. Interrelationship between affluence and household size on municipal solid waste arising: Evidence from selected residential areas of putrajaya. *Journal of Asian Scientific Research* 2:747–758.

Jain, S., Jain, S., Wolf, I.T., Lee, J., and Tong, Y.W. 2015. A comprehensive review on operating parameters and different pretreatment methodologies for anaerobic digestion of municipal solid waste. *Renewable & Sustainable Energy Reviews* 52:142–154.

Jiang, T., Ma, X., Yang, J., Tang, Q., Yi, Z., and Chen, M. 2016. Effect of different struvite crystallization methods on gaseous emission and the comprehensive comparison during the composting. *Bioresoure Technology* 217:219–226.

Jovanovic, S., Savic, S., Jovicic, N., Boskovic, G., and Djordjevic, Z. 2016. Using multicriteria decision making for selection of the optimal strategy for municipal solid waste management. *Waste Management & Research* 34:884–895.

Juarez, M.F., Prahauser, B., Walter, A., Insam, H., and Franke-Whittle, I.H. 2015. Co-composting of biowaste and wood ash, influence on a microbially driven-process. *Waste Management* 46:155–164.

Kolekar, K.A., Hazra, T., and Chakrabarty, S.N. 2016. A review on prediction of municipal solid waste generation models. *Procedia Environmental Sciences* 35:238–244.

Li, J., Liu, K., Yan, S., Li, Y., and Han, D. 2016. Application of thermal plasma technology for the treatment of solid wastes in China: An overview. *Waste Management* 58:260–269.

Li, J., Wang, C., Du, L., Lv, Z., Li, X., Hu, X., Niu, Z., and Zhang, Y. 2017. Did municipal solid waste landfill have obvious influence on polychlorinated dibenzo-p-
dioxins and polychlorinated dibenzofurans (PCDD/Fs) in ambient air: A case
study in East China. *Waste Management* 62:169–176.

Li, R., Wang, J.J., Zhang, Z., Shen, F., Zhang, G., Qin, R., Li, X., and Xiao, R. 2012.
Nutrient transformations during composting of pig manure with bentonite.
Bioresource Technology 121:362–368.

Li, Z., Lu, H., Ren, L., and He, L. 2013. Experimental and modeling approaches for
food waste composting: A review. *Chemosphere* 93:1247–1257.

Lim, S.L., Lee, L.H., and Wu, T.Y. 2016. Sustainability of using composting and vermicomposting technologies for organic solid waste biotransformation: Recent
overview, greenhouse gases emissions and economic analysis. *Journal of Cleaner
Production* 111:262–278.

Liu, Y., Xing, P., and Liu, J. 2017. Environmental performance evaluation of different
municipal solid waste management scenarios in china. *Resources, Conservation
and Recycling* 125:98–106.

Long, J.H., Aziz, T.N., de los Reyes, F.L., and Ducoste, J.J. 2012. Anaerobic co-digestion
of fat, oil, and grease (FOG): A review of gas production and process limitations. *Process Safety and Environment Protection* 90:231–245.

Ma, H., Cao, Y., Lu, X., Ding, Z., and Zhou, W. 2016. Review of typical municipal solid
waste disposal status and energy technology. *Energy Procedia* 88:589–594.

Manu, M.K., Kumar, R., and Garg, A. 2017. Performance assessment of improved
composting system for food waste with varying aeration and use of microbial
inoculum. *Bioresource Technology* 234:167–177.

Melikoglu, M., Lin, C.S.K., and Webb, C. 2013. Kinetic studies on the multi-enzyme
solution produced via solid state fermentation of waste bread by *Aspergillus
awamori*. *Biochemical Engineering Journal* 80:76–82.

Morita, M., and Sasaki, K. 2012. Factors influencing the degradation of garbage
in methanogenic bioreactors and impacts on biogas formation. *Applied
Microbiology & Biotechnology* 94:575–582.

Münster, M., Ravn, H., Hedegaard, K., Juul, N., and Ljunggren, S.M. 2015. Economic and
environmental optimization of waste treatment. *Waste Management* 38:486–495.

Pietronave, S., Fracchia, L., Rinaldi, M., and Martinotti, M.G. 2004. Influence of biotic
and abiotic factors on human pathogens in a finished compost. *Water Research*
38:1963–1970.

Rigamonti, L., Grosso, M., and Giugliano, M. 2010. Life cycle assessment of subunits composing a MSW management system. *Journal of Cleaner Production*
18:1652–1662.

Righi, S., Oliviero, L., Pedrini, M., Buscaroli, A., and Casa, C.D. 2013. Life cycle assessment of management systems for sewage sludge and food waste: Centralized
and decentralized approaches. *Journal of Cleaner Production* 44:8–17.

Sharholy, M., Ahmad, K., Mahmood, G., and Trivedi, R.C. 2008. Municipal solid
waste management in Indian cities: A review. *Waste Management* 28:459–467.

Soltani, A., Hewage, K., Reza, B., and Sadiq, R. 2015. Multiple stakeholders in multi-
criteria decision-making in the context of municipal solid waste management:
A review. *Waste Management* 35:318–328.

Sonle, H., and Louati, A. 2016. Modeling municipal solid waste collection: A gener-
alized vehicle routing model with multiple transfer stations, gather sites and
inhomogeneous vehicles in time windows. *Waste Management* 52:34–49.

Sukholthaman, P., and Sharp, A. 2016. A system dynamics model to evaluate effects of source separation of municipal solid waste management: A case of Bangkok, Thailand. *Waste Management* 52:50–61.

Thambirajah, J.J., Zulkali, M.D., and Hashim, M.A. 1995. Microbiological and biochemical changes during the composting of oil palm empty-fruit bunches. Effect of nitrogen supplementation on the substrate. *Bioresource Technology* 52:133–144.

Wang, H., Xu, J., Yu, H., Liu, X., Yin, W., Liu, Y., Liu, Z., and Zhang, T. 2015. Study of the application and methods for the comprehensive treatment of municipal solid waste in northeastern China. *Renewable and Sustainable Energy Reviews* 52:1881–1889.

Wang, Q., Awasthi, M.K., Ren, X., Zhao, J., Li, R., Shen, F., and Zhang Z. 2017. Effect of calcium bentonite on Zn and Cu mobility and their accumulation in vegetable growth in soil amended with compost during consecutive planting. *Environmental Science & Pollution Research* 24:1–10.

Wang, Q., Wang, Z., Awasthi, M.K., Jiang, Y., Li, R., Ren, X., Zhao, J., Shen, F., Wang, M., and Zhang, Z. 2016. Evaluation of medical stone amendment for the reduction of nitrogen loss and bioavailability of heavy metals during pig manure composting. *Bioresource Technology* 220:297–304.

Wu, D., Zhang, C., Lv, F., Shao, L., and He, P. 2014. The operation of cost-effective on-site process for the bio-treatment of mixed municipal solid waste in rural areas. *Waste Management* 34:999–1005.

Yang, L., Chen, Z., Liu, T., Jiang, J., Li, B., Cao, Y., and Yu, Y. 2013. Ecological effects of cow manure compost on soils contaminated by landfill leachate. *Ecological Indicators* 32:14–18.

Yap, H.Y., and Nixon, J.D. 2015. A multi-criteria analysis of options for energy recovery from municipal solid waste in India and the UK. *Waste Management* 46:265–277.

Yuan, J., Chadwick, D., Zhang, D., Li, G., Chen, S., Luo, W., Du, L., He, S., and Peng, S. 2016. Effects of aeration rate on maturity and gaseous emissions during sewage sludge composting. *Waste Management* 56:403–410.

Zhang, L., and Sun, X. 2016. Improving green waste composting by addition of sugarcane bagasse and exhausted grape marc. *Bioresource Technology* 218:335–343.

Zhao, X.G., Jiang, G.W., Li, A., and Li, Y. 2016. Technology, cost, a performance of waste-to-energy incineration industry in china. *Renewable and Sustainable Energy Reviews* 55:115–130.

Zhen, G., Lu, X., Kato, H., Zhao, Y., and Li, Y. 2017. Overview of pretreatment strategies for enhancing sewage sludge disintegration and subsequent anaerobic digestion: Current advances, full-scale application and future perspectives. *Renewable and Sustainable Energy Reviews* 69:559–577.

2

3R Processing for Different Types of Solid Waste Management

Ankaram Snehalata

CONTENTS

2.1 Introduction

2.1.1 Waste and Waste Management

Waste is the unused, discarded portion from human consumption. If no management of solid waste existed, the proportion of waste on earth would be increased. A balance between production and disposal of waste makes solid waste management sound. Solid waste is municipal, industrial, and hazardous, all emerging from various organic and inorganic sources of human activities.

Solid waste is a rich source of biomass which can be converted into fuel at waste to energy plants. Industrialization, urbanization, and rapid changes in human lifestyle contributed to the increase in solid waste. Solid waste management aims to reduce health hazards for inhabitants and to maintain a pollution-free environment. Waste management includes the three Rs: reducing, recycling, and reusing the trash produced around the globe. Solid waste management is associated with diverse techniques, including composting, dumping, incineration, pyrolysis, etc.

Waste management is the present immense needs for a sustainable environment and the avoidance of a rapid decline in natural resources. Special emphasis will be given to both technical and traditional methods of waste management. Solid waste management can be implemented through the 3 R strategies for recovering energy from the waste. The journey from genesis, compilation, transformation, and disposal of waste is destined to waste management. The three Rs serve as pioneer sequences of the processes to reduce waste and to shift from traditional to skilled waste management. The three Rs changed views toward waste.

A field investigation of the municipal solid waste management of around 59 cities was conducted. Strategies and guidelines for sound management were recommended by the author within the vicinity of Management and Handling Rules 2000 (www.indiaenvironmentportal.org).

2.2 Recycling

The cumulative processing of materials from refused waste into new products is recycling. The remedial process of recycling saves both the community and the environment. The experimental study was executed on industrial waste that included wood mixtures, refuse timber, and chromate copper arsenate mixed with poultry mixture and subjected for composting. The former ingredients were favorable as additives for nitrogen and moisture balance (McMahon et al., 2009). The construction and demolition scrap i.e. timber waste from landfill was recycled into compost, recyclates included are shredded chip board, hardboard, and melamine supplemented by eco-bio contents like poultry manure and green waste. It reflects the need of composting at centralized facilities (McMahon et al., 2008). The economic efficiency of recycling outs on 36 jobs for unemployed while merely dumping of 10,000 tons refuse in a landfill affords six recruitments (Banerjee, 2015).

During recycling, two facts to be considered are the amount of energy saved and the ratio of cost to benefit, which depends on the type of recyclate and the energy needed to recycle. According to the Environmental Protection Agency (EPA), recycling aluminum can save 95% of the energy needed to

produce similar quantity of aluminum from its original source, bauxite (http://www1.eere.energy.gov/industry/aluminum/pdfs/aluminum.pdf).
  Plastic recycling consists of three phases:

- The physical method creates new objects by remelting the old ones.
- The chemical method converts recyclate plastic into monomers; for example, Polyethylene terephthalate (PET) plastic is mixed with alcohol and a catalyst to form dialkyl terephthalate, which can be used with ethylene glycol to form a new polyester polymer.
- Pyrolysis generates fuel oil by thermal depolymerization process (http://www.wikiwand.com/en/Recycling).

2.2.1 Recyclates

Recyclate is a crude material resulting from waste debris and is used to construct new products. There is an array of diversity in recyclates comprising old refused rubber tires is a best recyclate. Scrap ship vessels are also a form as recyclates. Waste such as paper, glass, plastic, and metal are also the recyclates that can be extended further in the list of recyclates. Recyclate quality, its life shelf and cost effectiveness are also an important factors that are to be considered for green economy and zero waste environment. Figure 2.1 shows how a product is formed and its conversion into debris after its lifespan after which it enters the recycling process. This cycle from new product to recyclate included the following important stages: segregation, sorting, and transport to the recycling center.
  The extensive recyclates include paper, aluminium cans, batteries, etc. These recyclates are likely to be convertible materials. Reprocessing post-consumer waste helps in economic balance, approach for waste management,

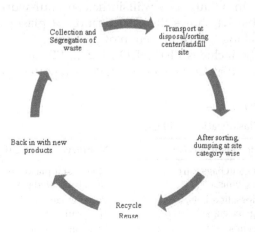

FIGURE 2.1
A product's journey.

and public awareness about the environment. A variety of policies need to be designed for recycling and available technologies.

During the recyclate processing, quality, shelf-life and cost approach of recyclates, supply and demand for reprocessed material should be examined. According to Baeyens and co-authors (2009), recycling has social, economic, and legal rationales. Demand for recyclable plastic includes low-density polyethylene (LDPE), of which 17% is reprocessed into shopping bags; 12% demand for high-density polyethylene (HDPE) to be converted into tables, benches, lorries, cargo liners, etc.; polypropylene (PP) claims around 19% reclaimed into small beans and battery boxes (Ferreira et al., 2012). In India, most wood is burned purposes; even the rejected wood chips from furniture shops are utilized as fuel. Some of it is valued as packaging support, animal bedding, garden mulches, etc. Recyclates come from both domestic and commercial uses, opted as second life. The important factors are funding, enhanced technologies, trained professionals, public awareness, and law enforcement for the successful outcome of the three Rs.

2.2.2 Recycling Consumer Waste

The refuse left after the intended consumer use may be discarded for disposal and then dumped at landfills or at any recycling center; all of this material is called post-consumer waste. This covers all sorts of discarded objects, for example, fruit peelings, meat bones, clothes, furniture, paper, electronics, industrial refuse, etc.

Construction activities in cities are increasing rapidly. In Delhi, 5000 tonnes of waste are generated per day of waste from building, demolition and land-clearing activities. Municipal Corporation of Delhi along with a Delhi-based recycling company infrastructure and financial services opted for the recycling of construction waste (facility center Burari). From the mixed waste collected from site, plastic and wood after segregation is sent to waste to energy plant in Okhla (www.ilfsindia.com/our-work/environment/construction). Table 2.1 shows the classification of plastics emerging from various goods packing and composition of packaging.

Recyclable plastic includes PET, HDPE, LDPE, PP, and Polyvinyl chloride (PVC). Central Pollution Control Board (CPCB) reported plastic waste as a

TABLE 2.1

List of Classification of Plastics

Recyclable Plastic Source	Nonrecyclable Plastic Source
Milk, detergent bags, carry bags Containers, films, etc.	Thermostat plastic, multilayer and laminated plastic
Cables, pipes, floorings, foam Packaging, tea cups	Polyurethane (PUF), bakelite, polycarbonate
Ice cream cups	Melamine, nylon

Source: www.cpcb. Indian Standards.

significant contributor of municipal solid waste, measuring 10,000 TPD of rejected plastic and 9% of 1.20 lakhs tons per day (TPD) of municipal solid waste. Major projects undertaken by CPCB include polymer coated bitumen road and plasma pyrolysis technology and conversion of plastic refuse into liquid fuel. The majority of these projects depend on the extraction of energy from waste.

To improve recycling efficiency, collection methods and avoiding mixed garbage should be acknowledged. In Japan, reclaiming waste collection methods depend on residents, who must separate recyclates at home, and group collection programs run and funded by municipal authorities (Matsumoto, 2012).

Waste paper reprocessing is provided by many private ventures in India. Recycling one ton of paper saves the cutting of 17 mature trees. To prevent deforestation and to boost paper recycling, ITC Paperboards and Specialty Business Company launched a program called "wealth out of waste" in India. It teamed up with many bigger Information Technology (IT) companies, residential welfare associations, and local bodies to expand the waste paper collection program (www.wealthywaste.com). Waste paper collection engages mostly informal sectors (95% waste collection): rag pickers, collector centers, and door-to-door collectors. Waste management can be augmented along with the active participation of is to the public, and providing them drop off centers, buyback centers and also citizen awareness by public education on recycling materials. In India, out of the total amount of consumed paper, only 27% of post-consumer paper is reintroduced into the system, which reflects that indigenous paper collection in India is to be promoted (www.sustanaibilityoutlook.in). A definite functional model needs to be designed and implemented for paper sorting, collection, and recycling; penalties must be charged for those who do not comply. Nongovernment organizations should also concurrently support in the waste management campaign. Up gradations in legislative rules for waste disposal is also to be redesigned.

In India, families (considered to include four persons) produce 2.5 kg garbage/day. This garbage consists of 30% or 750 g of recyclable material per day and 26 million ton recyclable material per annum. Paper is major constituent portion (dipp.gov.in).

In Germany, Green Dot System has the goal of developing schemes to collect, sort, and recycle used packaging (www.guiaenvase.com).

2.2.3 Recycling Industrial Waste

Factories, industry, mills, and mining operations produce industrial waste. Industrial waste is treated by one of two methods—it is reused and recycled where it was generated, or it is recycled by recycling sectors. Economic benefit from industrial waste incorporates extraction of products that carry commercial value. Biomass production from wastewater by cultivation of microalgae, especially filamentous cynobacteria, was employed (Markou and Georgakakis, 2011).

Bioelectrochemical systems (BESs) are unique systems capable of transforming the chemical energy of organic waste, including wastewater with lower stability and ligno-cellulosic biomass, into electropower or hydrogen/chemical products in microbial fuel cells (MFCs) or microbial electrolysis cells. BESs use organic substrates in place of precious metals as catalysts (Pant et al., 2012).

Thus, garbage is an important source of sustainable and reliable bioenergy. This tool is proved to be environmentally safe and reliable. In an experiment, industrial blast furnace slag and fly ash waste were used as supplementary cementation materials to reduce the burden and excessive use of cement (Portland). This co-product from industrial waste proved to be regulatory tool in concrete (Chen et al., 2010). Recycling of construction material such as asphalt, brick, concrete, ferrous metal, glass, masonry, nonferrous metal, paper and cardboard, plastic, and timber can create secondary products for future use (Tam and Tam, 2006).

2.2.4 Electronic Waste Recycling

Batteries and capacitors in Liquid-crystal displays (LCDs) are e-wastes that are harmful to the environment e-waste. After manual removal of batteries and capacitors from electronic waste (e-waste), the remaining parts are shredded to 10 mm particles and subjected to magnetic field for ferrous metal removal. While nonferrous metals are separated by centrifuge or vibrating plates by density, additional precious metals are isolated and dissolved in acid, sorted, and smelted into ingots. The leftover glass and plastic fractions extracted by density are sold to reprocessors (or diverted into reprocessing sector). Lead and Cathode Ray Tubes (CRTs) or backlight from TV sets, monitors, and LCDs are disassembled manually.

E-waste is an emerging issue because of the rapid increase in the amount of it being discarded but also because it is a business opportunity. Contents of e-waste include iron, copper, aluminum, gold, and other metals forming about 66% of e-waste and 2.70% pollutants. It is one of the fastest growing fractions of municipal waste. E-waste recycling is a market-based enterprise. China and India recycle through small sized enterprises in informal sectors (Widmer et al., 2005).

Carbonaceous adsorbent extracted from carbon slurry from the fertilizer industry succeeded in absorbing anionic dyes such as ethyl orange, metanil yellow, and acid blue. These adsorbents can be used as an affordable alternative for color removal from discharge (Jain et al., 2003).

Different types of industrial wastes as glass, fly ash, blast furnace slag, and silica fume can be used as pozzolanic supplement for partial restoration of Portland cement in cemented paste backfill. This additive enhances mechanical performance of cemented paste and reduces binder costs (Ercikdi et al., 2009). E-parisaraa private limited is another government-authorized e-waste recycling unit in Bangalore that recycles waste electrical and electronic equipments. To facilitate recycling, certain universal codes are used for the products from which an item is made. Some International Organization for Standardization (ISO) standards related to recycling include ISO 15270:2008 for plastic waste and ISO 14001:2004 for recycling practice (wikiwand.com).

Some e-waste is consumer-oriented electronics, that is, televisions, cell phones, ovens, irons, washing machines, etc. Commercial e-waste is derived from industrial engines, exhaust fans, generators, AC machines, computers, etc.

An approach for stabilization and applying guar gum industrial waste for vermicompost yield by employing earthworm species was studied. Earthworms can be used to manage the industrial waste by its bioconversion into vermicompost (Suthar, 2006).

2.2.5 Recycling Codes

Items to be sold display packaging labels and recycling symbols to give information about the contents of packaging that can be recycled. Table 2.2 shows some examples.

All consumer products, ranging from soft drink cans to PET bottles are coded to show whether they can be recycled or disposed of via other methods. Most post-consumer waste is recyclable. Recycling codes have been been allotted to diverse types of waste ranging from plastic, glass, paper, wood, metal, etc.

TABLE 2.2

Universal Recycling Codes of Various Recyclates

Code	Recyclate	Code	Recyclate
20 **PAP** #20 PAP	Corrugated cardboard paper	**70** GL #1 PET(E)	Polyethylene terephthalate Polyester fibers, beverage bottles
21 PAP #20 PAP	Other paper mixed paper magazines, mail	**71** GL #2 PEHD or HDPE	High-density polyethylene Plastic bottles, plastic bags, trash cans, oil cans, imitation wood
22 PAP #22 PAP	Paper	**72** GL #3 PVC	Polyvinyl chloride Window frameworks, tins for chemicals, flooring, plumbing ducts
01 PET #40 FE	Steel	**04** PE-LD #4 PELD or LDPE	Low-density polyethylene Plastic bags, buckets, soap dispenser bottles, milk cans, plastic tubes

(Continued)

TABLE 2.2 *(Continued)*

Universal Recycling Codes of Various Recyclates

Code	Recyclate	Code	Recyclate
⟳ **02** PE-HD #41 ALU	Aluminium	⟳ **05** PP #5 PP	Polypropylene Bumpers, car interior trim, industrial fibers, carry-out beverage cups
⟳ **03** PVC #41 ALU	Made-of-aluminium symbol	⟳ **06** PS #6 PS	Polystyrene Toys, flower vases, video cassettes, ashtrays, trunks, beverage/food coolers, beer cups, wine and champagne cups, carry-out food containers, styrofoam
⟳ **40** FE #70 GLS	Mixed glass container/multi-part container	⟳ **07** O #07(OTHER)	Polycarbonate (PC), polyamide (PA), styrene acrylonitrile (SAN), acrylic plastics/polyacrylonitrile (PAN), bioplastics All other plastics
⟳ **41** ALU #71 GLS	#71 GLS clear glass	⟳ **ABS** #9 or #ABS	Acrylonitrile butadiene styrene Monitor/TV coverings, coffee makers, cell phones, most computer plastic, most FFF 3D printed parts that are not bioplastic such as PLA
(alu) #72 GLS	#72 GLS green glass	⟳ **PA** PA	Polyamide nylon

Source: http://www.unicode.org/L2/L2001/01146-N2342-Recycling.pdf; https://en.wikipedia.org /wiki/Recycling_codes.

Among these materials, plastics are difficult to recycle because of the vast range of materials used to make them. Plastic containers should be equipped with labels including both plastic composition and recycling code. Plastic recycling symbol numbers different types of plastic. Various countries and industries have developed different recycling codes. Here are some examples:

Recycling code for plastics were designed by (1988) Society of Plastic Industry (plastic recycling codes archived, July 21, 2011).

Paper recycling codes were designed by American Forest and Paper Association (AF&PA).

Recycling codes for glass were planned by the Glass Packaging Institute (GPI).

Recycling codes for steel were created by the Steel Recycling Institute.

Universal recycling symbols were created by Gary Anderson (1970).

2.3 Reuse

Reuse is using a discarded item again for its originally intended use or repurposing the item. It does not involve reprocessing. Biogenic rejects are key factors for reprocessors in nature of biodegradation.

Biodegradable waste can be transformed into useful biomass by means of aerobic and anaerobic key processes. This degradable biomass is broken down biochemically by the action of microbes and various organisms. Composting is an important an option for resource recovery from waste and boosting soil health by applying organic matter. For optimum use of bio-oriented waste, biodegradable potentiality and microbial activities should be thoroughly studied.

From November 2015, it is mandatory for all road developers in India to use plastic with bituminous mixes for road construction. Installing waste-to-compost and bio-methanation plants reduces the amount of landfill states. Manual scavenging is the most common practice in India, but 76% of workers in informal recycling operations suffer from respiratory ailments like asthma, choking, coughing, irritation, and breathing difficulties (www.attero.in/e-waste facts).

Centre for Science and Environment reported that the city Moradabad receives 50% of India's total discarded circuit board e-waste, which affects 50,000 informal recycling workers. As a result of informal e-waste recycling, leachate of zinc and copper blended with soil beyond 5–15 times the safety limits. Ganga River has not escaped contamination with chromium and cadmium by up to 1.3–3 times; traces of mercury and arsenic were also reported (cseindia.org/content/cse-study-shows-e-waste-recycling).

2.3.1 Modeled on Nature: Biological Reprocessing in Waste Management

Biological reprocessing includes the decomposing of waste originating from organic source. The decomposition and breakdown may be initiated by means of microorganisms and annelids like earthworms. Waste refuses such as agriculture waste, vegetable waste, animal waste, paper waste, food industry waste, etc. can be converted into compost or vermicompost. The anaerobic digestion process of these wastes aided by microbes yields an energy based endproduct.

2.3.2 Composting

Composting is biological technically sound process applicable for all biomass. Soil amelioration, soil fertility, and soil health can be maintained by compost application.

2.3.2.1 Composting Process

The composting process involves the following steps: mesophile phase, thermophile phase, cooling stage, and maturing phase. The compost that results is a soft, moist humus that is also a biologically stable soil ready for field application as biofertilizer.

Compost combined with biochar applied to soil proved to have a positive synergic association by producing high yield and mineral fertilizer retention. Here the author also suggests the application of biochar in *terra preta* genesis technology that would preserve sustainability of natural resources (Fischer and Glaser, 2012). Composting methods consist of windrow composting, static pile composting, and in-vessel composting. The choice depends on different measures and organic raw materials. The different steps involved in the process are shown in Figures 2.2 and 2.3, respectively.

Table 2.3 shows different studies about composting studies. Many researchers have shown the potential use of different biomasses for waste stabilization, and volume and mass reduction, with a particular emphasis on microbial studies of compost, etc.

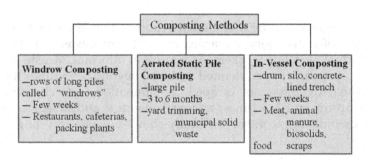

FIGURE 2.2
Different methods of composting.

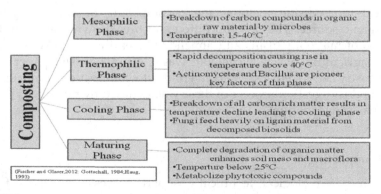

FIGURE 2.3
Steps involved in the process of composting.

TABLE 2.3

Studies on Composting from Different Types Waste

Authors	Title	Objectives	Findings
Cerda et al. (2017)	Composting of food wastes: Status and challenges	• Microbiology of food waste composting • Gaseous emissions and non-biodegradable impurities	• Role of refused food compost in soil bioremediation • Need in improvement of compost process monitoring
Singh and Kalamdhad et al. (2013)	Rotary drum composting of vegetable waste and tree leaves	• Higher temperature (60°C–70°C) at inlet zone and (50°C–60°C) at middle zone were observed with increased degradation in the drum • TOC, C/N ratio, CO_2 evolution, and coliforms were decreased significantly	• Compost parameters with total nitrogen (2.6%) and final total phosphorus (6 g/kg) was reported
Bhatia et al. (2013)	Diversity of bacterial isolates during full-scale rotary drum composting	• Different methods in rotary drum composting process was reviewed • Culture-dependent (indigenous population of bacterial isolates mainly total heterotrophic bacteria) • Culture-independent method were studied	• Fruitful composting technology providing turbulence and aeration during composting

(Continued)

TABLE 2.3 (*Continued*)
Studies on Composting from Different Types Waste

Authors	Title	Objectives	Findings
Jiwan Singh Ajay S. Kalamdhad (2013)	Assessment of bioavailability and leachability of heavy metals during rotary drum composting of green waste (water hyacinth)	• Total heavy metal concentration during water hyacinth composting was observed	• Cattle manure boosted organic matter degradation and humification process; reducing the toxicity of metals during rotary drum composting
Ajay S. Kalamdhad et al. (2008)	Stability evaluation of compost by respiration techniques in a rotary drum composter	• CO_2 evolution and oxygen uptake rate (OUR) were observed to assess the final compost	• Rotary drum composting of mixed organic waste can produce final compost within 20 days of composting • Solvita® maturity index displayed C/N 16, 30, and 38 has a of 7 considered as end product

2.3.3 Anaerobic Digestion

The waste originated from living material is acted upon by microorganisms in the absence of oxygen within a closed chamber. The materials that can be processed in an anaerobic digester are manures, food waste, fats, oils, greases, industrial residuals, and sewage sludge. Biogas is generated during the assimilation of organic waste by microbes. The contents of biogas are methane and carbon dioxide. The solid matter that is left after digestion is called digestate, and it can be separated as a solid or a liquid rich in nutrients that can be applied as a manure for crops. Thus, anaerobic digestion yields a renewable energy used in a variety of ways for human benefit. The leftover digestate is used as bedding for livestock, flower pots, soil amendments, and fertilizers (www.epa.gov). Different types of anaerobic digesters include standalone digesters, on-farm digesters, and wastewater treatment plant digesters.

Class A biosolids are produced in a thermophilic anaerobic digester and can be directly used for application per USEPA guidelines. Anaerobic digestion of waste can be processed through a series of steps like hydrolysis, acidogenesis, acetogenesis, and methanogenesis. This approach can be applied for waste management or for fuel production, as depicted in Figure 2.4. Biogas produced from anaerobic breakdown of biomass can be diverted for various energy uses such as in engines, fueling vehicles, and in domestic kitchens. See Table 2.4.

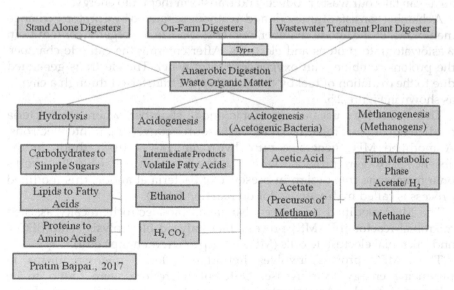

FIGURE 2.4
Process of anaerobic digestion.

TABLE 2.4

Use of Biogas

Resulting Biogas Product	Power engines, mechanical start over, heat/electricity systems
	Fuel boilers and furnaces, heating digesters, etc.
	Fuel vehicles
	Domestic natural gas

2.3.4 Microbial Fuel Cell (MFC)

This technological procedure has drawn the attention of waste management sectors due to its minimum operating conditions and a diversity of waste biological substrate is applicable in this method which is an alternative for bioremediation. It is a novel technology for processing waste and can replace the use of fossil fuels. In this biologically based fuel cell, the fuel is made by different sources of organic compounds. Landfill leachate, municipal and agro-industrial wastewaters and sediments, and solid organic wastes can be sources of electric power and commodity chemicals. MFCs can provide electricity, biohydrogen production, and wastewater treatment, and they can perform as a biosensor in pollution control. MFC technology can be used in breweries plant, wastewater treatment plants, and desalination plants for producing hydrogen, remote sensing and harvesting power source. Widespread use of MFCs in these areas can take our waste products and transform them into energy.

A chamber consists of an anode and a cathode separated by a proton exchange membrane. In the anode chamber, microbes oxidize the organic content from wastewater into protons and electrons. After entering the cathode chamber, the protons combine with oxygen and form water. The electrons generated due to the oxidation of fuel by the microbes are transferred through a circuit, as shown in Figure 2.5.

This technology uses an anaerobic anode chamber, where the bacteria oxidize the organic material from waste source converting it into electrons. A modified MFC generates pure hydrogen gas at the cathode in the absence of oxygen, provided with a little amount of voltage (0.25v) works on a membrane free design in presence of bacteria at anode, this modified process is called hydrogen evolution reaction (HER).

The MFC modified process is also called bio-electrochemically assisted microbial reactor (BEAMR) process, biocatalyzed electrolysis cells (BECs), and microbial electrolyte cells (MECs) (http://www.engr.ps.edu).

Thus, MFC process involves important role of microorganisms in producing energy from refuse. Different microflora were found to be effective, notably Actinobacteria, Firmicutes, Chloroflex, Rhodoferax, Ferruginibacter spp, Rhodopseudomonas, Ferribacterium, Clostridium,

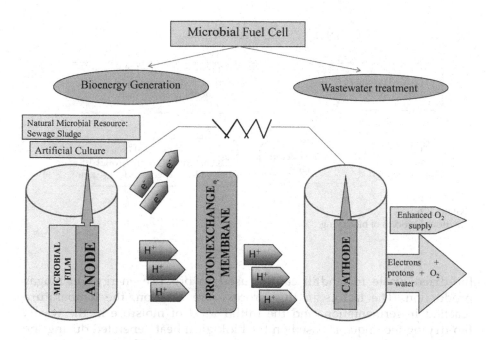

FIGURE 2.5
Microbial fuel cell.

Chlorobaculum, Rhodobacter, and Bradyrhizobium on anode biofilm, in generating an efficient electricity of 13.2 W/m³ from sewage sludge (Zhang et al., 2012). An experiment conducted by Moqsud and co-workers extracted bioelectricity by employing microbial fuel cell fueled by compost, organic matter holding grass cuttings, leaf mold, rice bran, oil cake (from mustard plants), and chicken droppings. Key microbes include a wide range, from *Shewanella oneidensis*, *R. ferrireducens*, *Geobacter sulfurreducens*, *C. butyricum*, *Pseudomonas aeruginosa*, *Saccharomyces cerevisiae*, and *Streptococcus lactis*. Marine sediment soil, active sludge water, and sewage are also enhanced power outputs of organisms that are used to generate power in MFCs (Rabaey et al., 2005; Ringeisen et al., 2006; Ray and Little, 2007; Najafpour et al., 2011).

2.3.5 Bio-Drying

Bio-drying, a convective evaporation process, is a novel tool for stabilization of municipal solid waste due to its high moisture content, by reducing water content facilitates odor free landfilling, impact of waste and self-ignition rate. It is a short-term mechanical bioconversion process including exothermic reactions in an aerobic environment, as shown in Figure 2.6.

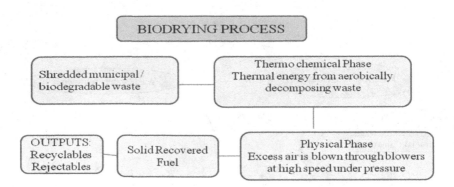

FIGURE 2.6
Functional process of biodrying.

Bio-dried waste in landfill can be used to produce energy via biogas production. The factors in this process are aeration, the temperature reached in fermentation, and the initial level of moisture in the waste. Bio-drying technique is based on the biological heat generated during the aerobic reactions of organic waste causing moisture and volume reduction of biomass (Tom et al., 2016).

Bio-drying supports the environment by decreasing landfill volume, increasing transport competency, condensing landfill emissions, and leaching of heavy metals-pollutants to soil layers. Bio-drying is a mechanical-biological process executed before landfill or combustion of solid waste. A number of studies are focused on aspects in bio-drying process, waste biomass reduction, the duration required for conversion of waste, etc. See Table 2.5.

TABLE 2.5

Contributions of Authors on Study of Biodrying Technology

Total Weight Loss %	Moisture Reduction %	Heating Value KJ/kg	Duration Required (days)	Authors
76.67	81.84	—	14	Ab Jalil et al. (2016)
30.00	—	—	30	Rada et al. (2014)
70.00	—	—	10	Zhang et al. (2011)
33.94	20.81	—	20	Tom et al. (2016)
32.00	50.00	—	13	Bilgin and Tulun (2008)
28.00	—	16779 KJ/kg	14	Tambone et al. (2011)
50.00	15	15,000 KJ/kg	30	Colomer et al. (2012)

Source: Ragazzi, M., and Rada, E.C., *Waste Manage. Environ.*, 10, 199–208, 2012.

2.4 Energy Recovery

Energy recovery can be done by waste-to-energy (WtE) technology yielding a value added end product like heat, electricity or fuel through an array of processes such as mass burning facilities, gasification, anaerobic digestion and landfilling. Around 86 energy restoration centers in United States are signed up in bioconversion of 28 million tons municipal solid waste to energy harvest yielding around 2,720 Megawatts of power per year. A waste-to-energy plant produces 550 Kw Hours of energy per ton of waste (www.epa.gov). Energy recovery from waste resources is a nonthreat process for waste management.

An analysis in India by Nixon et al. (2017) detailed the unsuccessful foundation of waste-to-energy strategy. Their survey was based on case studies of an incineration plant, a gasification plant, and coal-fueled plant. The results obtained were compared with the standards of European WtE plants. The authors recommended further developments in financial incentives, improvization in regulation on dumpsites, structuring criteria for ash disposal and maintaining the protocol of stack emission measurements.

Energy can be recovered from waste in the form of heat, electricity, and fuel by the processes such as thermal conversion, thermochemical conversion, and biochemical conversion. Waste in both its dry and wet forms can be utilized at various phases for energy recovery, as shown in Figure 2.7 and Table 2.6.

The emerging scientific field i.e. nanotechnology treasured its roots in boosting the reaction productivity and rate of reaction of electrochemical conversion of carbon dioxide into fuel. The synergic gain through mutual work of nanomaterials and reverse microbial fuel cells should be widely devised in electrochemical conversion of waste biomass (Ganesh, 2016).

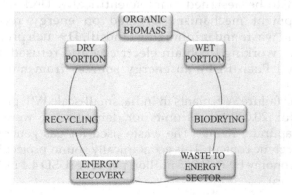

FIGURE 2.7
Type of waste and its course for recovery.

TABLE 2.6

Co-existing Forms of Energy Recovery Techniques

Thermal Conversion	Biochemical Conversion	Thermal Chemical Conversion	Electrochemical Conversion
Combustion	Hydrolysis	Combustion	Microbial fuel cells
Gasification	Fermentation	Gasification	Incineration
Pyrolysis	Catalysis	Pyrolysis	
Hydrothermal carbonization	Extraction	Carbonization	
	Anaerobic Digestion	Catalytic liquefaction	
	Torrefaction		
Form of energy			
Steam, electricity, gas, liquid, and liquid fuel	Hydrochar, biodiesel, methane, alcohol	Heat, gas, oil, char	Electricity, bio-hydrogen gas

Source: http://www.eai.in/ref/ae/wte/wte.html; fao.org; Chen, W.H., *Renew. Sust. Energ. Rev.*; Xiawei et al.

2.5 Pyrolysis

Pyrolysis is an endothermic process that thermochemically decomposes organic biomass at high temperatures in anaerobic conditions (see Figure 2.8). Types of waste subjected to pyrolysis are municipal solid waste, tires, wood, paper, kitchen waste, etc.

An energy recovery plan for waste may boost a nation's economy and environmental health by minimizing its dependence on landfilling of waste.

For effective extraction of carbon and methane from dumping sites, landfills need to be designed more scientifically. Under the projects, clean development mechanism targeted on energy recovery from municipal solid waste and minimizing landfill, 119 such projects around the globe were working to obtain electricity and refused derived fuel Refuse Derived Fuel (RDF), an energy sources from municipal solid waste.

In spite of the failure WtE plants in India, small-scale WtE projects proved to be successful; 20,000 biogas units for domestic use were installed in Thiruvananthapuram, Kerala. The waste used for gas generation ranged from 2.5% of organic content. This economically sound project had a saving in transport economy by INR 225 million per year (USD 4.4 million) (www.nswaienvis.nic.in).

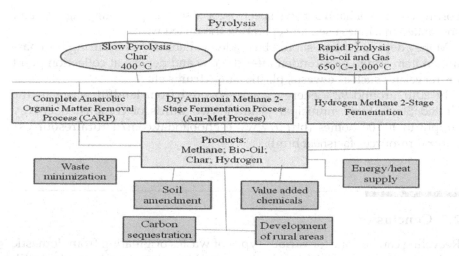

FIGURE 2.8
Pyrolysis process diagram.

2.6 Resource Recovery

Resource recovery has become an important concern for waste management sectors. It selectively abstracts from discarded refuse items for reuse or recycling with a goal to minimize landfilling. The resource recovery sector mainly incorporates waste management teams extensively working with environmental concern.

Resource recovery endeavors to monitor, control, and maintain the proper flow of product from manufacture to its disposal. Its attempts are mainly based on recovery of energy in many forms such as electricity, heat, steam, biogas, bio-oil char, etc. from discarded products. Maximum effort is made to prevent dumping either by recycling or reusing items. One of the greatest benefits is that natural resources for raw materials are not disturbed. For example, by reusing the bagasse and molasses from sugar factories or pulp from used paper, new paper product can be manufactured without cutting trees. In achieving zero waste, factors to be considered are quality of recyclate, sorting and storage facilities of recyclate for reprocessing product to be formed. Mixed waste collection including commingled recyclables must undergo strict sorting steps with enhanced technology. Desired recyclates must be sorted, separated, and cleaned from mixed waste spectrum, which requires extra efforts. For source separation at the consumer level, desired recyclates must be cleaned, sorted, and collected by the consumer, which requires extensive public education programs to avoid contamination in the recycled material.

Sorting consists of hand sorting, spectroscopic scanning, strong magnets, and automated machines used for separating the mixed waste.

Money deposit scheme should be made available for user during purchasing an item and to be refunded after the use and deposit at collection point for recyclates such as e-waste, plastic cans, drums etc.

In 2010, around 92% energy was recovered from landfill waste in the United States from municipal yielding two thirds of combusted energy supply to 100,00 homes (http://www.eschooltoday.com/naturalresources/natural-resources-factsheet.html).

2.7 Conclusion

Recycling can be done for various types of wastes originating from domestic and industrial regions. Waste can be reused by converting it into biofertilizer through composting, and into biofuel through anaerobic digestion and pyrolysis. Microbial fuel cell is another recent method to extract energy from waste with minimum infrastructure and input.

Improvement in regulations of solid waste and its disposal, management should be executed.

Initiatives for recycling needs to include local municipalities, the public, and voluntary organizations. Collaborative work must be encouraged between waste management sectors and recycling facilities. Also encouraging public participation in composting and various campaigns, awareness programs for waste generators can be structured. Public awareness and education; strict collection, sorting, and disposal legislation; donation sites for reusable items; collaboration between the public and the waste management sectors, communal collection awareness, etc., need to be revised and implemented.

Implementing the policy of the three Rs contributes toward sustainable zero waste and nullifies global problems such as depleting fossil fuels, environmental degradation, climate change, employment, etc.

References

Ab Jalil, N.A., H. Basri, N.E.A. Basri, M.F. Abushammala. (2016). Abushammala. Biodrying of municipal solid waste under different ventilation periods. *Environ Eng Res.* 2016. doi:10.4491/eer.2015.122.

Baeyens, J., A. Brems, R. Dewil. (2009). Recovery and recycling of post-consumer waste materials. Chemical engineering and chemical process technology – Vol. IV- *Recovery and Recycling of Post-Consumer Waste Materials*, J. Baeyens, A. Brems, R. Dewil (Eds.), ©Encyclopedia of Life Support Systems (EOLSS).

Banerjee, R. (2015). Importance of recycling. *IJIREEICE*. 3(6). doi:10.17148/IJIREEICE.2015.3611.

Bhatia, A., S. Madan, J. Sahooc, M. Ali, R. Pathania, A.A. Kazmi. (2013). Diversity of bacterial isolates during full scale rotary drum composting. *Waste Manage.* 33(7):1595–1601.

Bilgin, M., Ş. Tulun. (2008). Biodrying for municipal solid waste: Volume and weight reduction. *Resour Conserv Recycl.* 52(5):829–834.

Cerda, A., A. Artola, X. Font, R. Barrena, T. Gea, A. Sánchez. (2017). Composting of food wastes: Status and challenges. *Bioresour Technol.*

Chaudhuri, S.K., D.R. Lovley. (2003). Electricity generation by direct oxidation of glucose in mediatorless microbial fuel cells. *Nat Biotechnol.* 21(10):1229e32.

Chen, C., G. Habert, Y. Bouzidi, A. Jullien, A. Ventura. (2010). LCA allocation procedure used as an incitative method for waste recycling: An application to mineral additions in concrete. *Resour Conserv Recycl.* 54(12):1231–1240. doi:10.1016/j.resconrec.2010.04.001.

Chen, W.H. (2015). A state-of-the-art review of biomass torrefaction, densification and applications. *Renew Sust Energ Rev.* http://works.bepress.com/wei-hsin_chen/130/.

Colomer-Mendoza, F.J., F. Robles-Martinez, L. Herrera-Prats, et al. (2012). *Environ Dev Sustain.* 14: 1013. https://doi.org/10.1007/s10668-012-9369-1.

CPCB (2013). A Status Report on Municipal Solid Waste (MSW) Management in Mysuru City, Karnataka. http://cpcb.nic.in/cpcbold/zonaloffice/banglore/MSW_report_Mysuru.pdf.

Ercikdi, B., F. Cihangir, A. Kesimal, H. Deveci, İ. Alp. (2009). Utilization of industrial waste products as pozzolanic material in cemented paste backfill of high sulphide mill tailings. *J Hazard Mater.* 168(2–3):848–856.

Ferreira, B., J. Monedero, J.L. Martí, C. Aliaga, M. Hortal, A.D. López. (2012). The economic aspects of recycling, In *Post-Consumer Waste Recycling and Optimal Production*, E. Damanhuri (Ed.), InTech. http://www.intechopen.com/books/post-consumer-waste-recycling-and-optimal-production/the-economicaspects-of-recycling.

Fischer, D., B. Glaser. (2012) *Management of Organic Waste*, InTech, pp. 177–208.

Ganesh, I. (2016). Electrochemical conversion of carbon dioxide into renewable fuel chemicals: The role of nanomaterials and the commercialization. *Renew Sust Energy Rev.* 59:1269–1297.

Gottschall, R. (1984). *Kompostierung. Optimale Aufbereitung und Verwendung organischer Materialien im ökologischen Landbau*, Müller, Karlsruhe, Germany.

Haug, R.T. (1993). *The Practical Handbook of Compost Engineering*, Lewis Publishers, Boca Raton, FL.

https://cdn.cseindia.org/userfiles/moradabad-e-waste.pdf.

http://repositori.uji.es/xmlui/bitstream/handle/10234/62754/Registro_acceso_restringido.pdf?sequence=3.

http://www.eschooltoday.com/natural-resources/natural-resources-factsheet.html.

http://www.indiaenvironmentportal.org.in/content/257913/the-municipal-solid-wastes-management-and-handling-rules-2000/.

https://www.sciencedirect.com/topics/chemistry/bauxite.

http://www.unicode.org/L2/L2001/01146-N2342-Recycling.pdf; https://en.wikipedia.org/wiki/Recycling_codes.

Jain, A.K., V.K. Gupta, A. Bhatnagar Suhas. (2003). Utilization of industrial waste products as adsorbents for the removal of dyes. *J Hazard Mater.* 101(1):31–42.

Kalamdhad, A.S., M. Pasha, A.A. Kazmi. (2008). Stability evaluation of compost by respiration techniques in a rotary drum composter. *Resour Conserv Recycl.* 52(5):829–834. doi:10.1016/j.resconrec.2007.12.003.

Kalamdhad, A.S., Y.K. Singh, M. Ali, M. Khwairakpam, A.A. Kazmi. (2009). Rotary drum composting of vegetable waste and tree leaves. *Bioresour Technol.* 100(24):6442–6450. doi:10.1016/j.biortech.2017.06.133.

Markou, G., D. Georgakakis. (2011). Cultivation of filamentous cyanobacteria (blue-green algae) in agro-industrial wastes and wastewaters: A review. *Appl Energy.* 88(10):3389–3401.

Matsumoto, S. (2012). Group collection of recyclables in Japan. *Post-Consumer Waste Recycling and Optimal Production.* pp. 67–78. www.intechopen.com.

McMahon, V., A. Garg, D. Aldred, G. Hobbs, R. Smith, I.E. Tothill. (2008). Composting and bioremediation process evaluation of wood waste materials generated from the construction and demolition industry. *Chemosphere.* 71(9):1617–1628. doi:10.1016/j.chemosphere.2008.01.031.

McMahon, V., A. Garg, D. Aldred, G. Hobbs, R. Smith, I.E. Tothill. (2009). Evaluation of the potential of applying composting/bioremediation techniques to wastes generated within the construction industry. *Waste Manage.* 29(1):186–196. doi:10.1016/j.wasman.2008.02.025.

Moqsud, M.A., K. Omine, N. Yasufuku, M. Hyodo, Y. Nakata. (2013). Microbial fuel cell (MFC) for bioelectricity generation from organic wastes. *Waste Manage.* 33(11):2465–2469.

Najafpour, G., M. Rahimnejad, A. Ghoreshi. (2011). The enhancement of a microbial fuel cell for electrical output using mediators and oxidizing agents. *Energy Sour.* 33(24):2239e48.

Nixon, J.D., P.K. Dey, S.K. Ghosh. (2017). Energy recovery from waste in India: An evidence-based analysis. *Sustain Energy Technol Assess.* 21:23–32. https://doi.org/10.1016/j.seta.2017.04.003.

Pant, D., A. Singh, G. Van Bogaert, S.I. Olsen, P.S. Nigam, L. Diels, K. Vanbroekhoven. (2012). Bioelectrochemical systems (BES) for sustainable energy production and product recovery from organic wastes and industrial wastewaters. *RSC Adv.* 2:1248–1263. doi:10.1039/C1RA00839K.

Puna, J.F., M.T. Santos. (2010). Thermal conversion technologies for solid wastes: A new way to produce sustainable energy. *Waste Management,* E.S. Kumar (Ed.), p. 232, INTECH, Croatia, SCIYO.COM.

Rabaey, K., N. Boon, M. Hofte, W. Verstraete. (2005). Microbial phenazine production enhances electron transfer in biofuel cells. *Environ Sci Technol.* 39(9):3401e8.

Rada, E., L. Squazardo, G. Ionescu, A. Badea. (2014). Economic viability of SRF co-combustion in cement factory. *UPB Sci Bull Serie D.* 76:199–206.

Ragazzi, M., E.C. Rada. (2012). RDF/SRF evolution and MSW bio-drying. *Waste Manage Environ.* 10(1):199–208.

Rahimnejad, M., G.D. Najafpour. (2011). Microbial fuel cells: A new source of power. In *Biochemical Engineering and Biotechnology,* pp. 1–31.

Ray, R., B. Little. (2007). A miniature microbial fuel cell operating with an aerobic anode chamber. *J Power Sources.* 165(2):591e7.

Ringeisen, B.R., E. Henderson, P.K. Wu, J. Pietron, R. Ray, B. Little, J.C. Biffinger, J.M. Jones-Meehan. (2006). High power density from a miniature microbial fuel cell using *Shewanella oneidensis* DSP10. *Environ Sci Technol.* 40(8):2629e34.

Singh, J., A.S. Kalamdhad. (2013). Assessment of bioavailability and leachability of heavy metals during rotary drum composting of green waste (Water hyacinth). *Ecol Eng.* 52:59–69.

Suthar, S. (2006). Potential utilization of guar gum industrial waste in vermicompost production. *Bioresour Technol.* 97:2474–2477. doi:10.1016/j.biortech.2005.10.018.

Tambone, F., B. Scaglia, S. Scotti, F. Adani. (2011). Effects of biodrying process on municipal solid waste properties. *Bioresour Technol.* 102(16):7443–7450. doi:10.1016/j.biortech.2011.05.010.

Tom, A.P., R. Pawels, A. Haridas. (2016). Biodrying process: A sustainable technology for treatment of municipal solid waste with high moisture content. *Waste Manage.* 49:64–72. doi:10.1016/j.wasman.2016.01.004.

Vivian, W.Y.T., C.M. Tam. (2006). A review on the viable technology for construction waste recycling. *Resour Conserv Recycl.* 47(3):209–221.

Widmer, R., H. Oswald-Krapf, D. Sinha-Khetriwal, M. Schnellmann, H. Boni. (2005). Global perspectives on e-waste. *Environ Impact Assess Rev.* 25(5):436–458.

wikiwand.com.

www.attero.in/e-waste facts.

www.ilfsindia.com/our-work/environment/construction.

Zhang, G., Q. Zhao, Y. Jiao, K. Wang, D.L. Lee, N. Ren. (2012). Efficient electricity generation from sewage sludge using biocathode microbial fuel cell. *Water Res.* 46:43–52. doi:10.1016/j.watres.2011.10.036.

Zhang, D.-Q., H. Zhang, C.-L. Wu, L.-M. Shao, P.-J. He. (2011). Evolution of heavy metals in municipal solid waste during bio-drying and implications of their subsequent transfer during combustion. *Waste Manage.* 31(8):1790–1796.

Zhou, X., W. Li, R. Mabon, L.J. Broadbelt. (2016). A critical review on hemicellulose pyrolysis. *Energy Technol.* 5. doi:10.1002/ente.201600327.

3

Modern Energy Recovery from Renewable Landfill or Bio-Covers of Landfills

Rena, Pratibha Gautam, and Sunil Kumar

CONTENTS

3.1 Introduction

The stream of garbage generated from households and businesses and collected by the municipal corporation, the department of public works, or the sanitation service is known as municipal solid waste (MSW). MSW is highly heterogeneous. It consists of food waste, metals, paper, plastics, etc. Some of the generated waste is recycled and reused; the remainder is dumped directly into the landfill. The landfill receives other kinds of waste too, such as hazardous waste, construction and demolition (C&D) waste, e-waste, etc. It is expected that by 2025, approximately 2.2 billion MSW tonnes per year will be generated (Moya et al., 2017). The United States has 3581 landfills for disposal of MSW (Jain et al., 2013). In Australia, most MSW goes to the landfill (Dever et al., 2011). According to Ministry of Urban Development (MoUD) report (2016), 1.14 lakhs ton of waste is being generated in India, and only 82% of the waste is being collected; out of the total collected waste, only 22% is being treated. On average, nearly 4.38 pounds of trash is generated per person per day in the United States; collectively about 251 tonnes per year among the total 87 million tons of waste is being recycled (EPA 2015). Total recycling rate is around 34.5%, and the majority of the waste ends up in a landfill. Every year about 135 million tons of MSW are sent to landfill in counties across the country (National Association of Counties waste energy recovery: Renewable energy from county landfill 2015). The dumping ground after undergoing a long period of deposition and stabilization becomes an additional source of both energy and material recovery. In 2010, the global energy potential of waste can be estimated at 8–18 EJ/year and is expected to increase by 13–30 EJ in 2025 (Scarlet et al., 2015).

3.2 Importance of Landfill

The landfill is a reservoir of waste; it is a place where collected waste from all over a city is dumped. It serves different functions: It is helpful in the monitoring, placement, and compaction of waste, along with the installation of landfill environmental monitoring and control facilities. The importance of landfills is depicted in Figure 3.1.

3.3 Landfill Management

The management process for landfills starts right from the planning stage and ends at post-closure control of the landfill. Each and every step is crucial for managing the huge amount of waste that is dumped in the landfill every day. Proper management of a landfill includes proper planning, right design, smooth functioning, and an effective closure plan. The different steps of landfill management are shown in Figure 3.2.

FIGURE 3.1
The importance of landfill.

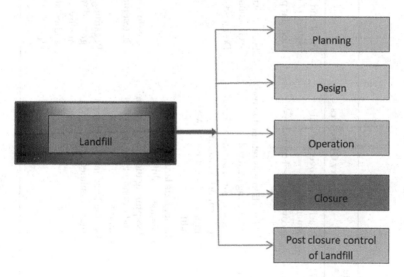

FIGURE 3.2
Different steps of landfill management.

3.4 History of Landfill Mining

The concept of the landfill is old; people have been dumping their waste into the environment whether or not they were aware of the concept of waste management. Table 3.1 shows the history of the landfill

TABLE 3.1

History of Landfill Mining

Year	Name of the Place	Primary Objectives	Equipment Used/Process Used
1953	Tel Aviv, Israel	To achieve recovery of soil amendment	Front-end loader and a clamp shell conveyers and rotating trammel screen (7 m long, 2 meters, in diameter, 13 RPM rotation)
1950–1980	The United States	Recovering steel containers	Modular processing and system
Mid 1950–1970	Barre, Massachusetts	Soil fraction was retained to use as a covering material	Section was lined prior to any additional filling. Test pits were dug, cells construction completely
Late 1960s–Early 1970	The United States	Assessment of technical feasibility of composting landfilled municipal solid waste (MSW) in situ	Designed cells in landfill
1982	Philippines	To upgrade landfill, Metro-Manila's Balut, Tondo	
1986	Thompson, Connecticut	Recapturing the landfill volume and extending the life of the landfill	Bull-dozers, pay-loaders, trucks, and a screen; excavation of about 20 test pits in the landfill
1989	Bethlehem, New Hampshire	Double-lined landfill adjacent to the old unlined one	Regarding daily cover, leachate management, testing the soil and ground water, odor management. Concerned were given to water quality management, concentration of oxygen, hydrogen, sulphide and volatile organics in the air, changes in the conductivity and pH Equipments: One front-end loader, four dump trucks, two bull dozers, one trammel screen, and one odor control sprayer

TABLE 3.2

Distribution of Landfill in Different Countries

Countries	Current Landfilling Rate
Sweden	9%
Finland	11%
Estonia	74%
Latvia	40%
Lithuania	58%
Norway	11%
Denmark	6%
United Kingdom	36%
Ireland	52%
Netherlands	3%
Belgium	8%
France	28%
Spain	43%
Portugal	38%
Germany	10%
Poland	30%
Czech Republic	30%
Austria	8%
Slovenia	12%
Slovakia	53%
Hungary	60%
Romania	49%
Croatia	63%
Italy	23%
Malta	93%
Turkey	82%
Bulgaria	88%
Serbia	88%
Cyprus	59%
Greece	79%

(http://www.enviroalternatives.com/landfill.html); Table 3.2 shows the current landfilling rate. (https://www.eurelco.org/infographic).

The rate of landfilling is quite dissimilar from one country to the next. Figure 3.3 shows the current rate of landfilling situation in the EU-28 (https://www.eurelco.org/infographic).

FIGURE 3.3
The current rate of landfilling in the EU-28.

3.5 Types of Landfill

Landfills fit in to three broad categories: municipal landfill, industrial landfill, and hazardous landfill. These categories are further divided into other subcategories (https://www.epa.gov/landfills/basic-information-about-landfills).

- **Municipal waste landfill:** This type of landfill is intended to store household waste such as paper, plastic, glass, rubber, along with hazardous materials.
- **Industrial landfill:** This type of landfill accepts commercial and industrial waste.
- **Construction and demolition (C&D) landfill:** This type of landfill is exclusively designed for C&D waste. The waste often contains bulky and weighty materials.
- **Coal combustion residual (CCR) landfill:** This type of landfill accepts coal combustion residual or coal ash. The regulation for disposal of CCR in landfills is published in federal registration 2015.
- **Polychlorinated biphenyl (PCB) landfills:** PCBs are strictly disposed of via Toxic Substance Control Act. Decontamination of PCB often does not require approval from the EPA. Figure 3.4 shows different types of landfill.

3.6 Types of Municipal Solid Landfill

There are different types of municipal solid landfill, including:

- Secure landfill: Secure landfills are designed for hazardous solid waste. A secure landfill is a type of engineered landfill that is constructed to reduce the amount of leachate and minimize the potential for environment damage.
- Mono-fill landfill: This is a type of landfill in which there is disposal of ash, C&D waste, and yard waste.
- Renewable landfill: This is a type of landfill from which materials like glass, plastics, and other combustible materials and

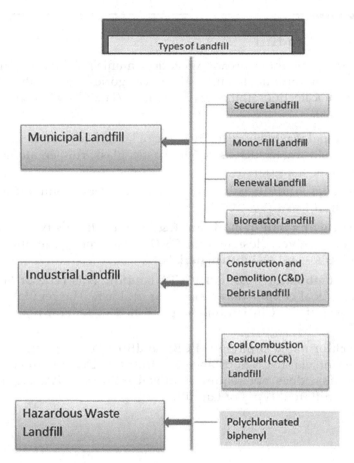

FIGURE 3.4
Types of landfill.

possible compost can be recovered. The site undergoes a lengthy stabilization period but it is very much different from the normal dried pile of waste.

- Bioreactor landfill: This is a type of waste treatment landfill that is equipped with technology that accelerates the decomposition of organic waste. It also reduces the leachate disposal cost.

The different type of municipal solid landfill is depicted in Figure 3.5.

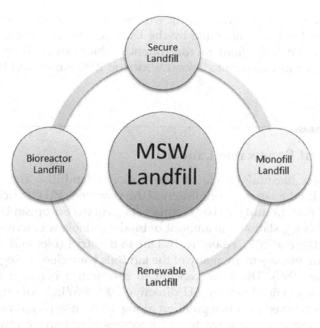

FIGURE 3.5
Types of municipal solid waste.

3.7 Renewable Landfill

MSW is highly heterogeneous in nature. The composition of MSW contains both organic and inorganic materials. The landfill in which MSW is disposed becomes a reservoir of valuables that can be used as secondary resources. Renewal landfill (RL) is a type of landfill from which material like glass, metal, potential compost, and combustible material can be recovered. Landfill mining saves space and lessens the burden of land encroachment; it is also very helpful in retrieving the material. Old landfills are unearthed and materials are segregated and categorized according to their types. The foundation of sustainable development is based on the concept of material recycling, This also helps to achieve the close loop economy. In order to achieve a closed loop economy, the RL concept can act as a solution which can reduce the pressure on the primary resources. The global population is increasing day by day; it is important for us to use our resources wisely and minimize the detrimental impacts that we are having on the world. Thus, waste material recycling is an important factor in maintaining the circular economy material loop. Land mining was introduced in 1953 at the Hiriya Landfill. This landfill is operated by the Dan Region Authority; since that time, the solution of the land encroachment problem and getting valuables from landfill mining came into force (Dickinson 1995; Krook and Baas, 2013).

3.8 Driver of Renewable Landfill

To reduce the harmful effects of conventional landfills, The regulation concerned with the same got published in the year 1999. EU Council directive 1999/31/EC was promulgated by member states of the European Union (EU). Because of this legislation, the amount of biodegradable was reduced to 50% and 35% in 2009 and 2016, respectively. Due to the strict rules and regulations for monitoring management, many of the landfills were closed (Regadio et al., 2012; Spencer 1990). The basic concept to distinguish between waste and by-products was laid down by EU directive 2008/98/EC. An end-of-waste criterion was defined in this regulation along with an explanation how and when waste ceases to be a waste and becomes a secondary raw material. Landfills in the United States are run under the guidelines of individual states' environmental agencies. With the advent of new techniques and processes, more and more recycling and material recovery are possible from heterogeneous stream of waste (Krook and Brass, 2103). Shifting the status of landfills from permanent to temporary prominent storage is achieved by implementing the concept of renewal landfill.

The concept of circular economy is satisfied as the waste degradation process occurs itself in the RL (Figure 3.6).

Waste is dumped in a landfill and is degraded under optimum conditions after stabilization of waste. It is then excavated, thus reducing the volume of the landfill and making it ready for incoming waste. Optimum biodegradation of biowaste is achieved by improvement in the water content in MSW and by adding nutrients. This has proved the most efficient option for landfill stabilization. In the initial stage, low-value material is recovered from the landfill through unearthing of land and land remediation. Unlike conventional landfill, renewable landfill assimilates material recycling and waste to energy. Thus, it increases the life span of landfill as it creates space and is helpful in reducing pollutants and land encroachment. RL is also very beneficial in recirculating landfill gases and leachate. It requires not only constant monitoring but also proper management. It is economically viable as it reduces the cost in purchasing new land for landfill.

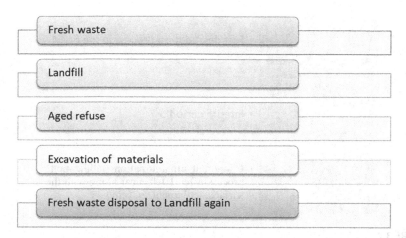

FIGURE 3.6
Waste degradation process.

3.9 Stabilization Process of RL

Stabilization time of the landfill is a critical point for RL. Waste in a landfill undergo complex degradation. The process o degradation in RL is investigated by: (i) Small scale lysimeter, (ii) Middle scale lysimeter, (iii) Field scale lysimeter (Kim and Townsend, 2012; Surmunen et al., 2008; Youcai et al., 2000, 2001, 2002; Youcai and Louchun, 2003). Both engineering and technological processes are required for stabilization of landfills. There are some materials that can bind the solid matrix, thus fixing the contaminants and not letting them percolate to the environment. Some of the bonding and filling agents are cement, fly ash, hydraulic lime, and hydrated lime. For organic contaminants like oil and PCB, hydraulic lime can be used.

Shanghai Laogang Landfill in South China was chosen for research. The landfill contained 400 million tons of waste from the past 15 years (Lou et al., 2009a, 2009b). The stabilization process was categorized into three steps based on reliable data about the leachate landfill gas and aged refuse from different years (Figure 3.7).

Phase three is ideal for material excavation. The basic idea of RL is to reuse the material and decrease the landfill volume to decrease the amount of waste at a landfill. This is also very helpful in speeding up the degradation process (Benson et al., 2007; Kai et al., 2008; Reinhart and Al-Yousfi, 1996;

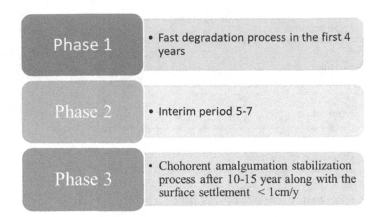

FIGURE 3.7
Different process of stabilization.

Sang et al., 2012; Valencia et al., 2009; White et al., 2011). Introduction of nutrient elements also enhances the microbial process (Qingshan et al., 1996). Injection of air into landfill maintains aeration (Marlies et al., 2013; Raga and Cossu, 2014; Rich et al., 2008; Ritzkowski and Stegmann, 2013; Pleasant et al., 2014). Addition of water and leachate circulation can balance the pH level and enhance the water content of waste. This ultimately boosts the physical, chemical, and biological degradation processes, which is essential for degradation process settlement (Benson et al., 2007; Valencia et al., 2009) NPK salts (0.24% w/w), i.e., KCl, KH_2PO_4, and NH_4CO_3, can also be used in order to stabilize the waste. They adjust and balance the nutrients level of C/N/P (100–150):5:1. Among all other compounds, KH_2PO_4 is the best one for speeding up biodegradation (Qingshan et al., 1996). Injection of air actively or passively leads to surface settlement in RL in comparison to CL under the same condition. It results in the availability of more space for the waste (Reinhart and Al-Yousfi, 1996; Sang et al., 2008). As the temperature increases due to the various major chemical activities, the leachate quantity decreases.

Additional flow of air enhances adequate aerobic decomposition of solid waste. For the growth of microorganisms, 0.06–0.94 l/min/kg of waste aeration rate is recommended. For aeration controlled amount of CO_2 is given as this gas is not economically viable. The amount of CO_2 is thus, kept below 15% (Vander Ghenyst et al., 1999).

Landfill mining or excavation is done only where the landfill is stabilized and has a stable methanogenic phase (IV) of CL (George et al., 2000) (Figure 3.8).

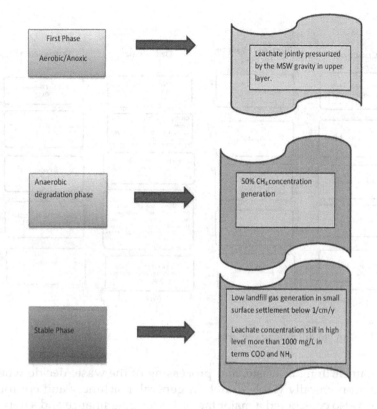

FIGURE 3.8
Different phases of stabilization.

3.10 In-situ and Ex-situ Landfill Mining Concepts

There are five methods which are suggested from different distinguished authors to manage and optimize the landfill mining process. They are shown in the Figure 3.9.

3.11 Methodology and Physical Mining Operation

The quantity and quality of waste are two of the most important factors that apply to RL. The complexity of the landfill site also decides the actual value of the waste (Fisher and Findlay, 1995; Van der Zee et al., 2004). The equipment

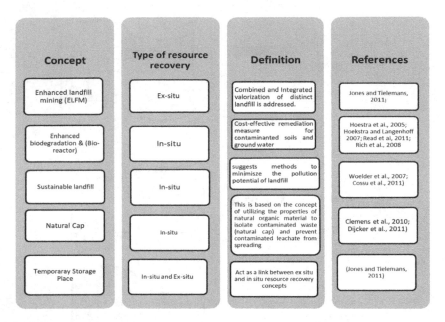

Concept	Type of resource recovery	Definition	References
Enhanced landfill mining (ELFM)	Ex-situ	Combined and Integrated valorization of distinct landfill is addressed.	Jones and Tielemans, 2011)
Enhanced biodegradation & (Bio-reactor)	In-situ	Cost-effective remediation measure for contaminanted soils and ground water	Hoestra et al., 2005; Hoekstra and Langenhoff 2007; Read et al, 2011; Rich et al., 2008
Sustainable landfill	In-situ	suggests methods to minimisze the pollution potential of landfill	Woelder et al., 2007; Cossu et al., 2011)
Natural Cap	In-situ	This is based on the concept of utilizing the properties of natural organic material to isolate contaminated waste (natural cap) and prevent contaminated leachate from spreading	Clemens et al., 2010; Dijcker et al., 2011)
Temporaray Storage Place	In-situ and Ex-situ	Act as a link between ex situ and in situ resource recovery concepts	(Jones and Tielemans, 2011)

FIGURE 3.9
In-situ and ex-situ landfill mining concepts.

selection, unearthing of waste, and processing of the waste decide whether the RL is economically viable or not. In general, machinery and equipment selection is also considered a major factor for judging finance and safety perspectives. In RL the major equipment used for the excavation process include:

- Equipment related to power generation
- Equipment related to transportation
- Conduction equipment
- Elevator conveyer belts
- Trommel screen
- Magnetic and eddy current separation machine
- Front end loader
- Equipment used to control odor (Kaartinen et al., 2013; USEPA, 1997).

3.12 Excavation Process

Landfill mining is the process where the solid wastes that were previously landfilled are excavated and processed. It is an integrated part of a renewable landfill's operational cycle. The excavated material is processed and recycled up to its maximum extent, and the land is again utilized for landfilling fresh waste. The process and techniques of excavation have not changed much for decades, and they resemble surface mining, colloquially called dig and haul. The following types of machinery are used in landfill mining:

- Cactus grab crane: It collects the waste samples up to maximum depth of landfill.
- Excavator: It uncovers the landfilled materials and places them on a conveyor belt, which takes them to sorting machinery.
- Trommel: It separates materials by size. Larger trommels separate materials like appliances and fabrics; smaller ones allow biodegradable soil fraction to pass through, leaving recyclable materials on screen to be collected.
- Front end loader: It moves sorted materials to trucks for further processing.
- Odor control sprayers: They are tractors with a cab having movable spray arm mounted on rotating platform.
- Air classifier: It separates light organic material from heavy organic material.

Recyclable materials, a combustible fraction, cover soil, and free landfill space can be recovered through this process. The recovering efficiency depends on the type of technology used for mining, the physical and chemical properties of waste, and the efficiency of applied technology. The limitations for excavation can be:

- Controlling fugitive emissions such as dust and particulates during operations.
- Opposition from communities during transportation of excavated material through populated areas.
- Selecting safer site for off-site disposal of excavated material.
- Worker safety and stewardship in case of partial excavation.

3.13 Separation

Many processes like screening, magnetic separation for recovery of ferrous and nonferrous metals, trommels, density separation, etc., are primarily used as separation techniques for excavated material. Other processes include shaker screens, tail-rotor magnets, belt magnets, eddy current separator, air knives and air classifiers, etc., but their application varies from case to case, and some of these processes have been rejected a few places due to the waste characteristics. The efficiency of separation is done through material characterization.

The first step in the separation process is to determine which materials are to be recovered. This needs a preliminary assessment of markets as well as characterization. On average, waste contains a maximum amount of fine particles that can pass through a 0.5 inch screen. The underrated organic material is about 10%, whereas unidentifiable larger items are about 18% of the waste. Notable is the presence of ferrous metals, aluminum, plastic, and glass, which are profitable to recycle. Ferrous and aluminum metals are particularly desirable because of their market costs. In comparison to ferrous metals, aluminum is difficult to separate from the mined material. The basic separation flow diagram is shown in Figure 3.10, which involves a trommel used for separation of fine particles from the complex mined material.

It is not easy to develop separation systems for waste from a newly excavated landfill without being able to custom-design, characterize, and modify the existing system and equipment. The maximum product quality cannot be known until a unit operation is designed for specific application. In a case study on enhanced landfill mining, the efficiency and sustainability of separation techniques used by two different contractors were studied on two different landfills that were partially excavated. The first landfill contained household wastes, plastics, ropes, etc.; the second had industrial materials like wires, glass, asbestos, etc. The separation technique used for the first landfill was based on separation in a wet environment consisting of floatation and wet sieving. The process started by sorting big objects manually, which was followed by metal separation by magnetic conveyors. Hydro cyclones and sieving were used to separate oil. The nonferrous fraction and the rest (glass and debris) were spitted using a mix of eddy currents and magnetic conveyors; 49.62% of combined soil fractions were derived as a result of this separation technique for this landfill. Moreover, 19.74% of organic waste and plastics, 15.23% mixed fraction of gravel and glass, 6.58% of organic waste (fine in nature), and 0.48% of metals was also derived in subsequent processes.

The separation technique for the second landfill started with manual presorting wherein the larger materials were shredded and the rest was sieved to get three different material sizes. At the later stage, refining was done by combining windshifting and sieving. At last the largest fraction was again sieved to remove smaller particles. This separation technique resulted in light (plastics) and heavy fraction around 48.37% (Weile et al., 2014).

In both cases, different separation schemes were selected based on characteristics of the excavated material.

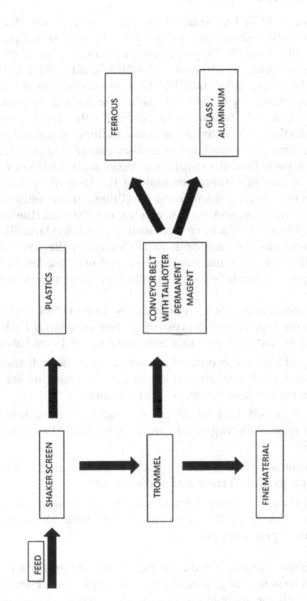

FIGURE 3.10

Separation of excavated material from landfill. (From Stessel, R.I. and Murphy, R.J., Processing of material mined from landfills, *Proceedings of the National Waste Processing Conference*, ASME, Detroit, MI, 101–111, 1991.)

3.14 Aerobic Landfill

A municipal waste landfill that transforms and stabilizes the decomposable organic waste within short time, typically 5–10 years, as compared to a conventional landfill taking 30–100 years using enhanced or accelerated biochemical processes is known as a bioreactor landfill (Reddy, 2006). Most of the time they are used as renewable landfills. Waste decomposition in renewable landfills can be accelerated by three main pathways: aerobic, anaerobic, and semi-aerobic. The pathway is determined according to the structures of landfill. All three can further be assisted by nutrient addition or leachate circulation in order to increase the overall decomposition rate (Ziyang et al., 2015).

The concept of aerobic landfills involves external recirculation of air into different heights of landfills, therefore making the landfill operate under aerobic conditions (the reverse of anaerobic conditions in conventional landfills). The waste stabilization and aerobic activity is enhanced due to the air circulation through horizontal and vertical wells. An aerobic landfill usually also involves recirculation of leachate in which leachate collected at the bottom of the landfill is collected and injected/sprayed over the landfill again. Aerobic landfills provide the following benefits over conventional landfills:

- Aerobic degradation of waste increases the rate of decomposition compared to the degradation process in conventional landfills; hence the total time period of a cycle in a renewable landfill is reduced.
- The quantity of leachate produced is comparatively less. If the process is assisted with leachate circulation, the amount of leachate actually needing treatment is many times reduced.

  It is also reported that simultaneous leachate circulation and aeration can reduce the organic loading in leachate (Hudgins and Harper, 1999.
- Aerobic decomposition of waste produces lesser quantities of methane gas than is produced from anaerobic process.
- Over the lifetime of a conventional landfill under anaerobic conditions, an aerobic landfill can be utilized multiple times due to enhanced decomposition rate.

As a result, environmental risk is reduced and the life of landfill is extended as more waste can be placed in gained airspace (Hudgins and Harper, 1999).

The limitation with the aerobic landfills is that, during forced aeration, the excess air can lead to land fires, which need to be extinguished. Slope failure may also result due to increased unit weight of the material if moisture is beyond limits during leachate recirculation. If aeration and leachate circulation is done in a controlled way, aerobic landfills can be a sustainable solution for MSW.

Similar to gas collection systems in anaerobic landfills, the air injection system in aerobic landfills is usually comprised of a network of vertical air injection wells placed in specific manner and connected to an electric blower or air compressor. "Leachate collection system" connected to a "leachate storage tank" from which it is pumped back to the landfills through injection wells (PVC well) or hoses (generally used in drip irrigation). First, injected leachate percolates through waste and gets mixed with the air (which is forced inside the waste cells) and then starts influencing the waste decomposition process inside the landfill.

In an experimental study by Hudgins and Harper (1999), two aerobic landfill systems were installed in Georgia. Prior to start-up of the air injection system, leachate was injected into waste to achieve mass moisture contents of above 60%. Leachate generated in the cell or remaining unutilized after re-injection was drained back to respective leachate collection system. Air was injected through vertical air injection wells, and rates for leachate and air injection were adjusted based on field data for adequate moisture and adequately aerated waste. It is reported that after 3–4 months after the start-up phase, aerobic conditions were stabilized. Aerobic conditions were ensured by primary and secondary data collection. Primary data includes inspection of waste visually, and measurement of landfill gases (CO_2, O_2, CH_4, and Volatile organic compounds [VOCs]) and waste mass temperatures, whereas secondary data includes leachate production and analysis (pH, Biochemical oxygen demand [BOD], Chemical oxygen demand [COD], and VOCs) (Hudgins and March, 1998) and landfill cover subsidence.

It was reported in a study period of approximately one year that methane generation was reduced drastically, in the range of 50%–90%, after installation of aerobic landfills. The leachate quantity was reduced by 86% and its BOD_5 was found to be reduced from 1100 to 500 ppm, and leachate VOCs were reduced by 75%–99%. After 11 months, the comparison between the quality of decomposed waste from aerobic and anaerobic landfills shows that, in aerobic systems, the waste was decomposed very rapidly and the quality of compost produced was comparable to the compost produced from food/vegetable waste; in anaerobic conditions, there was almost no decomposition of waste. It has been concluded in the study that under controlled aerobic conditions, aerobic landfills can effectively treat MSW sustainably.

3.15 Resource Recovery from Fine Soil-Like Materials

The waste dumped in landfills undergoes several physiochemical and biochemical changes because the composition of this waste is completely heterogeneous in nature and contains a considerable amount of trapped moisture. After few months of dumping, these landfills are comparable to biochemical reactors where all the biodegradable fraction of waste is decomposed in aerobic or anaerobic conditions.

The waste dumped in conventional landfills takes a few decades to become stabilized, whereas the same can be reduced drastically by installing an aerobic system. After stabilization, the landfills are excavated and the stabilized output is recovered. The composition of this stabilized output may vary to a great extent depending on the quality of the waste dumped. It mostly includes soil-like material (produced from degradation of organic waste), plastics, stones, glasses, metals, and other miscellaneous waste.

The key issue in assessing the economic feasibility of a landfill mining project is the possibility of recycling soil-like materials obtained by excavating and screening of old landfills. These materials include soil, garbage, sand, and other fine fractions accounting for 50%–60% of total stored wastes. The remaining includes combustible fraction (20%–30%) and approximately 10% of inorganic materials such as concrete, stone, glass, and metal (Hull and Krogmann, 2005). Dumped materials and the length of storage are responsible for the difference in composition of stored wastes. Stored wastes from landfills biodegraded for more than six years contain more soil-like materials (Hogland et al., 2004). The soil-like material acts like a biofertilizer because it includes organic matters, moisture, mineral particles, organic matters, air, and microorganisms. It is a loose material and holds fertilizer effects resembling natural black soils (Zhou et al., 2015). For a bioreactor landfill, around 50% of the stabilized waste is soil/compost mixture that can be used as cover material. A small fraction of around 10% consists of metals, which may be recovered using screening and magnetic separator for recycling. The remaining fraction contains plastics and other miscellaneous inert materials. The plastic material can be used as a fuel source or as feedstock for low-grade plastic products.

3.16 Recovery of Soil-Like Material

The excavated material of a renewable landfill can be processed to separate different materials such as plastic, metal, glass, etc., and the remaining material can be screened (by sieving) to recover the soil-like material, which is also commonly known as aged refuse of landfills. This soil-like end product of an excavated renewable landfill is a processed product of biodegradable fraction of municipal waste dumped into landfills over the years. It possesses high organic carbon content, necessary nutrient (macro and micro) value (as required for good quality compost), and high active biomass concentration. It also contains some undesirable fraction such as heavy metals.

3.17 Resource Recovery from Soil-Like Material

The soil-like material recovered from excavated renewable landfill is nutrient rich and very active in terms of microbial activity. It can be considered as potential means for resource recovery.

The stockpile of the mined waste contains different valuable materials. The fine fraction materials that are less than 40 mm in size are categorized into soil-like fine particles as mentioned earlier. Each serves its different purposes. The stockpile also contains coarser materials like light material stone and bricks. Light materials are further categorized into high- and low-grade plastic. Wood, rubber, and fiber also come under the same category and can be used as residual derived fuel (RDF). High-grade plastic granules are melted and transformed into regenerative granules. Metals are sent to recycling centers. Other coarser materials like stones, bricks, etc., are used for re-landfilling or re-construction for liner system. A detailed characterization of this soil-like material should be performed to understand its physiochemical properties; it should also be checked for its valuable applications that would make it renewable.

3.17.1 Application as a Cover Material for Other Landfills

As per standard practice for sanitary landfill, compacted waste should be covered with a daily cover of the local soil. This soil-like material can be used for daily covering of compacted waste; it not only works as local soil but also enhances the microbial activity in the dumped waste because it contains a high concentration of active mass (Xiaoli and Youcai, 2006).

3.17.2 Application in Soil Amendment/Soil Conditioning

Being rich in nutrient value, this soil-like material can be used to amend the quality of barren land or land with low nutrient value. It is also reported that, due to its structural stability, enhanced porosity, and appropriate moisture availability, this aged refuse is capable of improving soil properties, enhancing nutrient transport, and reducing soil erosion (Ziyang et al., 2015).

3.17.3 Application in Wastewater Treatment

Soil-like material recovered from renewable landfill has been reported to show good potential for wastewater treatment. The high active biomass concentration of this aged refuse shows good results as a biofilter in treating wastewater. Two mechanisms are responsible for its ability to treat wastewater: biodegradation of pollutant (by a microorganism) and adsorption of pollutants on its surface (various heavy metals can be removed by using aged refuse as adsorbent) (Xiaoli and Youcai, 2006).

3.17.4 Application in Agriculture as an Organic Fertilizer

A study done on the characterization of aged refuse reveals that Total Nitrogen (TN) concentrations in this soil-like material was found to vary between 2.30 and 5.50 mg/g, whereas the concentrations of Total Phosphorus (TP) and Total Potassium (TK) were observed as 2.13–7.43 mg/g and 23.07–38.75 mg/g, respectively. Organic matter in the aged refuse was reported between 6.32% and 11.57% (Lou, 2009). With such high nutrient value, this aged refuse can be a potential substitute for chemical fertilizer. However, a detailed characterization of this soil-like material is highly recommended before using it as a fertilizer or for any other purpose.

3.18 Conclusion

One of the major methods of MSW disposal is landfilling. By using available innovative and transformation technology, both energy and material can be excavated from the landfill. According to Krook et al. (2012), landfill mining helps to solve the issues related to landfill management. The problem linked to land encroachment and pollution is mitigated by applying landfill mining techniques. Surprisingly, little attention is given to integrated resource recovery from landfills. Modern energy recovery from landfills promotes the concept of the circular economy by producing post-consumer end-of-life products, thus closing the loop by direct recycling of pre-consumer manufacturing.

Landfill mining is beneficial in so many other ways: It helps to enhance human health and also provides safety benefits by reducing the concentration of other non-methane organic compounds (including hazardous air pollution and other VOCs). It ultimately helps to save the environment from further degradation. Landfill mining produces secondary resources and thus helps to reduce the enormous pressure on primary resources.

References

Basic Information about Landfills. https://www.epa.gov/landfills/basic-information-about-landfills, accessed May 15, 2018.

Benson, C.H., Barlaz, M.A., Lane, D.T., Rawe, J.M. 2007. Practice review of five bioreactor/recirculation landfills. *Waste Management*, 27 (1), 13–29.

Data Launched on the Landfill Situation in the EU-28. https://www.eurelco.org/infographic, accessed April 25, 2018.

Dickinson, W., 1995. Landfill mining comes of age. *Solid Waste Technology.* 9 (2), 42–47.

Dever, S.A., Swarbrick, G.E., Stuetz, R,M. 2011. Passive drainage and biofiltration of landfill gas: Results of Australian field trial. *Waste Management,* 31 (5), 1029–1048.

EPA. 2015. Municipal solid waste generation, recycling, and disposal in the United States: Facts and figures for 2012. https://www.epa.gov/sites/production/files/2015-09/documents/2012_msw_fs.pdf.

Fisher, H., Findlay, D., 1995. Exploring the economics of mining landfills. *World Waste,* 38, 50–54.

George, T., Hilary, T., Samuel, V. 2000. *Integrated Solid Waste Management,* Springer, pp. 384–387.

Hogland, W., Marques, M., Nimmermark, S. 2004. Landfill mining and waste characterization: A strategy for remediation of contaminated areas. *Journal of Material Cycles Waste Management,* 6, 119–124. doi:10.1007/s10163-003-0110-x.

Hudgins, M., Harper, S. 1999. Operational characteristics of two aerobic landfill systems, *The Seventh International Waste Management and Landfill Symposium in Sardinia,* Italy, October 4.

Hudgins, M., March, J. 1998. In-situ municipal solid waste composting using an aerobic landfill, *Composting in the Southeast, Proceedings of the 1998 Conference,* 144–158.

Hull, M.R., Krogmann, U., ASCE, M., Strom, F.P. 2005. Composition and characteristics of excavated materials from a New Jersey Landfill, *Journal of Environmental Engineering,* 131:3, 478–490, doi:10.1061/(ASCE)0733-9372(2005)131:3(478).

Jain, P., Townsend, T.G., Johnson, P. 2013. Case study of landfill reclamation at a Florida landfill site. *Waste Management,* 33 (1), 109–116.

Jones, T.P., Geysen, D., Tielemans, Y., Van Passel, S.T., Pontikes, Y., Blainpain, B., Quaghebeur, M., Hoeskstra, N. 2013. Enhanced landfill mining in view of multiple resource recovery: A critical review. *Journal of Cleaner Production,* 55 (15), 45–55.

Kaartinen, T., Sormunen, K., Rintala, J. 2013. Case study on sampling, processing and characterization of landfilled municipal solid waste in the view of landfill mining. *Journal of Cleaner Production,* 55, 56–66.

Kai, S., Matti, E., Jukka, R. 2008. Internal leachate quality in a municipal solid waste landfill: Vertical, horizontal and temporal variation and impacts of leachate recirculation. *Journal of Hazardous Materials,* 160 (2–3), 601–607.

Kim, H., Townsend, T, G., 2012. Wet landfill decomposition rate determination using methane yield results for excavated waste samples. *Waste Management,* 32 (7), 1427–1433.

Krook, J., Baas, L., 2013. Getting serious about mining the technosphere: A review of recent landfill mining and urban mining research. *Journal of Cleaner Production,* 55, 1–9.

Landfill Mining. http://www.enviroalternatives.com/landfill.html, accessed March 28, 2018.

Lou, Z. 2009. Landfill refuse stabilization process characterized by nutrient change. *Environmental Engineering Science,* 26 (11), 26–31.

Lou, Z., Zhao, Y., Chai, X., Yuan, T., Song, Y., Niu, D. 2009b. Landfill refuse stabilization process characterized by nutrient change. *Environmental Engineering Science,* 26 (11), 1655–1660.

Lou, Z., Zhao, Y., Yuan, T., Song, Y., Chen, H., Zhu, N., Huan, R. 2009a. Natural attenuation and characterization of contaminants composition in landfill leachate under different disposing ages. *Science of the Total Environment*, 407 (10), 3385–3391.

Marlies, H., Oliver, G., Marion, H.-H. 2013. Comparison between lab- and full-scale applications of in situ aeration of an old landfill and assessment of long-term emission development after completion. *Waste Management*, 33 (10), 2061–2073.

Moya, D., Aldas, C.,Lopez, G., Kaparaju, P. 2017. Municipal solid waste as a valuable renewable energy resource: A worldwide opportunity of energy recovery by using waste-to-energy technologies. *Energy Procedia*, (134), 286–295.

Municipal Solid Waste Generation, Recycling, and Disposal in the United States: Facts and Figures for 2012. https://www.epa.gov/sites/production/files/2015-09/documents/2012_msw_fs.pdf, accessed March 27, 2018.

Municipal Solid Waste Management Manual Part 1 An Overview. 2016. Central Public Health and Environmental Engineering Organization (CPHEEO). http://www.smmurban.com/uploads/swm/SWM%20Manual/Book%201.pdf.

National Association of countries waste energy recovery: Renewable energy from county landfill. https://www.naco.org/sites/default/files/documents/WasteEnergy_FINAL.pdf.

Pleasant, S., O'Donnell, A., Jon, P., Jain, P., Townsend, T. 2014 Evaluation of air sparging and vadose zone aeration for remediation of iron and manganese-impacted groundwater at a closed municipal landfill. *Science of the Total Environment*, 485–486, 31–40.

Qingshan, Z., Youcai, Z., Dimin, X. 1996. Refuse degradation and stabilizing process in municipal refuse test lysimeters. *Journal of Tongji University*, 24 (5), 596–600.

Raga, R., Cossu, R. 2014. Landfill aeration in the framework of a reclamation project in Northern Italy. *Waste Management*, 34 (3), 683–691.

Reddy, K. 2006. *Geotechnical Aspects of Bioreactor Landfills*, IGC 2006, Chennai, India, 79–94.

Regadío, M., Ruiz, A.I., De Soto, I.S., Rodriguezrastrero, M., Sanchez, N., Gismera, M.J., da Sevilla, P., Rodríguez Procopio, J., Cuevas, J. 2012. Pollution profiles and physicochemical parameters in old uncontrolled landfills. *Waste Management*, 32, 482–497.

Reinhart, D.R., Al-Yousfi, A.B. 1996. The impact of leachate recirculation on municipal solid waste landfill operating characteristics. *Waste Management & Research*, (14), 337–346.

Rich, C., Gronow, J., Voulvoulis, N. 2008. The potential for aeration of MSW landfills to accelerate completion. *Waste Management*, 28 (6) 1039–1048.

Ritzkowski, M., Stegmann, R. 2013. Landfill aeration within the scope of post-closure care and its completion. *Waste Management*, 33 (10), 2074–2082.

Sang, N, N., Soda, S., Ishigaki, T., Ike, M. 2012. Microorganisms in landfill bioreactors for accelerated stabilization of solid wastes. *Journal of Bioscience and Bioengineering*, 114 (3), 243–250.

Scarlet, N., Motola, V., Dallemand, J.F., Monforti-Ferrario, F., Mofor, L. 2015. Evaluation of energy potential of municipal solid waste from African urban areas renewable and sustainable. *Energy Reviews*, (50). 1269–1286.

Sormunen, K., Ettala, M., Rintala, J. 2008. Internal leachate quality in a municipal solid waste landfill: Vertical, horizontal and temporal variation and impacts of leachate recirculation. *Journal of Hazardous Materials*, 160 (2–3), 601–607.

Spencer, R. 1990. Landfill space reuse. *Biocycle*, 31 (2), 30–33.

Stessel, R.I., Murphy, R.J. 1991. Processing of material mined from landfills. *Proceedings of the National Waste Processing Conference*, ASME, Detroit, MI, 101–111.

Swachh Bharat Urban, Ministry of Housing and Urban Affairs, Government of India. www.swachhbharaturban.gov.in, accessed January 10, 2018.

United States Environmental Protection. 1997. Landfill reclamation, Solid Waste and Emergency Response, EPA 530-F-97-001.

Valencia, R., van der Zon, W., Woelders, H., Lubberding, H.J., Gijzen, H.J. 2009. The effect of hydraulic conditions on waste stabilisation in bioreactor landfill simulators. *Bioresource Technology*, 100 (5), 1754–1761.

VanderGheynst, J.S., Gossett, J.M., Walke, L.P. 1999. High-solids aerobic decomposition: Pilot-scale reactor development and experimentation. *Process Biochemistry*, 32, 361–375.

Van der Zee, D.J., Achterkamp, M.C., De Visser, B.J. 2004. Assessing the market opportunities of landfill mining. *Waste Management*, 24, 795–804.

Weile, K. V. D., Moerenhout, T., Marien, E. 2014. *Review on Sustainable Innovative Separation Techniques for Enhanced Landfill Mining (ELFM)*, EGU General Assembly 2014, Vienna, Austria, id. 10720.

White, J.K., Beaven, R.P., Powrie, W., Knox, K. 2011. Leachate recirculation in a landfill: Some insights obtained from the development of a simple 1-D model. *Waste Management*, 31 (6), 1210–1221.

Xiaoli, C., Youcai, Z. 2006. Adsorption of phenolic compound by aged-refuse," *Journal of Hazardous Materials*, 137, 410–417.

Youcai, Z., Liyan, S., Renhua, H., Song, L., Li, X. 2007. Recycling of aged refuse from a closed landfill. *Waste Management & Research*, 25, 130–138.

Youcai, Z., Luochun, W., Renhua, H., Dimin, X., Guowei, G. 2002. A comparison of refuse attenuation in laboratory and field scale lysimeters. *Waste Management*, 22, 29–35.

Youcai, Z., Zhugen, C., Qingwen, S., Renhua, H. 2001. Monitoring and long-term prediction for the refuse compositions and settlement in large-scale landfill. *Waste Management & Research*, 19 (2), 160–168.

Youcai, Z., Jiangying, L., Renhua, H., Guowei, G. 2000. Long-term monitoring and prediction for leachate concentrations in Shanghai refuse landfill. *Water Air Soil Pollution*, 122 (3–4), 281–297.

Zhou, C., Xu, W., Gong, Z., Fang, W., Cao, A. 2015. Characteristics and fertilizer effects of soil-like materials from landfill mining. *Clean—Soil, Air, Water*, 43 (6), 940–947.

Ziyang, L., Luochun, W., Nanwen, Z., Youcai, Z. 2015. Material recycling from renewable landfill and associated risks: A review. *Chemosphere*, 131, 91–103.

4

Composting and Environmental Benefits of Resource Recovery

Ran Xiao, Quan Wang, Junchao Zhao, Xiuna Ren, and Zengqiang Zhang

CONTENTS

4.1 Introduction

For the past few decades, dealing with the rising amount of generated organic waste has become a pressing issue due to population increases and rising living standards (Hoornweg et al., 2013). Various treatment approaches, such as landfill, incineration, pyrolysis, composting, and anaerobic digestion, have been applied for the treatment of organic waste. Among them, composting has gained increased attention due to the relatively lower investment, less technical requirements, and the production of organic fertilizer (Xiao et al., 2017). The disadvantages associated with composting are the requirements of land and time as well as potential environmental risks due to the emission of odors and greenhouse gases. Therefore, most recent studies focused on accelerating the composting process, increasing the quality of the final product (compost), and reducing the impact on the local environment.

Composting is a microorganism-mediated process that uses microbes to artificially promote the conversion of biodegradable organic matter to humus-like products (Lim et al., 2016). During this process, plant nutrients, including both macronutrients (e.g., nitrogen, phosphorus, potassium, calcium, etc.) and micronutrients (e.g., iron, boron, chlorine, zinc, etc.), were recycled and enriched in the final compost (Larney et al., 2006; Zhang et al., 2016). Due to the high organic matter and nutrient contents, compost is widely used as soil amendment or organic fertilizer in horticultural and agricultural production (Lim et al., 2016; Onwosi et al., 2017). In addition, land application of compost can also improve soil structure, which makes it easier for plants to absorb nutrients (Sánchez et al., 2017). For these reasons, composting is highly recommended for organic waste management and resource recovery worldwide (Wang et al., 2015; Onwosi et al., 2017). Furthermore, the deficiencies of plant nutrients, especially manganese, phosphorus, sulfur, zinc, boron, and molybdenum, required the recovery and recycling of such elements. Thus, a comprehensive understanding of the composting process and its function as a resource recovery method is urgently needed for the further development and implementation of composting in organic waste management.

4.2 Development of Compost Mixtures Formulation

Composting is a natural process, with the involvement of various microorganisms, including bacteria, actinobacteria, fungi, protozoa, and rotifers; when composting is done in unfavorable conditions, the process includes inadequate C:N ratio and moisture content as well as air flow (Onwosi et al., 2017). However, several drawbacks are associated with the aforementioned unfavorable conditions, including low efficiency, loss of nitrogen, odorous gas emissions, and reduced compost quality due to residual inorganic/organic pollutants or immaturity (Li et al., 2017; Onwosi et al., 2017; Xiao et al., 2017). Composting practice is driven by various microbes, so a scientific formulation of the start compost mixture or supplementation with nutrients and microorganisms can lead to enhanced composting processes as well as a high-quality compost product (Wu et al., 2016; Sánchez et al., 2017).

4.2.1 Components of the Composting Mixture

Primary substrates and additives (including bulking agents) are two main constituents of a composting mix, and the key to making great compost is the right mixture of ingredients. The primary substrate is organic waste that needs to be treated, for example, poultry manure, sewage sludge, and food waste (Onwosi et al., 2017). The primary substrates can be classified into green and brown materials according to the elemental constitution (http://www.compost-info-guide.com/index.htm). Green materials include vegetable, manure, grass clippings; they are higher in the nitrogen content. By comparison, brown materials, such as leaves, straw, and sawdust, are higher in carbon content.

A wide variety of additives have been utilized in composting mixture to obtain more efficient compost processes or a higher-quality final product (Barthod et al., 2018). Basic characteristics (i.e., nutrient contents, particle size, and moisture content, etc.) of different primary substrates are critical to microbial activities, and substrate C:N ratio and moisture content were two main characteristics that needed to be evaluated when preparing the starting material. The most common additives include zeolite, bentonite, medical stone, and lime. Biochar, the solid material produced from organic biomass through pyrolysis, has proved to be a promising compost additive (Xiao et al., 2017). The selection of the compost additive should be based on the basic physicochemical properties of the substrate. Furthermore, multiple additives can be utilized at the same time in a mixed compost, depending on the availability of additives and the properties of substrates. Furthermore, bulking agents, a special type of additive, are generally added in the composting mixture to improve the matrix physical structure so that it provides enhanced porosity and facilitates aeration, although most bulking agents are not involved in the biological reactions (Zhou et al., 2017; Rich et al., 2018). Previous studies have demonstrated that the desired air space for composting is approximately 30% (Kulcu and Yaldiz, 2007; Iqbal et al., 2010). Bulking agents promote the decomposition of compost as well as improve the quality of the end product (Bernal et al., 2009; Onwosi et al., 2017). Apart from the facilitated aeration with the addition of bulking agents. Zhou et al. (2014, 2015) demonstrated that the addition of straw materials as a bulking agent facilitated the formation of humic substance during composting due to the high amount of lignin. During the past few years, composting bulking agents have become a hot hit. Agents from various sources were analyzed for their influence on composting efficiency (Wang et al., 2016a, 2016b; Onwosi et al., 2017; Wang et al., 2017). Material availability and cost effectiveness should also be considered for industrial application.

The additives can be generally categorized into mineral additives, chemical agents, and microorganism additives.

4.2.1.1 Mineral Additives

Mineral additives have proved to be a practical method to deal with disadvantages associated with traditional composting. So far, multiple additives like zeolite, bentonite, lime, medical stone, and clay have been applied in composting to control nitrogen loss, reduce GHG emissions, and reduce the mobility of heavy metals during composting (Li et al., 2012; Wang et al., 2016a; Awasthi et al., 2016b). For example, Wang et al. (2016b) found the introduction of medical stone led to an improvement in compost quality by reducing nitrogen loss (27.9% to 48.8%) during the composting process and the leachability of Cu/Zn in the compost. Awasthi et al. (2017) found biochar addition from 2% to 18% accelerated composting process, increased the water-soluble nutrient contents and the availabilities of heavy metals (i.e., Cu, Zn, Ni, and Pb) were reduced. They also found that the mixture of biochar-zeolite addition in DFSS composting led to a reduction of CH_4 and N_2O emission by 92.85% to 95.34% and 95.14% to 97.3%, respectively (Awasthi et al., 2016c). Moreover, the addition of lime and magnesium and phosphorus salts can reduce odorous gas emissions (e.g., VFA and ammonia) as well as improve the variation of total bacteria, which in turn enhanced organic matter decomposition (Wang et al., 2018). Several studies reported that the involvement of mineral additives could control the antibiotic resistance genes and reduce nitrogen loss during composting (Chan et al., 2016; Zhang et al., 2016; Lim et al., 2017).

Overall, the addition of mineral additives is relatively cheaper and more practical compared to other methods for improving the compost, and a recent study found the addition of clay that contained over 70% montmorillonite and kaolinite amendment could prolong the thermophilic phase, reduced the maturity period of composting, and reduced N_2O by 25.3% to 63.4%, CH_4 26.01% to 50.24%, and NH_3 85% to 70.5% (Chen et al., 2018).

4.2.1.2 Chemical Agents

The introduction of chemical agents not only facilitates the composting process but also improves nutrient contents of the final product. For example, struvite crystallization process (Equation 4.1), which was normally applied in wastewater treatment for the recycling of Mg and P, was utilized in composting to form the optimal slow release N/P fertilizer (i.e., struvite, $MgNH_4PO_4 \cdot 6H_2O$) (Jeong and Kim, 2001; Lee et al., 2009).

$$Mg^{2+} + NH_4^+ + PO_4^{3-} \rightarrow MgNH_4PO_4 \cdot 6H_2O \qquad (4.1)$$

According to Equation 4.1, both NH_4^+ and PO_4^{3-} ions were reclaimed with a molar ratio of 1:1, which in turn resulted in a decreased emission of NH_3 gas and recycling of nonrenewable P resources. So far, various magnesium salts [e.g., $Mg(OH)_2$, $MgCl_2$, $MgSO_4$] and phosphorus salts (e.g., H_3PO_4, KH_2PO_4, NaH_2PO_4, and Na_2HPO_4) were used to facilitate formation of the struvite. For example, Wang et al. (2016a) found that the addition of MgO and K_2HPO_4 reduced the nitrogen loss by 27.4%–44.3% during composting. They also found that the addition of lime reduced the salinity and improved the maturity of compost. Luo et al. (2013) found that the addition of phosphogypsum and dicyandiamide resulted in a reduction in NH_3, CH_4, and N_2O emissions. Jang et al. (2002) found that the addition of glucose at 5% wt. increased the abundance of bacteria and actinomycetes by 10–100 times, and an enhanced degradation of plastic during food waste composting process was observed.

4.2.1.3 Microbial Inoculum

Composting is indeed a biological degradation of organic matters, so the diversity and abundance of microorganisms are crucial for successful composting. Even though composting can occur with naturally developed microbial biota (Manu et al., 2017), the indigenous microorganisms of natural composting are generally inadequate and easily affected by environmental factors (Tran et al., 2015). Thus, composting at natural conditions is time-consuming, and the produced compost is lower in quality due to lower nutrient contents and immaturity. In order to resolve this problem, scientists and engineers started to introduce external microbial inoculum to obtain accelerated composting. As demonstrated in various studies, the involvement of external microbial inoculum can enhance composting efficiency by rapidly decomposing stable organic components (Sarkar et al., 2010) and thus improve the maturity of compost (Wei et al., 2007). For example, Manu et al. (2017) found that the use of microbial inoculum

(EM1), which was a mixture of lactic acid bacteria, yeast, and phototrophic bacteria, reduced the whole composting period from 54 days (control) to 36 days. In addition, lignocellulose compounds were degraded at a faster rate with the involvement of inoculum compared with the control (Manu et al., 2017). Tran et al. (2015) found that the inoculation of lactic acid bacterium *Pediococcus acidilactici* (TM14) achieved faster composting by reducing the production of the detrimental acetic acid and enhancing the proliferation of thermophilic bacteria.

4.3 Critical Conditions on Monitoring and Parameter Adjustments

The physiochemical parameters of compost matrix are crucial factors that affect composting processes as well as the quality of final products. In order to accelerate composting processes, improve compost quality, and reduce the potential environmental risks, various parameters should be adjusted to optimal conditions and monitored during the composting process.

4.3.1 Temperature

Pile temperature is a parameter for monitoring composting processes (Onwosi et al., 2017). In addition, a period of high pile temperature is helpful for eliminating the existing pathogens or seeds in the substrates (Godlewska et al., 2017). Due to the activities of microorganisms, heat was generated with the decomposition of organic materials, which turned the composting process into an exothermic process. Based on the dynamic of pile temperature, composting can be categorized into four stages (i.e., mesophilic, thermophilic, cooling, and maturation stages) with the involvement of different microbes (Onwosi et al., 2017; Xiao et al., 2017). As microbial activities that are driven by mesophilic microorganisms were increased soon after compost piles are formed, the temperature of the compost piles increased rapidly to over 40°C within 24–72 hours. After that, the mesophilic microorganisms become less competitive and are replaced by others that are thermophilic, or heat-loving. The pile temperature continues to increase, and pathogen and weed seeds can be killed within three days when the pile temperature is over 55°C (Wang et al., 2016a). Then, the temperature is gradually reduced due to the consumption of liable organic materials, and the composting process goes into the maturation phase. The temperature of traditional composting cannot reach the standard in reality, which means prolonged composting processes are necessary. In addition, pathogenic bacteria cannot be eliminated at the lower temperature. Therefore, multiple additives, as illustrated below, were used to regulate the properties of compost (Xi et al., 2012; Gabhane et al., 2012; Wang et al., 2016a, 2016b; Awasthi et al., 2017; Yuan et al., 2017) in order to obtain a high temperature of composting. These aforementioned treatments could increase pile temperature with an acceleration of organic matter degradation. A high pile temperature may lead to a higher loss of nitrogen due to the volatilization of NH_3. Temperature is a critical parameter that indicates the process of organic matter degradation, so the measurement of pile temperature is important when monitoring the composting process.

4.3.2 Moisture Content

The moisture of the compost pile is another factor that significantly affects microorganism activities. The moisture content of a compost pile is related to several pile properties such as the oxygen uptake rate, free air space, microbial activities, and the temperature (Petric et al., 2012). Generally, a moisture content ranging from 50% to 60% is suitable for composting (Bernal et al., 2009). Microorganism activities are significantly restricted with low moisture content (Li et al., 2012), whereas a higher moisture content in the compost pile leads to an insufficient oxygen supply, which in turn leads to anaerobic zones and anaerobic fermentation (Mohammad et al., 2012; Wang et al., 2016b).

Due to the high moisture content, over 80%–90%, feedstocks like sewage sludge, animal manure, and olive waste are not suitable for composting. Dewater pretreatment or conditioner are thus needed before starting composting. To maintain a suitable moisture content (mainly for aeration purposes) for composting, bulk agents (woodchips, sawdust, stalks, etc.) or mineral additives (bentonite, biochar, and zeolite) are generally added to the starting materials (Awasthi et al., 2016a). Adjusting the aeration rate and the frequency of turning piles might also help to control the moisture of the composting pile (Petric et al., 2012).

4.3.3 C:N Ratio

Carbon and nitrogen are the most critical elements for composting since all living organisms require a certain amount of C and N. On the one hand, C serves as both energy source and the basic structure that accounts for about 50% of microbial cells. On the other hand, N is a necessary element for cell growth and function via the formation of proteins, nucleic acids, and enzymes (Onwosi et al., 2017). A C:N ratio ranged from 25 to 30 for the starting materials is generally suitable for successful composting (Huang et al., 2004). At a low C:N ratio, nitrogen is abundant, and the majority of nitrogen in the starting materials will be lost as ammonia gas, especially when pile pH and temperature are high, which leads to the reduction of compost quality and an increase in air pollution (Awasthi et al., 2016a). By comparison, insufficient nitrogen supply reduces the growth and proliferation of microorganisms, which significantly prolongs the composting cycle and reduces the peak temperature of piles (Petric and Mustafiće, 2015). For those reasons, the adjustment the C:N ratio in the starting material is quite important for active composting. Straws from various feedstocks, sawdust, and green wastes were commonly used as agents in adjusting compost C:N ratio in both lab-scale and industrial-scale studies (Li et al., 2012; Yang et al., 2015; Wu et al., 2016). Furthermore, a suitable C:N ratio has a positive effect on reducing the release of GHGs and ammonia (Jiang et al., 2011) as well as heavy metal availability (Wu et al., 2016). The C:N ratio gradually reduced during composting, and the investigation of the compost C:N ratio is helping in the understanding of mineralization and maturity of compost substrates (Awasthi et al., 2017).

4.4 Analysis of Raw Materials and Compost

Chemical analysis of the starting material is essential for the starting of composting since the characteristics of the raw material are closely related to the successful practice of composting. Compared with the background information from various references, laboratory analysis is more helpful to guide compost practice. Chemical analysis of the raw material and compost in the laboratory helps to guarantee the quality of the final compost. For example, investigating various pollutants in the starting material or compost products is helpful to avoid associated environmental risks. It is necessary to evaluate the fertility of compost by analyzing the nitrogen, phosphorus, potassium, and organic matter content in the compost products (Wang et al., 2016a, 2017). However, a routine analysis of compost material is generally expensive and time-consuming. Therefore, rapid tests that can be conducted on a farm scale are needed. More specifically, such tests include the analysis of moisture, pH, EC, soluble salts, and particle size distribution. Analysis of the pH value of a compost is the most common due to the simplicity of the testing procedure and low cost for the analytical equipment. The moisture content of compost can also be quickly characterized by the difference of sample weight. Detailed information about the testing procedures is provided in Table 4.1.

TABLE 4.1

Detailed Information about the Composting Index Testing

Index	Method	Note	References
Temperature	Using a probe that reaches deep into the compost, and waits until the reading is stabilized		
Moisture Content	Drying a certain amount of sample at 105°C for 24 h, and record the percent of the mass loss	Fresh sample	Chan et al. (2016)
pH and EC	Analyzing 1:10 aqueous extract (w/v) using pH meter and EC probe	Fresh sample	Waqas et al. (2018)
NH_4^+-N, NO_3^--N	Extracting sample with 2 M KCl at 1:5 (w/v) ratio, and analyzing the samples by a flow analyzer	Fresh sample	Li et al. (2012)
Organic matter	Incineration samples at 550°C in a muffle furnace for 3 h, and record the mass loss	Dried sample	FCQAO (1993)
Total P	Calorimetrically at 690 nm as molybdovanadate phosphoric acid after concentrated HNO_3–$HClO_4$ digestion	Dried sample	Li et al. (2012)
Total metal	Acid digested with a mixture of HNO_3–HCl–HF and analyzed using a flame atomic absorption spectrophotometer	Dried sample	Li et al. (2012)

4.5 Mathematical Modeling of Solid Waste Composting Process

Composting, as a biological treatment method, has been widely utilized to deal with and recycle various organic waste (Bernal et al., 2009). In order to accelerate the composting and obtain high-quality final products, composting conditions and processes need to be controlled and designed. Composting is a labor-intensive and time-consuming process, and it is also complex, involving physical, chemical, and biological processes (Petric and Selimbašić, 2008; Gabhane et al., 2012). Several factors, including the nature of starting material, composting reactor, and the ambient conditions, influence both composting reaction and product quality (Vasiliadou et al., 2015). Optimizing the composting process and evaluating the quality of compost through physiochemical and biochemical experimentations are both challenging and costly. Accordingly, mathematical modeling is a useful tool for understanding the behavior of science and engineering systems such as landfills, anaerobic digestion, and wastewater treatment (Mason, 2006; Hettiarachchi et al., 2007;

Xie et al., 2016). In the last few decades, many different kinds of mathematical models have been proposed and improved in regard to the composting process to help researchers understand the dynamics of the nutrients transformation, organic matter degradation, biomass growth, CO_2 emission, oxygen uptake rate, and so on. Mason (2006) examined different mathematical models and found that most of them could predict the temperature profiles. By analyzing organic matter decomposition in different compartments, Zhang et al. (2012) developed a novel model for the analysis of overall decomposing of organic substrates. Utilizing a simplified mathematical model, Bari and Koenig (2012) investigated the change of composting efficiency under different aeration rates. Meanwhile, some integrated mathematical models were also developed and validated for the composting process (Petric and Selimbašić, 2008; Vlyssides et al., 2009; Vidriales-Escobar et al., 2017). However, the mathematical modeling is related to the temperature, moisture content, substrate characteristics, composting reactor, and oxygen content. The different initial materials, composting process, and conditions influence the sensitivity and veracity of the mathematical modeling. Therefore, the composting conditions and relative parameters should be considered and then calibrated before simulating the model.

4.6 Composting Methods and Operational Costs

Composting methods can be generally divided into anaerobic compost and aerobic compost. Compared to anaerobic composting, aerobic composting is widely used in developing countries because it has the advantages of fast decomposition of organic matter, reduced time, and less odor emission (Villasenor et al., 2011; Awasthi et al., 2016a). Normally, the composting system can be divided into different categories according to the reactor type, material flow characteristics, type of turning, and way of oxygen supply. The most widely used past and present composting methods are shown in Table 4.2.

The choice of composting method hinges on the properties of the raw material, area, cost, and the requirement of the end product. According to the complexity of technology and usage, composting methods are divided into windrow, static pile, and reactor system.

4.6.1 Windrow Composting

Windrow composting involves pilling the mixture of starting materials into a long row (windrows), which is periodically turned over by mechanical equipment and given natural ventilation (Shi et al., 1999). The height, width, and shape of the windrow varies with the nature of the raw material and the type

TABLE 4.2

Main Composting System Classification

Openness	Stir	Air Blow	Compost Type
Open	Do not stir	No aeration	Traditional compost
		Aeration	Static compost
	Stir	No aeration	Windrow composting (natural ventilation)
		Aeration	Windrow composting (forced ventilation)
Confined	Material flow direction	Means of intervention	Compost type
		Static	Tunnel compost
	Horizontal	Stir	Stirred tank compost
		Rollover	Drum compost
	Vertical	Stir	Tower compost
		Fill	Silo compost

of dump equipment used. The stacking section can be a trapezoid, irregular quadrilateral, or triangle. Typically, the height of these piles is 2 meters and the width is 5 meters (Beck-Friis et al., 2000; Andersen et al., 2010). The length of stacking is determined according to the quantity of composting material and the actual location of the composting site. Smaller windrows mean more heat loss; thus, it would be hard for the compost to reach the hygienization requirement. However, bigger stacking generally has problems of weaker oxygen supply and odor emissions (Figure 4.1).

Windrows composting can be aerated by either natural ventilation or forced aeration. High porosity of the composting material is necessary for an effective aeration, which can be adjusted by periodically turning of the

FIGURE 4.1
Windrow composting.

compost piles or the addition of bulking agent at the start of composting. Turning the composting matrix can also control the temperature at a suitable range due to release of excessive heat from the windrow. Therefore, the temperature of windrows is the main factor in the scheduling of the turning during composting. Generally, turning is mostly performed at the early stage of composting when the active microorganism consumes organic matter and much is heat generated. The windrow composting period ranges from 3 to 9 weeks, depending on the raw materials and the frequency of turning.

4.6.2 Aerated Static Composting

Aerated static composting has a ventilation layer that contains small pieces of wood, crushed straw, or other superior air permeability materials. The raw material also keeps a proper porosity by mixing with bulking agents. Blowers are used to blow air into the composting pile to supply enough oxygen or to suction air from the pile through the ventilation layer. Aerated static composting does not need turning equipment; thus, the frequency of aeration is important for the aerated static composting. Two methods can be utilized for ventilation control. One is the time control method, which provides enough air for regular aeration to meet the need of oxygen during the composting process. The other is called the temperature control method, which controls the air blower to work or stop according to the range of set temperatures. The use of a blower for aeration also requires more calculations about suitable size, number, and type to provide enough oxygen supply. Aerated static composting usually has a top layer, which is mainly mature compost or wood dust, to reduce odor emission and to reduce water and heat loss (Figure 4.2).

FIGURE 4.2
Aerated static pile.

4.6.3 Reactor System

Since the 1980s, many countries have developed a large number of reactor composting systems. The composting reactor equipment is designed to promote microbial metabolism, improve fermentation rate, and achieve the purpose of mechanized production (Jarvis et al., 2009). There are different kinds of common reactor composting systems, some of which are described below.

4.6.3.1 Silo Compost Reactor

This reactor is a silo in which materials are fed from the top into the silo and then are discharged from the bottom. Air flows through the piles from the bottom of the silo, and waste gas is treated at the top. The typical composting cycle is 10 days, so one-tenth of the material must be collected from the top of the reactor every day. This composting reactor covers a small area, but the temperature and ventilation are not easy to control.

4.6.3.2 Drum Type Composting Reactor

The drum type composting reactor is a composting system that uses a horizontal drum to mix, ventilate, and transport materials. The drum usually has more than two warehouses, which can be divided into single-chambered or combined type. The flow direction of air is opposite to the movement of the raw material. The residence time of composting is determined by drum rotation speed and tilt angle. Additional treatment of the compost is still necessary to achieve further degradation of the organic substance. It is a proper method for dealing with some regular produced organic waste.

4.6.3.3 Stirred Tank Composting Reactor

The reactor is a horizontal stirred tank reactor that is not capped and is equipped with forced ventilation and mechanical agitation for the aim of flexible operation. The movement of the material is stirred by an agitation device, which is installed in the top of the box. The raw material in the reactor is circulated once a day, and the residence time is nearly 30 days.

4.6.3.4 Stirred Bed Composting Reactor

This blender is a composting device that moves the material from the circular perimeter to the center by agitation. Fresh material is delivered into the reactor and then mixed with the original material. Beds are constructed to facilitate aeration. Manure compost is moved to the end of the bed and collected by the transfer device. Even though the size of the reactor and the frequency of tuning determine the duration of composting, the recommended periods for commercial stirred bed composting reactors generally range from 14 to 30 days.

4.6.4 Operational Costs

The common raw materials for compost production are livestock and poultry excrement and agricultural waste. These raw materials are relatively inexpensive. However, the most important factors are the transportation cost and the difficulty of obtaining the raw materials. Many composting plants usually run some breeding plants, with the aim of reducing the cost of obtaining raw materials. However, the transportation cost of raw materials must be taken into account when evaluating operation costs. Second, composting is a time-consuming practice; thus, a piece of broad land is generally required for both feedstock stocking and composting. The area and value of land significantly affect the selection of composting methods, which should be most effectively adjusted based on the equipment and land available prior to operation. After determining the basic costs, uncontrollable factors such as labor costs, fuel, electricity, operating equipment, and maintenance used in the compost production should also be calculated, as should the compost profits based on the market conditions. Prior to implementation, on-site preparation should also include the planning costs to obtain the necessary permits, as well as the equipment installation and shipping routes. If the final product is focused on the market, additional handling such as screening and packaging, as well as compost quality testing, should also be calculated.

4.7 Quality and Use of Compost End Product

As a source of plant nutrient, the fertility of compost is an essential factor in determining compost quality. The stability and maturity of the compost are also important indexes of compost quality. Some experimental methods can be utilized to evaluate the quality, or more precisely, the maturity, of compost. Maturity characteristics include the odor of compost, temperature of compost mixture, and color of the final product. Laboratory analysis can be more precise in evaluating compost quality.

The quality of compost can be evaluated with several physiochemical parameters such as pH, organic carbon/organic nitrogen (C:N) ratio, organic matter (OM) content, humification ratio, cation exchange capacity (CEC), etc. The pH value of the compost can be used to evaluate the maturity of compost. During composting, the pH value is dynamic and generally increases with the prolonged duration, which is caused by the decomposition of organic acid and the release of ammonium. A decrease in compost pH is due to the generation of mini-molecule volatile fatty acids or the formation of carbonate (Li et al., 2012; Awasthi et al., 2016b). Normally, the pH of the final compost ranges from 5.5 to 8.5, which seems suitable for land utilization and the activity of microorganism in soil (Test Methods for the Examination of Composting and Compost (TMECC), 2002). As for the C:N ratio of a compost, a ratio within the range from 25 to 35 is believed to be suitable for the development of microorganisms (Bishop and Godfrey, 1993). The C:N ratio usually decreases as the composting process proceeds and can be a good indicator of nitrogen availability for the microorganisms and the maturity of the compost product (Bernal et al., 2009). In contrast, an increase in the C:N ratio can also be detected in the early stage of composting process, with the rapid volatilization of ammonia and the slow mineralization of organic carbon (Awasthi et al., 2016b). Satisfying mature compost requirements requires a lower C:N ratio, which is considered below 20 (Fang and Wong, 1999). The organic matter content is required to exceed 40% according to TMECC (2002).

The humus in soil is an important factor of soil fertility, so compost rich in humic substances is considered a high-quality organic fertilizer. The generation of humic-like substances during the composting process indicates the maturation of compost and the formation of stable organic matter, which can also enhance the utilization of the compost in agricultural production. During composting, the content of humic acid increased, but the amount of fulvic acid and dissolved organic carbon decreases due to microbial degradation. The parameters used for evaluation of the humification level in compost materials are shown in Table 4.3.

The increase in these parameters during composting indicates that the degree of humification is enhanced. These humic compounds have a high adsorption to positive ion compounds due to the formation of functional groups like carboxyl and phenolic. Thus, studies have shown that compost with a CEC value greater than 60 meq. 100 g^{-1} is considered manure (Iglesias-Jimenez and Perez-Garcia, 1992).

TABLE 4.3

Index for Evaluation of Compost Humification Level

Parameters	
Humification ratio (HR)	Humic substance (alkali-extractable organic carbon)/organic carbon
Humification index (HI)	Humic acid/Organic carbon
Percent of humic acids (PHA)	Humic acid/Humic substance (alkali-extractable organic carbon)
Polymerization index (PI)	Humic acid/Fulvic acid

Source: Bernal, M.P., et al., *Bioresour. Technol.*, 100, 5444–5453, 2009.

The quality of compost is quite important for its land application and agricultural performance. High-quality compost can improve soil fertility and physical structure as well as promote crop growth. However, the application of immature compost may lead to an inhibition of crop growth and increase in the risk of pests and diseases. Therefore, effective engineering and a technical evaluation system need to be established to ensure the application value of compost.

4.8 Disease Agents and Environmental Considerations

Organic wastes from various sources may contain different kinds of microorganism that might be infectious to human beings (Böhm, 2007). Health concerns related to airborne particles emitted from composting operation have increased dramatically (Harrison, 2007). Bioaerosols that consist of various microorganisms and biomolecules have become major concerns due to the potential risks to nearby residents (Wéry 2014; Douglas et al., 2017). For example, Gutarowska et al. (2016) found that the number of mesophilic microorganisms in the air around composting plants can be over $2.5 \times 10^4 \, CFU/m^3$. Meanwhile, the total amount of endotoxins in the dust ranged from 0.006 to 0.014 nmol/mg. A higher proportion of the dust with smaller particle diameter is contaminated with a microorganism (Gutarowska et al., 2016). Even worse, the contact with these airborne particles may lead to a wide range of health problems, including throat and eye irritation and even respiratory symptoms when those tiny particles penetrate deep into the lung and the alveolar sac (Pearson et al., 2015; Douglas et al., 2017). For example, Shen et al. (2004) found that exposure to *Aspergillus fumigatus*, a typical fungal pathogen from composting, resulted in invasive aspergillosis for some people.

Limited attention from researchers and governments is focused on exposure and the subsequent health problems related to bioaerosols from composting process. Therefore, most countries do not have legal regulations defining the limits of bioaerosol contents. The United Kingdom was among the first nations to regulate the atmospheric bioaerosol concentration around composting plants. The acceptable levels within 250 m from processing places were 300, 500, and 100 (CFU/m^3) for gram-negative bacteria, *Aspergillus fumigatus*, and total bacteria, respectively (Environment Agency, 2010). A similar regulation is also enforced in Germany, but no specific threshold of bioaerosol content is provided (BUNR, 2002). Apart from the lack of legal regulation, the monitoring or bioaerosols is a great challenge. The residual pathogens in the finished compost might also be related to some health problems. Even though composting can inactivate various pathogens, health risks might exist especially when the temperature or exposure time are high enough or long enough for insufficient activation of the targeted organisms

(Currie et al., 2014). Even worse, there were no specific hygienic standards so far for most countries. Therefore, practices like the validation of treatment of disinfection, continuous recording of the relevant process parameter, monitoring the change of microbiological community, and restricting the utilization of the final product should be enforced for the safe utilization of the compost (Böhm, 2007).

Apart from the health considerations, environmental impacts should also be considered. Such environmental consideration include:

1. Preventing water contamination.
2. Restricting the potential of air pollution due to odor emissions.
3. Preventing nuisances such as scavenging animals, vermin, and flies.

4.9 Overview of the Composting Practice

As demonstrated below, the enhanced biological reaction process is feasible for the management of organic wastes. Both substrate properties and environmental factors can influence the efficiency of composting (Xiao et al., 2017). Therefore, different practices, including mitigating compost mixture formulation and adapting different composting techniques, were conducted in composting to boost its efficiency. As the main product, composts can be used in different quantities and forms according to their final destination, including soil amendment and organic fertilizer. Land application of compost requires careful evaluation of the physiochemical properties of the compost by the analysis of its nutrient content, pH, salinity, and water-holding capacity.

During the beginning stages of composting, pile temperature increases with the decomposition of organic subtracts. Pathogenic bacteria and seeds are killed under the high temperature. Despite that, health risks still exist due to the residual pathogenic bacteria for the reduced duration of high temperature or maturity. The formation of bioaerosols might also impair human health, especially for those living around composting factories.

4.10 Economic Considerations

Compost is widely used as the alternative to inorganic fertilizer, and it is rich in nutrient elements for plant growth (Doan et al., 2013). Compared with other organic waste recycling treatments, the cost of compost is cheaper and the revenue is comparable with other techniques (Dimitris and Robert, 2004).

The main costs of composting were from manure collection and transportation, construction, processing, and the demands of the site. Therefore, an accelerated composting process can dramatically reduce the cost of composting. Enhanced composting efficiency and the high-quality product can also increase the benefit of composting.

In Italy, compost products are categorized into three classes: well-prepared compost, normal fertilizer, and soil conditioner. More specifically, 34%, 64% and 2% of the well-prepared composts were utilized in garden centers, as organic fertilizer, and soil amendment, respectively (Rigamonti et al., 2010). Different composting plants have different market requirements, and the kinds of goods produced and their costs are diverse. In the state of South Dakota in the United States, the shipments of compost are around 3,000–3,500 tons per year, which accounted for 30% of the annual income (Emerson, 2005). More specifically, the price of fine composting (<1 cm) was $30/ton, and the price for 0.5 cm size composting product was $25.

Kan et al. (2008) reported that the regulations set by the Israeli authorities encouraged the disposal of organic wastes by composting, which created an opportunity for additional benefits for the reuse of these wastes. The application of high-quality compost increases crop yields, reduces the cost of chemical fertilizer and pesticide applications, and significantly improves the economic benefits of agricultural production. Composting is a sustainable means of improving soil fertility in poor soil areas. It also uses irrigation water more efficiently because it increases the retention of soil moisture. Organic waste composting is an environmentally friendly engineering and technical means for agricultural production, which not only can improve resource utilization but also increase income.

4.11 Critical Research and Advanced Training Needs

Composting is becoming increasingly popular as a waste management method for treating organic waste (Li et al., 2012; Wang et al., 2016b). Through composting, organic waste is effectively disposed of and recycled with the production of value-added by-product (compost) (Bernal et al., 2009). Soil application of compost could enrich soil organic matter content and improve soil structure and other agricultural-related properties (Huang et al., 2006; Wang et al., 2017). However, several drawbacks were associated with composting, such as greenhouse gas emissions, nitrogen loss, and low efficiency of organic matter transformation. The greenhouse gas emissions and ammonia volatilization, as well as the low degree of humification, causes secondary pollution, and reduces the agricultural value and quality of the final products (Chen et al., 2010; Awasthi et al., 2016c). Therefore, mitigating greenhouse gas emissions and nitrogen loss and improving the organic matter humification have become the critical focus of research about composting. In the last decade, many practical methods such as adjusting the C:N ratio; adding mineral, chemical, and microbial additives; and controlling the aeration rate have been proposed (Gabhane et al., 2012; Yang et al., 2015; Wang et al., 2017). Doing more research to explore the mechanism of GHGs and ammonia emissions and organic matter transformation could be useful for improving composting. On the other hand, composting is also an engineering technology. Many large-scale composting plants have been set up to dispose of organic waste and produce the commercial compost (Farrell and Jones, 2009; Gutiérrez et al., 2017). The professional quality and knowledge reserves of the technicians are very important factors for maintaining the production and ensuring the product quality as well as reducing the potential environmental risks. With societal, economic, and technological advances, the problems in composting will also change. The operation skills of workers will also be higher. In this case, advanced training for an employee who works in composting plant is essential.

4.12 Current and Future Trends of Composting

So far, composting has proved to be a reliable method for organic waste management. With the rising amount of generated organic waste due to the increases in population and urbanization, composting has gained wide popularity around the world.

Because of the strict restrictions on organic waste disposal and land application, the composting industry is growing rapidly in the United States. According to Brenda and Goldstein (2014), over 19 million tons of organic waste, which included yard trimmings, food scraps, bio-solids, and other agricultural waste, were diverted to composting. Among the investigated 44 states, 4914 composting operations were operating in 2014. However, most of them were running on a small scale, with a capacity of less than 5000 tons/year, which lags far behind the rising waste volume and quantity.

With the enforcement of the European Union (EU) Landfill Directive (1999/31/EC), composting has gradually taken over landfilling as an alternative strategy for the management of biodegradable waste. Today, annual compost production has reached 12 to 15 million tons in Europe. In the United Kingdom, for example, 362 permitted operational compost sites were operating in 2014 (310 in England, 29 in Scotland, 17 in Wales, and six in Northern Ireland), with compost production at 3.21 million tons, according to the latest Waste and Resources Action Programme (WRAP) organic recycling industry status report (http://www.wrap.org.uk/content/organics-recycling-industry-status-report-2015-0, 2015). Both the number of sites and the processing capacities were significantly increased from 80 and 0.62 million tons in 1999 (Slater and Frederickson, 2001). A significant change happened in Italy: From 1993 to 1999, the compost plants have increased from 10 to 114, and the number has risen to 261 in 2015. Meanwhile, the production capacity has reached 8.1 million tons in 2015 according to European compost network country report (https://www.compostnetwork.info/).

The composting industry in China has steadily increased during the past few years. Preliminary statistics show that the number of factories related to organic fertilizer production has reached over 5506 in 2016. The industry increased at a rapid rate of about 36.8%, and the market is expected to be over $US23 billion. Despite that, most of the composting factories are running at a small scale, with annual production of about 6000 tons.

Although the composting industry is flourishing worldwide, capacity still lags far behind the generation of organic waste. Several practical methods are available for enhancing the development of this industry, and they will be discussed next.

4.12.1 Enhanced Composting Process

An accelerated composting process is required for future composting practice since composting is time-consuming as it is practiced today (Onwosi et al., 2017). The scale of a compost plant could be enlarged with the enhanced efficiency of composting, and thus more organic waste can be processed and transferred to the value-added final product. The efficiency of composting can be enhanced through the enhancement of aeration, the mitigation of feedstock C:N ratio, the addition of chemical agents, and the use of bulk materials. However, some of these methods are costly, which hinders their application beyond the laboratory. Thus, cost-effective methods that can improve the efficiency of the composting process should be developed in the future.

4.12.2 Combined Commercial Composting with On-Farm Composting

For many countries, the commercial large-scale composting factory is a pivotal strategy for the management of solid organic waste. For example, about 3.9 million tons of waste, accounting for over 40% of the total household waste in 2014 in the United Kingdom, was processed in commercial composting factories based on the 2016 European Compost Network (ECN) country reported. Recently, a shortage of composting facilities has become a major barrier to the increasing demand for the processing of increasing amounts of organic waste. Several factors should be considered for the setup of a composting factory, including the anticipated type/accessibility of raw material, the technique and equipment utilized, and the projections for growth. The desirable composting plant should locate close to the sources of organic waste, which can significantly reduce transportation costs. Another benefit of the increased size of a facility can dilute the operation cost, and larger facilities are generally better equipped to handle a wide variety of feedstocks and create a diverse number of products. More and more people are making their own compost using inexpensive compost equipment. This practice not only provides an outlet for organic waste but also generates another source of revenue. From the aspect of the ecosystem, the on-site management of organic waste and utilization of compost facilitate the balancing of plant nutrients and soil organic carbon since most of the compost product is used today as organic fertilizer or soil conditioner. As a whole, the combined commercial large-scale and on-farming processes would facilitate the management of organic waste.

4.12.3 Energy Reclaimed from Composting Processes

Even though energy recovery from composting processes dates back thousands of years, heat recovery is often overlooked in commercial composting plants. Heat recycling from the composting process, especially from larger-scale commercial composting plants, provides an alternative source for localized heating needs since the highest temperature of a compost pile is 70°C, and it is maintained for a long period of time. Heat can be collected form composting factories via several technologies (Smith et al., 2016). A recent study by Smith and Aber (2018) found that energy recovery rates can be as high as 17,700 to 32,940 kJ/h when the vapor temperature ranges between 51°C and 66°C. Generally speaking, energy recovery from the composting process depends on the heat exchanger type and facility scale. The heat recovery system is generally customized specifically to the operation's scale and site-specific characteristics, so energy recovery rates varied dramatically from site to site. Energy recovery rates are higher for commercial-scale systems and can be as high as 7084 kJ/kg DM compared with that of 1159 kJ/kg DM for lab-scale systems, according to Smith et al. (2016). In the future, more research should explore the strategy of energy recovery from the composting process.

4.13 Conclusion

Composting would be the main disposal method for the management of the increasing amount of generated organic waste in the future. Therefore, enhancing composting processes, improving compost quality, and eliminating the associated potential environmental and health risks are becoming issues to be addressed. The integrated commercial–home scale composting practice and energy reclaiming during composting would significantly reduce the cost of composting and facilitate the development of the composting industry.

References

Andersen, J.K., Boldrin, A., Samuelsson, J., Christensen, T.H., and Scheutz, C. 2010. Quantification of greenhouse gas emissions from windrow composting of garden waste. *Journal of Environmental Quality*, 39, 713–724.

Awasthi, M.K., Wang, Q., Chen, H., Wang, M., Ren, X., Zhao, J., Li, J., et al. 2017. Evaluation of biochar amended biosolids co-composting to improve the nutrient transformation and its correlation as a function for the production of nutrient-rich compost. *Bioresource Technology*, 237, 156–166.

Awasthi, M.K., Wang, Q., Huang, H., Li, R., Shen, F., Lahori, A.H., Wang, P., Guo, D., Guo, Z., Jiang, S., and Zhang, Z.Q. 2016c. Effect of biochar amendment on greenhouse gas emission and bio-availability of heavy metals during sewage sludge co-composting. *Journal of Cleaner Production*, 135, 829–835.

Awasthi, M.K., Wang, Q., Huang, H., Ren, X., Lahori, A.H., and Mahar, A. 2016a. Influence of zeolite and lime as additives on greenhouse gas emissions and maturity evolution during sewage sludge composting. *Bioresource Technology*, 216, 172–181.

Awasthi, M.K., Wang, Q., Ren, X., Zhao, J., Huang, H., Awasthi, K.S., Lahori, H.A., Li, R., Zhou, L., and Zhang, Z.Q. 2016b. Role of biochar amendment in mitigation of nitrogen loss and greenhouse gas emission during sewage sludge composting. *Bioresource Technology*, 219, 270–280.

Bari, Q. H., and Koenig, A. 2012. Application of a simplified mathematical model to estimate the effect of forced aeration on composting in a closed system. *Waste Management*, 32, 2037–2045.

Barthod, J., Rumpel, C., and Dignac, M.F. 2018. Composting with additives to improve organic amendments. A review. *Agronomy for Sustainable Development*, 38(2), 17.

Beck-Friis, B., Pell, M., Sonesson, U., Jönsson, H., and Kirchmann, H. 2000. Formation and emission of N_2O and CH_4 from compost heaps of organic household waste. *Environmental Monitoring and Assessment*, 62, 317–331.

Bernal, M.P., Alburquerque, J.A., and Moral, R. 2009. Composting of animal manures and chemical criteria for compost maturity assessment. A review. *Bioresource Technology*, 100, 5444–5453.

Bishop, P.L., and Godfrey, C. 1983. Nitrogen transformation during sewage composting. *Biocycle*, 24, 34–39.

Böhm, R. 2007. Chapter 9 Pathogenic agents. In Diaz L.F., Bertoldi de M., Bidlingmaier W., Stentiford E. (Eds.), Waste *Management Series*, 8:177–200.

Breda, P., and Nora, G. 2014. State of composting in the U.S. *Biocycle*, 55, 19.

BUNR (Federal Ministry for Environment, Nature Conservation and NuclearSafety), 2002. Erste Allgemeine VerwaltungsvorschriftzumBundes— Immissionsschutzgesetz, Technische Anleitung zur Reinhaltung der Luft— TA Luft (First General Administrative. Regulation Pertaining the Federal Mission Control Act, Technical Instructions on Air. Quality Control—TA Luft). English version, http://www.bmub.bund.de/fileadmin/bmuimport/files/pdfs/allgemein/application/pdf/taluft engl.pdf.

Chan, M.T., Selvam, A., and Wong, J.W.C. 2016. Reducing nitrogen loss and salinity during 'struvite' food waste composting by zeolite amendment. *Bioresource Technology*, 200, 838–844.

Chen, H., Awasthi, M.K., Liu, T., Zhao, J.C., Ren, X.L., Wang, M.J., Duan, Y., Awasthi, S.K., Zhang, Z.Q. 2018. Influence of clay as additive on greenhouse gases emission and maturity evaluation during chicken manure composting. *Bioresource Technology*, 266: 82–88.

Chen, Y.X., Huang, X.D., Han, Z.Y., Huang, X., Hu, B., Shi, D.Z., and Wu, W.X. 2010. Effects of bamboo charcoal and bamboo vinegar on nitrogen conservation and heavy metals immobility during pig manure composting. *Chemosphere*, 78, 1177–1181.

Currie, S.L., Beattie, T.K., Knapp, C.W., and Lindsay, D.S. 2014. Legionella spp. in UK composts: A potential public health issue? *Clinical Microbiology & Infection*, 20(4), O224–O229.

Dimitris, P.K., and Robert, K.H. 2004. Life-cycle inventory of municipal solid waste and yard waste windrow composting in the United States. *Journal of Environmental Engineering*, 130, 1390–1400.

Doan, T.T., Ngo, P.T., Rumpel, C., Nguyen, B.V., and Jouquet, P. 2013. Interactions between compost, vermicompost and earthworms influence plant growth and yield: A one-year greenhouse experiment. *Scientia Horticulturae*, 160, 148–154.

Douglas, P., Hayes, E.T., Williams, W.B., Tyrrel, S.F., Kinnersley, R.P., and Walsh, K. 2017. Use of dispersion modelling for environmental impact assessment of biological air pollution from composting: Progress, problems and prospects. *Waste Management*, 70, 22–29.

Emerson, D. 2005. Latest trends in yard trimmings composting. *Biocycle*, 46, 22.

Environment Agency. 2010. Composting and potential health effects frombioaerosols: Our interim guidance for permit applicants. Position Statement, http://organics-recycling.org.uk/uploads/article1822/Composting & bioaerosols position statement-fina 2010l%5B1%5D.pdf.

Fang, M., and Wong, J.W.C. 1999. Effects of lime amendment on availability of heavy metals and maturation in sewage sludge composting. *Environmental Pollution*, 106, 83–89.

Farrell, M., and Jones, D.L. 2009. Critical evaluation of municipal solid waste composting and potential compost markets. *Bioresource Technology*, 100, 4301–4310.

Federal Compost Quality Assurance Organization (FCQAO), 1993. Methods book for the analysis of compost-in addition with the results of the parallel interlaboratory test, Stuttgart, pp. 1–8.

Gabhane, J., William, S.P., Bidyadhar, R., Bhilawe, P., Anand, D., Vaidya, A.N., and Wate, S.R. 2012. Additives aided composting of green waste: effects on organic matter degradation, compost maturity, and quality of the finished compost. *Bioresource Technology*, 114, 382–388.

Godlewska, P., Schmidt, H.P., Yong, S.O., and Oleszczuk, P. 2017. Biochar for composting improvement and contaminants reduction. A review. *Bioresource Technology*, 246, 193–202.

Gutarowska, B., Skóra, J., Łukasz, Stępień, Szpinar, B., Otlewaka, A., and Pielech-Przybylska, K. 2016. Assessment of microbial contamination within working environments of different types of composting plants. *Journal of the Air and Waste Management Association*, 65(4), 466–478.

Gutiérrez, M.C., Serrano, A., Siles, J. A., Chica, A.F., and Martín, M.A. 2017. Centralized management of sewage sludge and agro-industrial waste through co-composting. *Journal of Environmental Management*, 196, 387–393.

Harrison E.Z. 2007. Health impact of composting air emissions. *BioCycle*, 48(11): 44.

Hettiarachchi, C.H., Meegoda, J.N., Tavantzis, J., and Hettiaratchi, P. 2007. Numerical model to predict settlements coupled with landfill gas pressure in bioreactor landfills. *Journal of Hazardous Material*, 139, 514–522.

Hoornweg, D., BhadaTata, P., and Kennedy, C. 2013. Environment: Waste production must peak this century. *Nature*, 502(7473), 615–617.

Huang, G. F., Wong, J. W. C., Wu, Q. T., Nagar, B. B. 2004. Effect of C/N on composting of pig manure with sawdust. *Waste Management*, 24(8), 805–813.

Huang, G.F., Wu, Q.T., Wong, J.W., and Nagar, B.B. 2006. Transformation of organic matter during co-composting of pig manure with sawdust. *Bioresource Technology*, 97, 1834–1842.

Iglesias-Jimenez, E., and Perez-Garcia, V. 1992. Composting of domestic refuse and sewage sludge. II. Evolution of carbon and some "humification" indexes. *Resource Conservation and Recycling*, 6, 243–257.

Iqbal, M.K., Shafiq, T., and Ahmed, K. 2010. Characterization of bulking agents and its effects on physical properties of compost. *Bioresource Technology*, 101(6), 1913–1919.

Jang, J.C., Shin P.K., Yoon, J.S., Lee, I.M., Lee, S.H., and Kim, M.N. 2002. Glucose effect on the biodegradation of plastics by compost from food garbage. *Polymer Degradation and Stability*, 76, 155–159.

Jarvis, Å., Sundberg, C., Milenkovski, S., Pell, M., Smårs, S., Lindgren, P.E., and Hallin, S. 2009. Activity and composition of ammonia oxidizing bacterial communities and emission dynamics of NH_3 and N_2O in a compost reactor treating organic household waste. *Journal of Applied Microbiology*, 106, 1502–1511.

Jeong, Y.K., and Kim, J.S. 2001. A new method for conservation of nitrogen in aerobic composting processes. *Bioresource Technology*, 79, 129–133.

Jiang, T., Schuchardt, F., Li, G., Guo, R., and Zhao, Y. 2011. Effect of C/N ratio, aeration rate and moisture content on ammonia and greenhouse gas emission during the composting. *Journal of Environmental Sciences*, 23, 1754–1760.

Kan, I., Ayalon, O., and Federman, R. 2008. Economic efficiency of compost production: The case of Israel. Hebrew University of Jerusalem, Department of Agricultural Economics and Management, 2008. https://ideas.repec.org/p/ags/huaedp/42831.html (accessed April 19, 2018).

Kulcu, R., and Yaldiz, O. 2007. Composting of goat manure and wheat straw using pine cones as a bulking agent. *Bioresource Technology*, 98(14), 2700–2704.

Larney, F.J., Sullivan, D.M., and Buckley, K.E. 2006. The role of composting in recycling manure nutrients. *Canadian Journal of Soil Science* 86(4), 97–611.

Lee, J.E., Rahman, M.M., and Ra, C.S., 2009. Dose effects of Mg and PO_4^{3-}, sources on the composting of swine manure. *Journal of Hazardous Material*, 169, 801–807.

Li, H., Duan, M., Gu, J., Zhang, Y., Qian, X., and Ma, J. 2017. Effects of bamboo charcoal on antibiotic resistance genes during chicken manure composting. *Ecotoxicology and Environmental Safety*, 140, 1–6.

Li, R., Wang, J.J., Zhang, Z.Q., Shen, F., Zhang, G., Qin, R., Li, X., and Xiao, R. 2012. Nutrient transformations during composting of pig manure with bentonite. *Bioresource Technology*, 121, 362–368.

Lim, S.L., Lee, L.H., and Wu, T.Y. 2016. Sustainability of using composting and vermicomposting technologies for organic solid waste biotransformation: Recent overview, greenhouse gases emissions and economic analysis. *Journal of Cleaner Production*, 111: 262–278.

Lim, S.S., Park, H.J., Hao, X., Lee, S.I., Jeon, B.J., Kwak, J.H., and Choi, W.J. 2017. Nitrogen, carbon, and dry matter losses during composting of livestock manure with two bulking agents as affected by co-amendments of phosphogypsum and zeolite. *Ecological Engineering*, 102, 280–290.

Luo, Y.M., Li, G.X., Luo, W.H., Schuchardt, F., Jiang, T., and Xu, D. 2013. Effect of phosphogypsum and dicyandiamide as additives on NH3, N2O and CH4 emissions during composting. *Journal of Environmental Science-China*, 25, 1338–1345.

Manu, M.K., Kumar, R., and Garg, A. 2017. Performance assessment of improved composting system for food waste with varying aeration and use of microbial inoculum. *Bioresource Technology*, 234: 167–177.

Mason, I.G. 2006. Mathematical modelling of the composting process: A review. *Waste Management*. 26, 3.

Mohammad, N., Alam, M.Z., Kabbashi, N.A., and Ahsan, A. 2012, Effective composting of oil palm industrial waste by filamentous fungi: A review. *Resources, Conservation and Recycling*, 58, 69–78.

Onwosi, C.O., Igbokwe, V.C., Odimba, J.N., Eke, I.E., Nwankwoala, M.O., and Iroh, I.N. 2017. Composting technology in waste stabilization: On the methods, challenges and future prospects. *Journal of Environmental Management*, 190, 140–157.

Pearson, C., Littlewood, E., Douglas, P., Robertson, S., Gant, T.W., Hansell, and A.L. 2015. Exposures and health outcomes in relation to bioaerosol emissions from composting facilities: A systematic review of occupational and community studies. *Journal of Toxicology and Environmental Health Part B*, 18, 43–69.

Petric, I., Helić, A., and Avdić, E.A. 2012. Evolution of process parameters and determination of kinetics for co-composting of organic fraction of municipal solid waste with poultry manure. *Bioresource Technology*, 117, 107–116.

Petric, I., and Mustafić, N. 2015. Dynamic modeling the composting process of the mixture of poultry manure and wheat straw. *Journal of Environmental Management*, 161, 392–401.

Petric, I., and Selimbašić, V. 2008. Development and validation of mathematical model for aerobic composting process. *Chemical Engineering Journal*, 139, 304–317.

Rich, N., Bharti, A., and Kumar, S. 2017. Effect of bulking agents and cow dung as inoculant on vegetable waste compost quality. *Bioresource Technology*, 252, 83.

Rigamonti, L., Grosso, M., and Giugliano, M. 2010. Life cycle assessment of subunits composing a MSW management system. *Journal of Cleaner Production*, 18, 1652–1662.

Sánchez, Ó.J., Ospina, D.A., and Montoya, S. 2017. Compost supplementation with nutrients and microorganisms in composting process. *Waste Management*, 69, 136–153.

Sarkar, S, Banerjee, R, Chanda, S, Das, P, Ganguly, S, and Pal, S. 2010. Effectiveness of inoculation with isolated Geobacillus, strains in the thermophilic stage of vegetable waste composting. *Bioresource Technology*, 101, 2892–2895.

Shen, D.K., Noodeh, A.D, Kazemi, A, Grillot, R., Roboson, G., and Brugère, J.F. 2004. Characterisation and expression of phospholipases B from the opportunistic fungus *Aspergillus fumigatus*. *FEMS Microbiology Letters*, 239(1), 87–93.

Shi, W., Norton, J.M., Miller, B.E., and Pace, M.G. 1999. Effects of aeration and moisture during windrow composting on the nitrogen fertilizer values of dairy waste composts. *Applied Soil Ecology*, 11, 17–28.

Slater, R.A., and Frederickson, J. 2001. Composting municipal waste in the UK: Some lessons from Europe. *Resources Conservation and Recycling*, 32(3), 359–374.

Smith, M.M., and Aber, J.D. 2018. Energy recovery from commercial-scale composting as a novel waste management strategy. *Applied Energy*, 211, 194–199.

Smith, M. M., Aber, J. D., and Rynk, R. 2016. Heat recovery from composting: A comprehensive review of system design, recovery rate, and utilization. *Compost Science and Utilization*, 1–12.

Test Methods for the Examination of Composts and Composting (TMECC), 2002. In Thompson, W., Leege, P., Millner, P., Watson, M.E. (Eds.), The US Composting Council, US Government Printing Office, http://tmecc.org/tmecc/index.html (accessed April 19, 2018).

Tran, Q.N.M., Mimoto, H., and Nakasaki, K. 2015. Inoculation of lactic acid bacterium accelerates organic matter degradation during composting. *Intentional Biodeterioration and Biodegradation*, 104, 377–383.

Vasiliadou, I.A., Chowdhury, A.K., Akratos, C.S., Tekerlekopoulou, A.G., Pavlou, S., and Vayenas, D.V. 2015. Mathematical modeling of olive mill waste composting process. *Waste Management*, 43, 61.

Vidriales-Escobar, G., Rentería-Tamayo, R., Alatriste-Mondragón, F., and González-Ortega, O. 2017. Mathematical modeling of a composting process in a small-scale tubular bioreactor. *Chemical Engineering Research and Design*, 120, 360–371.

Villasenor, J., Rodriguez, L., and Fernandez, F.J. 2011. Composting of domestic sewage sludge with natural zeolites in a rotary drum reactor. *Bioresource Technology*, 102, 1447–1454.

Vlyssides, A., Mai, S., and Barampouti, E. M. 2009. An integrated mathematical model for co-composting of agricultural solid wastes with industrial wastewater. *Bioresource Technology*, 100, 4797–4806.

Wang, Q., Awasthi, M. K., Zhao, J., Ren, X., Li, R., Wang, Z., Wang, M.J., and Zhang, Z.Q. 2017. Improvement of pig manure compost lignocellulose degradation, organic matter humification and compost quality with medical stone. *Bioresource Technology*, 243, 771–777.

Wang, K., He, C., You, S., Liu, W., Wang, W., Zhang, R., Qi, H., and Ren, N. 2015. Transformation of organic matters in animal wastes during composting. *Journal of Hazardous Materials*, 300, 745–753.

Wang, Q., Li, R., Cai, H., Awasthi, M.K., Zhang, Z.Q., Wang, J.J., Ali, A., and Amanulah, M. 2016a. Improving pig manure composting efficiency employing Ca-bentonite. *Ecological Engineering*, 87, 157–161.

Wang, X., Selvam, A., Lau, S.S.S., and Wong, J. 2018. Influence of lime and struvite on microbial community succession and odour emission during food waste composting. *Bioresource Technology*, 247, 652–659.

Wang, Q., Wang, Z., Awasthi, M.K., Jiang, Y., Li, R., Ren, X., Zhao, J., Shen, F., Wang, M., and Zhang, Z.Q. 2016b. Evaluation of medical stone amendment for the reduction of nitrogen loss and bioavailability of heavy metals during pig manure composting. *Bioresource Technology*, 220, 297–304.

Waqas, M., Nizami, A.S., Aburiazaiza, A.S., Aburiazaiza, A.S., Barakat, M.A., Ismail, I.M.I., and Rashid, M.I. 2018. Optimization of food waste compost with the use of biochar. *Journal of Environmental Management*, 216, 70–81.

Wei, Z., Xi, B., Zhao Y., Wang, S., Liu, H., and Jiang, Y. 2007. Effect of inoculating microbes in municipal solid waste composting on characteristics of humic acid. *Chemosphere*, 68, 368–374.

Wéry, N. 2014. Bioaerosols from composting facilities: A review. *Frontiers in Cellular and Infection Microbiology*, 4, 42.

Wu, S., Shen, Z., Yang, C., Zhou, Y., Li, X., Zeng, G., Ai, S., and He, H. 2016. Effects of C/N ratio and bulking agent on speciation of Zn and Cu and enzymatic activity during pig manure composting. *International Biodeterioration and Biodegradation*, 119, 429–436.

Xi, B.D., He, X.S., Wei, Z.M., Jiang, Y.H., Li, M.X., Li, D., Li, Y., and Dang, Q.L. 2012. Effect of inoculation methods on the composting efficiency of municipal solid wastes. *Chemosphere*, 88, 744–750.

Xiao, R., Awasthi, M.K., Li, R, Park, J.H., Scott, M.P., Wang, Q., Jim, J.W., and Zeng, Q.Z. 2017. Recent developments in biochar utilization as an additive in organic solid waste composting: A review. *Bioresource Technology*, 246, 203–213.

Xie, S., Zhan, X., Hai, F.I., Guo, W., Hao, H.N., Price, W.E., and Nghiem, L.D. 2016. Anaerobic co-digestion: A critical review of mathematical modelling for performance optimization. *Bioresource Technology*, 222, 498–512.

Yang, F., Li, G., Shi, H., and Wang, Y. 2015. Effects of phosphogypsum and superphosphate on compost maturity and gaseous emissions during kitchen waste composting. *Waste Management*, 36, 70–76.

Yuan, J., Zhang, D., Li, Y., Chadwick, D., Li, G., Li, Y., and Du, L. 2017. Effects of adding bulking agents on biostabilization and drying of municipal solid waste. *Waste Management*, 62, 52–60.

Zhang, Y., Lashermes, G., Houot, S., Doublet, J., Steyer, J.P., Zhu, Y.G., Barriuso, E., and Garnier, P. 2012. Modelling of organic matter dynamics during the composting process. *Waste Management*, 32, 19–30.

Zhang, H., Li, G., Gu, J., Wang, G., Li, Y., and Zhang, D. 2016. Influence of aeration on volatile sulfur compounds (VSCs) and NH3 emissions during aerobic composting of kitchen waste. *Waste Management*, 58, 369–375.

Zhou, C., Liu, Z., and Huang, Z.L. 2015. A new strategy for co-composting dairy manure with rice straw: Addition of different inoculate at three stages of composting. *Waste Management*, 40, 38–43.

Zhou, Y., Selvam, A., and Wong, J.W.C. 2014. Evaluation of humic substances during co-composting of food waste, sawdust and Chinese medicinal herbal residues. *Bioresource Technology*, 168, 229–234.

Zhou, Y., Selvam, A., and Wong, J. 2017. Chinese medicinal herbal residues as a bulking agent for food waste composting. *Bioresource Technology*, 249, 182–188.

5

Modern Anaerobic Digestion and Mechanism of Energy Recovery from Solid Organic Waste

Zilin Song and Gaihe Yang

CONTENTS

5.1 Introduction

The production of biogas by the anaerobic digestion (AD) of solid organic wastes is a promising alternative to fossil fuels. AD technology can generate clean energy from organic waste, it is energy efficient and environmentally beneficial, and it has great potential for development and application. Its development has some limitations, however, such as the lack of governmental support, the high cost of digestion feedstocks, lack of available technology for economical biogas production, inadequate management of biomass supply chains, and low energy efficiency. An integrated effort to comprehensively elucidate AD technology is thus urgently needed. This chapter discusses AD development, current status, biogas industries, developmental barriers, microbial functions, and mechanism. This information should improve our understanding of the history, current development, technologies, and functional mechanism of AD.

5.2 Concept of AD Development

AD is a method commonly used to treat agricultural wastes and yield renewable energy. It is widely used in the treatment of wastewater, the management of municipal waste, and the treatment of livestock manure. AD is a biological process under anaerobic conditions in which three physiological groups of microorganisms (fermenting, acidogenic, and methanogenic bacteria) decompose complex biodegradable lignocellulose to CH_4 and CO_2. AD has been extensively applied in practice to alleviate energy emergencies, but it is not a new technology. Table 5.1 outlines the development of AD. Briefly, Jan Baptist van Helmont (1580–1644) first observed in 1630 that the degradation of organic material yielded ignitable gas. Alessandro Volta (1745–1827) subsequently discovered methane in 1776 by gathering gas from Lake Maggiore in Switzerland, and John Dalton (1766–1844) confirmed its chemical composition in 1804 (Abbasi et al. 2012). Jean Louis Mouras then first defined AD in 1870, and Louis Pasteur (1822–1895) found that biogas can be utilized for heating, which was further confirmed by Donald Cameron in 1895, and for street lighting (Sárvári et al. 2016).

There was an energy shortage in the 1970s due to the petroleum crisis, and biogas development provided another choice for alternative energy sources. AD has since been widely investigated and has been developed into a technology. AD has developed rapidly in recent decades, particularly for anaerobic

TABLE 5.1

Development of Anaerobic Digestion

Year	Event	Person
1630	Observed that degradation of organic material yielded gas	Jan Baptist van Helmont
1776	Discovered methane by collecting gas	Alessandro Volta
1804	Confirmed the chemical composition of methane	John Dalton
1870	Defined anaerobic digestion	Jean Louis Mouras
1895	Stated that biogas could be used for heating	Louis Pasteur

co-digestion (ACOD). ACOD is the simultaneous digestion of two or more organic substrates. ACOD has a better C:N ratio than mono-digestion and relieves the inhibition caused by excessive free N (Mata-Alvarez et al. 2011; Xie et al. 2011). AD is now being used for the treatment of both solid organic waste and wastewater. AD has thus become an interesting alternative source of energy production due to its environmental advantages of using waste as raw material to produce biogas and producing high-quality fertilizer as its main product (Sárvári et al. 2016).

5.3 Potential of Biogas Production and Recovery

Biogas comes from the anaerobic decomposition of organic compounds and can be a substitute for natural gas and fossil fuels. Instead of treating the discarded material as waste, the solid organic matter is considered as having great potential as an energy resource in waste-to-energy technologies, such as AD fuel recovered from landfill gas. Yi et al. (2018) reported that the city of Daejeon in South Korea produced 184,686 tons/year of solid waste, with a potential recovery of methane gas of 3,105,749 Nm^3/year, equivalent to 16.8 Nm^3/ton of waste. The rate of energy recovery for methane gas is 0.177 GJ/ton for solid waste and 0.0175 GJ/ton for industrial waste.

The application of biogas also plays important roles in the reduction of emissions of greenhouse gases (GHGs). The use of straw AD can save an estimated 311.4–312.4 Yuan/year/household (Song et al. 2014) compared to the use of raw coal for fuel. In addition to the economic benefits, household savings can have large social and ecological benefits for farmers. An evaluation of the potential production of biogas from straw and the estimation of GHG reduction indicated that the direct burning of straw would produce 8.85×10^8 t CO_2/year in China. If all straw was used for AD, 12.97×10^{11} m^3 biogas/year could be produced, reducing CO_2 emissions by 6.70×10^8 tons/year (Song 2013).

High emissions of GHGs from the combustion of fossil fuels have led to a growing concern about environmental issues, so renewable fuels are urgently needed (Hosseini and Wahid 2013) as substitutes for fossil fuels. Biogas production in Europe was estimated at 6 million tons of oil equivalents in 2007 and is expected to increase to 23 million tons of oil equivalents by 2020 (Eriksson and Olsson 2007). A reduction in GHG emissions of >50% is consequently expected after 2017 (Yang et al. 2014). The European Union has set a target for the use of renewable energy systems of 20% of the European energy provision by 2020, and the proportion of vehicles using biocompressed natural gas is expected to increase globally from the current level of <2% to >25% by 2045 (Arto et al. 2016). Green and economical energy substitutes for fossil fuels are thus urgently needed (Sun et al. 2015).

5.4 Industries Using Anaerobic Digesters

Growing concerns around the world about environmental issues and the exhaustion of fossil-fuel energy are promoting the development of bio-energy industries. Most countries have proposed a series of commercial and marketing policies for the development of AD technology (Song et al. 2015). The production of biogas from organic waste by AD is cost-effective because it not only alleviates the pollution caused by excessive agricultural residues (e.g., crop straw, livestock manure, and kitchen waste) but also allows livestock and poultry farms to be self-supporting in the provision of electrical and heat energy (Insam et al. 2014). Industries using anaerobic digesters mainly include the use of the digested residues and bioenergy from biomass C during digestion. Digested residues are usually used as organic fertilizer because they contain large amounts of organic C and nutrients that are necessary for plants. The use of digestates from AD of various wastes is popular in some countries and has benefited soil quality (Juarez et al. 2013). Digestates provide components containing N, K, and P and improve soil structure by the addition of organic matter (Nkoa 2014). Digestate attributes and application frequency and soil texture determine the extent of nutrient accumulation and distribution in the soil (Stumborg et al. 2007). The use of organic waste by AD is generally an economical and environmentally friendly way to reduce the potential production of GHGs; its residues can also be used as an efficient agricultural fertilizer. Studies have demonstrated that the loss of soil nutrients (water-soluble C and N fractions) is substantially lower for digested than undigested matter, and soil organic C probably does not suffer from reduced C input compared to mineral fertilization (Insam et al. 2015).

The largest contribution of AD is the production of biogas, an alternative renewable energy to natural gas and fossil fuels. Biogas can play an important role in the developing market for renewable energy, and the global use of biogas is estimated to increase twofold in the coming years, growing from 13.0 GW in 2010 to 28.0 GW by 2020 (Maroneze et al. 2014). Biogas has high potential and a bright future for four reasons: (1) H_2 can be produced from biogas; (2) electricity can be generated, with combined heat and power production; (3) heat and steam can be produced; and (4) biogas can be used as a vehicular fuel. Biogas emits less dangerous and less hazardous material than fossil fuels into the atmosphere, such as S, Pb, and other heavy hydrocarbons (e.g., Cu and Zn). Table 5.2 shows the number of biogas plants and biomethane filling stations in European countries. Sweden uses 65% of its biogas in transport vehicles (Farzaneh-Gord and Branch 2011). Germany has allocated 46% of its biogas produced by the digestion of municipal waste into the production of heat and steam (Subramanian et al. 2013). These countries use biogas in their own ways, which could be used as examples for developing countries such as Pakistan, Malaysia, and Indonesia.

TABLE 5.2

Number of Biogas Plants and Biomethane Filling Stations in Selected European Countries

Country	Biogas Plants	Gas Filling Stations	Biogas Filling Stations
Germany	178	203	308
Italy	5	903	—
Sweden	59	190	218
Netherlands	21	150	60
UK	37	80	—
Switzerland	24	137	130

Source: Khan, I.U., et al., *Energy Convers. Manag.*, 150, 277–294, 2017.

5.5 Current Barriers to AD Deployment

Biogas has received much attention as a renewable and sustainable energy source in recent years. Biogas technology may seem to be relatively mature, but it is not yet completely feasible economically and needs considerable financial support to compete with common energy technologies based on fossil fuels (Budzianowski 2012). Four limitations are preventing the development and application of biogas technology: (1) lack of governmental support, (2) high cost of digestion feedstocks, (3) lack of available technology for economic viability, and (4) inadequate management of biomass supply chains.

Most countries and local governments provide fewer policies and less financial support for the development of, promotion of, and marketing for biogas industries compared to the energy technologies based on fossil fuels, on which governments greatly depend. Bioenergy has thus been considered noncommercial energy. Most important, some countries do not directly use existing biogas technologies from developed countries, where the technologies usually use material from crops with high costs, because such biogas systems would lead to net economic losses. Available innovation and advanced technology that could promote the cost-effective production, conditioning, and use of biogas is thus very important (Mao et al. 2015), especially for undeveloped and developing countries (Aggarangsi et al. 2013; Dong and Lu 2013; Maghanaki et al. 2013) where low-cost feedstocks are not available. The biomass supply chain for AD is another barrier to the development of the biogas industry. Agricultural residues from harvested crops may have high water content (>50%) and cannot be stored directly until properly dried, which could increase transportation costs (Holm-Nielsen et al. 2009). In addition, the origin of the biomass, particularly agricultural crops, is usually rural and can be managed solely by the local people. Most local residents, however, seem unwilling to manage agricultural residues due to the low profits. Local residents should thus be encouraged to participate in the treatment of agricultural biomass for biogas feedstocks by taking full advantage of all resources, such as sophisticated instruments, policy decisions, and business and marketing (Bacenetti et al. 2013). The biomass provision chain generally plays an important role in the continuous supply of feedstocks to the AD industry.

The low production rate of biogas is thus also a barrier to AD deployment. Many studies have been conducted to improve the energy efficiency of AD for methane generation. Mixing N- and C-rich material can optimize the C:N ratio; for example, mixing straw and livestock manure can promote microbial activity, thereby increasing biogas production. Pretreatment of cellulose-rich material is another way to improve methane production; for example, $Ca(OH)_2$ and H_2O_2 can be used as chemical agents for pretreating crop straw. These two procedures can increase AD efficiency and biogas production (Song et al. 2012). Other physical or biological methods can also be used for increasing methane yields, such as microwave treatment and the addition of white-rot fungi. These methods can stimulate biogas production to some extent.

5.6 Evolution of the Use of Ruminal Microorganisms and Other Microbes for AD

AD of lignocellulosic biomass for producing biogas has been widely applied as a promising alternative to fossil fuels. The efficiency of this technology is limited, however, due to the difficulty in degrading the components (lignin, cellulose, and hemicellulose); less available soluble compounds with low molecular weights can thus be used by anaerobic microbes (Song et al. 2012). Various pretreatments have been used to improve the efficiency of energy conversion in AD, including physical, chemical, and biological, and combinations thereof (Song et al. 2014). Biological pretreatments are considered to be more environmentally friendly and less energy intensive. The lack of a continuous supply of microbial additives and the high costs of screening render the direct use of microbial populations unfeasible. Microorganisms with high decomposition efficiencies, low costs, and environmental friendliness are needed.

Ruminal fluid in the fore-stomachs of cattle can decompose cellulose, hemicellulose, and lignin better than other anaerobic microbes (Creevey et al. 2014). The ruminal fluid has a complex microbial composition and contains a variety of microbes, such as bacteria, archaea, and fungi. The rates of decomposition of cellulose, hemicellulose, and lignin are higher for ruminal microbes than for microbes from other anaerobic digesters

(Zhang et al. 2016). Hu et al. (2006) found that the degradation of volatile solids was faster, and the yield of volatile fatty acids (VFAs) was higher, for ruminal microorganisms than for bacteria from conventional anaerobic digesters. The efficient decomposition of lignocellulosic biomass by the ruminal microbial community could thus greatly increase the rate of conversion of VFAs to methane. Furthermore, ruminal fluid is a waste product of slaughterhouses, so the use of ruminal microbes to pretreat the lignocellulosic biomass would not only eliminate the cost of disposal of ruminal fluid but also improve the energy efficiency of AD systems.

Microbes other than ruminal microorganisms, such as white-rot fungi (Zhao et al. 2014), mesophilic lignocellulolytic microbes (BYND-5) (Yan et al. 2012), and thermophilic microorganisms (MG1) (Yuan et al. 2014), have been utilized in the anaerobic fermentation of agricultural crops, kitchen wastes, and woody substances. These functional microbes can effectively increase the efficiency of decomposition of biomass and the rate of conversion of substrates to methane.

5.7 Modeling Microbial Structure in AD

Digestion involves various biochemical processes in which microbes with different functions cooperate with each other to maintain digestion. For example, acidogens responsible for the uptake of sugars and amino acids, and acetogens that degrade long-chain fatty acids (e.g., valerate, butyrate, and propionate) cooperate to produce hydrogen and acetate. Methanogens that use acetate and hydrogen as resources subsequently transform them into methane (Ramirez et al. 2009). Excessive amounts of free ammonia N could cause AD to fail (Borja et al. 1996). Instability caused by ammonia always leads to a large accumulation of VFAs, which decrease the pH and thereby influence the functioning of microbial groups. A population of acetoclastic methanogens in one study shifted from *Methanosaeta* to *Methanosarcina* during the start-up of five up-flow anaerobic reactors as ammonia concentrations increased (Calli et al. 2005).

Anaerobic Digestion Model No. 1 (ADM1) is a mathematical model, established by the International Water Association (IWA) Task Group, that focuses on the function and structure of various constituents during AD. This model has been widely applied in various digestion systems to simulate fermentation, including the co-digestion of livestock manure and agricultural crops, organic waste and activated sludge, and olive-mill wastewater and solid waste (Boubaker and Ridha 2008; Derbal et al. 2009; Sun et al. 2013). ADM1 identified interactions between the substrates, microorganisms, and products. Lübken et al. (2007) and Schoen et al. (2009) calculated the number of bacteria and archaea in the co-digestion of manure and renewable agricultural crops using ADM1 and found that *Methanosarcina* was the dominant methanogen during methanogenesis. Most studies have indicated that ADM1 could potentially simulate digestion accurately. ADM1 does not distinguish between microbes performing the same reaction, however, so all microorganisms are presumed to have the same attributes. ADM1 therefore cannot fully represent or predict experimental results due to the diversity of microbial species (Rivas-García et al. 2013).

An extended model, ADM10, was proposed by the Ramirez team to overcome the drawbacks and improve the accuracy of ADM1 by incorporating microbial diversity under both normal conditions, which do not lead to imbalances and unusual conditions. The 10 designation in this model represents a microbial diversity of 10 species. The original ADM1 contains seven biomass species (seven functional groups, one species per group) and 24 state parameters, so ADM10 includes 70 biomass species (seven functional groups, 10 species per group), for a total of 87 state parameters. The number of associated reactions has been increased from 19 to 154. The microbial groups are thus completely different between the two models, although they belong to the same genera.

Substrate characterization is the largest challenge for the application of the ADMs, despite the advantages in simulating the interactions between the substrates, microorganisms, and products in digestion. Substrate characterization requires more intensive analysis of the substrates than during normal plant operation. Biernacki et al. (2013) proposed using degradation kinetics or analyzing characteristics other than those required by ADM1. Indeed, ADM could be more easily applied in plant operations if input parameters could be found online and linked to the model.

5.8 Kinetic Modeling and the Evolution of Synergistic Impact during AD

The available kinetic models for AD can be grouped into five types: basic kinetic models, ADM1, statistical models, computational fluid dynamic (CFD) models, and other algorithmic models (Xie et al. 2016). Table 5.3 presents the advantages and disadvantages of different models. The characteristics of models probably overlap to some extent because all the mathematical models were established based on the balance of specific parameters.

5.8.1 Basic Kinetic Models

The establishment of kinetic models must depend on the growth of microbes and the rate of substrate consumption. Nutrient substrates are

TABLE 5.3

Advantages and Disadvantages of the Five Models

Model Category	Advantages	Disadvantages
Basic kinetic models	• Easy to use • Confirm the hydrolysis rate constant	• Oversimplify changes of process • Cannot provide direct practice knowledge for full-scale implementation
ADM1	• Most widely applied and recognized models in the research area • Allows the expansion and broadening of its application scope	• Need an enhanced practicality • Substrate characterization is redundant • Conversion and distribution of S, P, and N are underdeveloped
Statistical models	• Qualitative analysis to aid the design of initial conditions and parameters for optimal AD output	• Unstable reaction kinetics and inaccurate for full-scale predictions
CFD models	• Intuitive visual analysis of results • Identify the impact of mixing on AD performance • Offer optimal hydraulic design for AD systems	• Complicate numerical simulation • Unsteadiness when coupling physical process with biological process
Other algorithmic models	• Minimum requirements for the knowledge of reaction mechanisms and measurement of experimental parameters during AD	• Lack of flexibility for reactor design and scale-up • Need a input-output association to train the method for real-world scenarios

Source: Xie, S.H., et al., *Bioresour. Technol.*, **222**, 498–512, 2016.

presumed to be adequate in kinetic models, and inhibition reactions are also taken into account. Table 5.4 shows the common kinetic models for simulating AD. Kinetic models probably oversimplify rate-limiting steps because these steps can be affected by the operating parameters and are thus more likely to be inconsistent (Yu et al. 2013). The influence of inhibition by intermediates produced in the digestion on the stability of the digester is also usually neglected in basic kinetic models. Basic kinetic models are usually applied to laboratory-scale batches and cannot offer direct practical information for large-scale anaerobic plants. These disadvantages become barriers to the further practical application of such models.

TABLE 5.4

Kinetic Models of Anaerobic Co-Digestion

Model	Expression
First-order kinetic model	$\dfrac{dS}{dt} = -Ks, maxS \; M(t) = P[1-exp(-kt)]$
Monod kinetic model	$\dfrac{dS}{dt} = \dfrac{\mu_m}{Y} \dfrac{SB}{K_s + S}$
Contois kinetic model	$\dfrac{dS}{dt} = \dfrac{\mu_m}{Y} \dfrac{SB}{K_x B + S}$
Haldane kinetic model	$\dfrac{dS}{dt} = \dfrac{\mu_m}{Y} \dfrac{SB}{K_s + S + S(S/K_I)^n}$
Chen and Hashimoto model	$M(t) = P \cdot \left(1 - \dfrac{K_{CH}}{HRT \times \mu_m + K_{CH} - 1}\right)$
Modified Gompertz model	$M(t) = P \cdot exp\left\{-exp\left[\dfrac{R_{max} \cdot e}{B_0}(\lambda - t) + 1\right]\right\}$
Dual pooled first-order kinetic model	$M(t) = P \cdot \left[1 - \alpha \cdot exp(-K_f t) - (1 - \alpha) \cdot exp(-K_L t)\right]$

Note: S, substrate concentration (g/L); B, microorganism concentration (g/L); K_s, max, maximum rate of specific substrate uptake (d^{-1}); μ_m, maximum specific growth rate (h^{-1}); Y, growth yield coefficient (dimensionless); K_x, Contois kinetic constant (dimensionless); K_s, half-saturation coefficient (g/L); n, Haldane index ($n = 1$ or 2); K_i, inhibition constant (g/L); M, cumulative methane production (mL/g volatile solid); P, ultimate methane production (mL/g VS); HRT, digestion time (d); KCH, Chen and Hashimoto kinetic constant (dimensionless); k, first-order rate constant (1/d); t, digestion time (d); Rmax, maximum rate of methane yield (mL/g VS/d); k, lag time (d); e, α constant (2.7183); K_f and K_L, rate constants for rapidly degradable substrates and total degradable substrates, respectively, (d^{-1}) and α, proportion of rapidly degradable substrate to total substrate.

5.8.2 ADM1

The IWA Task Group for Mathematical Modeling of Anaerobic Digestion Processes established ADM1 to meet the demand for a generic model depicting AD (Batstone et al. 2002). This model includes five periods: disintegration, hydrolysis, acidogenesis, acetogenesis, and methanogenesis. It contains (1) 19 biochemical processes (four equations representing the degradation of particulate matter, eight equations describing the degradation of soluble matter, and seven equations representing biomass concentrations); (2) six acid/base equilibria linked with pH calculation; (3) three gas-liquid transfer processes (CH_4, CO_2, and H_2); (4) inhibitions; and (5) a number of variables, 12 of which represent particulates, 24 of which represent solute, and three of which represent gases. A large number of variables render ADM1 more comprehensive, but ADM1 neglects some processes and species associated with more specific applications, and some variables and identifiability difficulties render ADM1 more complex and difficult to operate.

5.8.3 Statistical Models

Statistical models mainly focus on the relationships between key factors (e.g., feedstocks, inoculation, C:N ratio, and solid content) and the products (CH_4 contents and reduction of volatile solids) (Xie et al. 2012). The simplex-centroid mixture design (SCMD) and the central composite design (CCD) are the two commonly used designs for statistical models. The former includes various operational parameters, and the latter contains several factors. The SCMD can establish a surface model of continuous variables to optimize the proportions of all components for a target response variable. The CCD similarly optimizes the operational parameters based on surface methodology. Functional relationships between responses (Y) and factors (X1 and X2) can be expressed by the coefficients of a second-order polynomial model using experimental data (Equation 5.1). Statistical models can design the conditions and parameters to reach optimal production for full-scale AD operation, but the accuracy of these models is sensitive to variable performance. The predicted result usually varies greatly (McLeod et al. 2015). The similarity of the reaction kinetics, design parameters, and operational additions among different digesters determines their predictive accuracy.

$$Y = \beta_0 + \beta_1 X_1 + \beta_2 X_2 + \beta_{11} X_1^2 + \beta_{22} X_2^2 + \beta_{12} X_1 X_2 \qquad (5.1)$$

5.8.4 CFD Models

CFD models provide an effective way to evaluate the volumes of the intensity of mixture based on digester structure and parameter performance. These models consist of five steps: (1) establishing the digester by the computer design program; (2) dividing the whole into smaller fractions; (3) establishing the boundary variables, including input, output, and intermediate parameters; (4) defining the attributes of the digestion phase; and (5) choosing the models to evaluate how the phases (gas, liquid, and solid) are influenced by the operational parameters and boundary variables (Lindmark et al. 2014). CFD has been widely applied to evaluate the performance of AD systems, but some limitations remain that prevent its extensive application. The complicated relationship between numerical simulation and model stability leads to an unclear relationship between the feedstocks and CH_4 production.

5.8.5 Other Algorithmic Models

Algorithmic models mainly include artificial neural networks and ant-colony-optimization (ACO). Artificial neural networks are an effective approach to investigate AD performance. This model does not require information about the relationships among critical parameters and has been used to simulate AD and optimize the biogas yield in various systems. ACO was developed based on the use of pheromones by ants to trace. Several researchers have successfully used ACO to optimize the biogas yield in systems of organic-waste digestion. Fang et al. (2009) analyzed the dynamics of kinetic parameters of digested sludge based on weighted nonlinear least-squares and an accelerating genetic algorithm. The algorithmic models need fewer parameters and only a little information about mechanisms. Their disadvantages are inflexibility and sensitivity to reactor design, requiring information on the relationship between inputs and outputs.

5.9 Conclusions

The production of biogas by the anaerobic digestion of agricultural wastes is a promising alternative to fossil fuels. This chapter discussed the development, developmental barriers, current status, biogas industries, microbial functions, and mechanism of AD. People are generally becoming more aware of the importance of AD in the alleviation of energy shortages and the reduction of environmental pollution. Most national governments have provided some financial support to promote the AD industry, such as funding for scientific experimentation, subsidies for using AD, and policy preference. Most countries and local governments offer less policy support, however, for the development, promotion, and marketing of biogas industries compared to the energy technologies based on fossil fuels, on which governments greatly depend. Furthermore, most studies of the technology and mechanism of AD have been conducted on the laboratory scale, which differs from the intermediate and large scales of biogas plants. We thus recommend greater governmental support and scientific experimentation at the intermediate and the large scale to develop further and strengthen the AD industry.

References

Abbasi, T., Tauseef, S.M., and Abbasi, S.A. 2012. *Biogas Energy*. New York: Springer, 11–23.

Aggarangsi, P., Tippayawong, N., Moran, J.C., and Rerkkriangkrai, P. 2013. Overview of livestock biogas technology development and implementation in Thailand. *Energy for Sustainable Development*, 17(4), 371–377.

Arto, I., Capellan–Perez, I., Lago, R., Bueno, G., and Bermejo, R. 2016. The energy requirements of a developed world. *Energy for Sustainable Development*, 33, 1–13.

Bacenetti, J., Negri, M., Fiala, M., and Gonzalez–Garcia, S. 2013. Anaerobic digestion of different feed stocks: Impact on energetic and environmental balances of biogas process. *Science of the Total Environment*, 463–464, 541–551.

Batstone, D.J., Keller, J., Angelidaki, I., Kalyuzhnyi, S., Pavlostathis, S., Rozzi, A., Sanders, W., Siegrist, H., and Vavilin, V. 2002. The IWA anaerobic digestion model no 1(ADM 1). *Water Science and Technology A Journal of the International Association on Water Pollution Research*, 45(10), 65–73.

Biernacki, P., Steinigeweg, S., Borchert, A., Uhlenhut, F., and Brehm, A. 2013. Application of anaerobic digestion model no. 1 for describing an existing biogas power plant. *Biomass and Bioenergy*, 59, 441–447.

Borja, R., Sanchez, E., and Weiland, P. 1996. Influence of ammonia concentration on thermophilic anaerobic digestion of cattle manure in upflow anaerobic sludge blanket (UASB) reactors. *Process Biochemistry*, 31(5), 477–483.

Boubaker, F. and Ridha, B.C. 2008. Modelling of the mesophilic anaerobic co-digestion of olive mill wastewater with olive mill solid waste using anaerobic digestion model no. 1 (ADM1), *Bioresource Technology*, 99, 6565–6577.

Budzianowski, W.M. 2012. Sustainable biogas energy in Poland: Prospects and challenges. *Renewable and Sustainable Energy Reviews*, 16(1), 342–349.

Calli, B., Mertoglu, B., Inanc, B., and Yenigun, O. 2005. Community changes during start–up in methanogenic bioreactors exposed to increasing levels of ammonia. *Environmental Technology*, 26, 85–91.

Creevey, C.J., Kelly, W.J., Henderson, G., and Leahy, S. 2014. Determining the culturabilityof the rumen bacterial microbiome. *Microbial Biotechnology*, 10, 1751–7915.

Derbal, K., Bencheikh–Iehocine, M., Cecchi, F. Meniai, A.H., and Pavan, P. 2009. Application of the IWA ADM1 model to simulate anaerobic co–digestion of organic waste with waste activated sludge in mesophilic condition. *Bioresource Technology*, 100, 1539–1543.

Dong, F., and Lu, J. 2013. Using solar energy to enhance biogas production from livestock residue–A case study of the Tongren biogas engineering pig farm in South China. *Energy*, 57, 759–765.

Eriksson, P., and Olsson, M. 2007. The potential of biogas as vehicle fuel in Europe: A technological innovation systems analysis of the emerging bio–methane technology. Chalmers University of Technology, Göteborg, Sweden, 1, 1–60.

Fang, F., Ni, B.-J., and Yu, H.Q. 2009. Estimating the kinetic parameters of activated sludge storage using weighted non–linear least–squares and accelerating genetic algorithm. *Water Research*, 43(10), 2595–2604.

Farzaneh–Gord, M., and Branch, S. 2011. Real and ideal gas thermodynamic analysis of single reservoir filling process of natural gas vehicle cylinders. *Theoretical and Applied Genetics*, 41(12), 21–36.

Holm-Nielsen, J.B., Al Seadi, T., and Oleskowicz–Popiel, P. 2009. The future of anaerobicdigestion and biogas utilization. *Bioresource Technology*, 100(22), 5478–5484.

Hosseini, S., and Wahid, M. 2013. Feasibility study of biogas production and utilization as a source of renewable energy in Malaysia. *Renewable and Sustainable Energy Reviews*, 19, 454–462.

Hu, Z.H., Yu, H.Q., and Zheng, J.C. 2006. Application of response surface methodology for optimization of acidogenesis of cattail by rumen cultures. *Bioresource Technology*, 97(16), 2103–2109.

Insam, H., Franke-Whittle, I.H., and Podmirseg, S.M. 2014. Agricultural waste management in Europe, with an emphasis on anaerobic digestion. *Journal of Integrated Field Science*, 11, 13–17.

Insam, H., Gomez–Brandon, M., and Ascher, J. 2015. Manure-based biogas fermentation residues–Friend or foe of soil fertility? *Soil Biology and Biochemistry*, 84, 1–14.

Juarez, M.F.D., Waldhuber, S., Knapp, A., Partl, C., Gomez–Brandon, M., and Insam, H. 2013. Wood ash effects on chemical and microbiological properties of digestate–and manure–amended soils. *Biology and Fertility of Soils*, 49, 575–585.

Khan, I.U., Mohd Othman, H.D., Hashim, H., Matsuura, T., Ismail, A.F., Rezaei-Dasht Arzhandi, M., and Azelee, I.W. 2017. Biogas as a renewable energy fuel: A review of biogas upgrading, utilization and storage. *Energy Conversion and Management*, 150, 277–294.

Lindmark, J., Thorin, E., BelFdhila, R., and Dahlquist, E. 2014. Effects of mixing on the result of anaerobic digestion: Review. *Renewable and Sustainable Energy Reviews*, 40, 1030–1047.

Lübken, M., Wichern, M., Schlattmann, M., and Gronauer, A. 2007. Modelling the energy balance of an anaerobic digester fed with cattle manure and renewable energy crops. *Water Research*, 41, 4085–4096.

Maghanaki, M.M., Ghobadian, B., Najafi, G., and Janzadeh Galogah, R. 2013. Potential ofbiogas production in Iran. *Renewable and Sustainable Energy Reviews*, 28, 702–714.

Mao, C.L., Feng, Y.Z., Wang, X.J., and Ren, G.X. 2015. Review on research achievements of biogas from anaerobic digestion. *Renewable and Sustainable Energy Reviews*, 45, 540–555.

Maroneze, M.M., Zepka, L.Q., Vieira, J.G., Queiroz, M.I., and Jacob–Lopes, E. 2014. Production and use of biogas in Europe: A survey of current status and perspectives. *RevistaAmbiente E Água An Interdisciplinary Journal of Applied Science*, 9(3), 445–458.

Mata–Alvarez, J., Dosta, J., Macé, S., and Astals, S. 2011. Co-digestion of solid wastes: A review of its uses and perspectives including modeling. *Critical Reviews in Biotechnology*, 31(2), 99–111.

McLeod, J.D., Othman, M.Z., Beale, D.J., and Joshi, D. 2015. The use of laboratory scale reactors to predict sensitivity to changes in operating conditions for full-scale anaerobic digestion treating municipal sewage sludge. *Bioresource Technology*, 189, 384–390.

Nkoa, R. 2014. Agricultural benefits and environmental risks of soil fertilization with anaerobic digestates: A review. *Agronomy for Sustainable Development*, 34, 473–492.

Ramirez, I., Volcke, E.I.P., Rajinikanth, R., and Steyer, J.P. 2009. Modeling microbial diversity in anaerobic digestion through an extended ADM1 model. *Water Research*, 43, 2787–2800.

Rivas-García, P., Botello-Álvarez, J.E., Estrada-Baltazar, A., and Navarrete-Bolaños, J.L. 2013. Numerical study of microbial population dynamics in anaerobic digestion through the anaerobic digestion model no. 1 (ADM1). *Chemical Engineering Journal*, 228, 87–92.

Sárvári, H.I., Tabatabaei, M., Karimi, K., and Kumar, R. 2016. Recent updates on biogas production: A review. *Biofuel Research Journal*, 10, 394–402.

Schoen, M.A., Sperl, D., Gadermaier, M., Goberna, M., Franke-Whittle, I., Insam, H., Ablinger, J., and Wett, B. 2009. Population dynamics at digester overload conditions. *BioresourceTechnology*, 100, 5648–5655.

Song, Z.L. 2013. Optimization of straw pretreatment for anaerobic fermentation and estimation of greenhouse gas emission reduction. Dissertation for Doctor Degree. Northwest A&F University, Yangling, China.

Song, Z.L., Yang, G.H., Guo, Y., and Zhang, T. 2012. Comparison of two chemical pretreatments of rice straw for biogas production by anaerobic digestion. *Bioresources*, 7, 3223–3236.

Song, Z.L., Yang, G.H., Liu, X., Yan, Z., Yuan, Y., and Liao, Y.Z. 2014. Comparison of seven chemical pretreatments of corn straw for improving methane yield by anaerobic digestion. *PLoS One*, 9(4), e93801.

Song, Z.L., Zhang, C., Yang, G.H., Feng, Y.Z., Ren, G.X., and Han, X. 2014. Comparison of biogas development from households and medium and large–scale biogas plants in rural China. *Renewable and Sustainable Energy Reviews*, 33, 204–213.

Stumborg, C., Schoenau, J.J., and Malhi, S.S. 2007. Nitrogen balance and accumulation pattern in three contrasting prairie soils receiving repeated applications of liquid swine and solid cattle manure. *Nutrient Cycling in Agroecosystems*, 78, 15–25.

Subramanian, K.A., Mathad, V.C., Vijay, V.K., and Subbarao, P.M.V. 2013. Comparative evaluation of emission and fuel economy of an automotive spark ignition vehicle fuelled with methane enriched biogas and CNG using chassis dynamometer. *Applied Energy*, 105, 17–29.

Sun, Q., Li, H., Yan, J., Liu, L., Yu, Z., and Yu, X. 2015. Selection of appropriate biogas upgrading technology–a review of biogas cleaning, upgrading and utilisation. *Renewable and Sustainable Energy Reviews*, 51, 521–532.

Sun, Y., Wang, D., and Qiao, W., Wang, W., and Zhu, T.L. 2013. Anaerobic co-digestion of municipal biomass wastes and waste activated sludge: dynamic model and material balances. *Journal of Environmental Sciences*, 25, 2112–2122.

Xie, S., Lawlor, P.G., Frost, J.P., Hu, Z., and Zhan, X. 2011. Effect of pig manure to grass silage ratio on methane production in batch anaerobic co–digestion of concentrated pig manure and grass silage. *Bioresource Technology*, 102(10), 5728–5733.

Xie, S., Wu, G., Lawlor, P.G., Frost, J.P., and Zhan, X. 2012. Methane production from anaerobic co–digestion of the separated solid fraction of pig manure with dried grass silage. *Bioresource Technology*, 104, 289–297.

Xie, S.H., Hai, F.I., Zhan, X.M., Guo, W.S., Ngo, H.H., Price, W.E., and Nghiem, L.D. 2016. Anaerobic co–digestion: A critical review of mathematical modeling for performance optimization. *Bioresource Technology*, 222, 498–512.

Yan, L., Gao, Y.M., Wang, Y.J., Liu, Q., Sun, Z.Y., Fu, B., Wen, X., Cui, Z.J., and Wang, W.D. 2012. Diversity of a mesophiliclignocellulolytic microbial consortium which isuseful for enhancement of biogas production. *Bioresource Technology*, 111, 49–54.

Yang, L., Ge, X., Wan, C., Yu, F., and Li, Y. 2014. Progress and perspectives in converting biogas to transportation fuels. *Renewable and Sustainable Energy Reviews*, 40, 1133–1152.

Yi, S., Jang, Y.-C., and An, A.K. 2018. Potential for energy recovery and greenhouse gas reduction through waste-to-energy technologies. *Journal of Cleaner Production*, 176, 503–511.

Yu, L., Ma, J., Frear, C., Zhao, Q., Dillon, R., Li, X., and Chen, S. 2013. Multiphase modeling of settling and suspension in anaerobic digester. *Applied Energy*, 111, 28–39.

Yuan, X.F., Wen, B.T., Ma, X.G., Zhu, W.B., Wang, X.F., Chen, S.J., and Cui, Z.J. 2014. Enhancing the anaerobic digestion of lignocellulose of municipal solid wasteusing a microbial pretreatment method. *Bioresource Technology*, 154, 1–9.

Zhang, H.B., Zhang, P.Y., Ye, J., Wu, Y., Fang, W., Gou, X.Y., and Zeng, G.M. 2016. Improvement of methane production from rice straw with rumenfluid pretreatment: A feasibility study. *International Biodeterioration and Biodegradation*, 113, 9–16.

Zhao, J., Zheng, Y., and Li, Y.B. 2014. Fungal pretreatment of yard trimmings forenhancement of methane yiled from solid–state anaerobic digestion. *Bioresource Technology*, 156, 176–181.

6

Greenhouse Gas Emissions through Biological Processing of Solid Waste and Their Global Warming Potential

Mukesh Kumar Awasthi, Hongyu Chen, Sanjeev Kumar Awasthi, Tao Liu, Meijing Wang, Yumin Duan, and Jiao Li

CONTENTS

6.1 Introduction

With the rapid development of industry and improvements in living standards around the world, the amount of solid waste has increased year by year and has become an urgent problem to be solved around the world. Solid waste usually refers to the solid, semisolid, or muddy material that is human generated and is discarded at a certain time duration and place. It is estimated that the global annual increase is about 10 billion tons of solid waste, or 2 tons per capita (Dong et al. 2003). At present, China's industrial solid waste accumulation has reached 5.92 billion tons, covering 55000 ha, of which 3700 ha is farmland, showing that the increase in solid waste has brought great harm to the environment and become one of the culprits of environmental pollution (Li et al. 2015). Solid waste accumulates in large amounts and covers a wide area. Because of the migration of harmful components and its residual ingredients of waste water and waste gas in the treatment are often converted into solid form, which causing that solid waste is in a sense of the existence of harmful components of final state. Hazardous substances in solid waste are not easily decomposed and causes many harmful effects. During the past decades, we have paid attention only to the prevention of water and gas pollution, coupled with the imperfect legal system and management as well as free disposal of these solid waste, some even directly into river, lake and sea which caused serious pollution.

Solid waste treatment usually refers to the process of converting solid waste into suitable form like waste to energy and fertilizer to make it easier to transport, store, use, or dispose of through the physical, chemical, biological, and biochemical methods. The goal of solid waste treatment is to conversion of biomass into harmless end product. General techniques for handling solid waste pollution include sorting, compaction, curing, biological treatment, incineration, pyrolysis, etc. These treatments produce varying degrees of greenhouse gases (GHGs).

The most of gases emit from solid waste treatment and that have the most influence on the global warming effect are CH_4, CO_2, water vapor, CO, and ozone, etc., where carbon dioxide, methane, and chlorofluorocarbons (CFCs) constitute the important contributors, and carbon dioxide is the major component. Methane emissions are gradually increasing; it is expected that the 20-year methane concentration will reach 2.34 ppm V, at which time, methane is anticipated to be the ultimate GHG (Yang et al. 2013). Meanwhile, the input of CO_2 and CH_4 to the GHGs effect is more than 75%. The harmful effect of CH_4 to GHGs is 21 times of that of CO_2, and its amount in the atmosphere is significantly less than that of CO_2, but the increasing rate is modestly greater than other gases. Landfill is one of the major sources of methane emissions. Other biological treatment methods also produce GHGs. The influence of biological processing of solid waste on GHG emissions reduction is introduced in this chapter.

6.2 Methodologies for Biogenic Emissions from Selected Source Categories

6.2.1 Landfills

Landfills are a garbage disposal sites where all kinds of waste are deposited directly without prior any treatment. Normally they are placed far from human living spaces. Recent landfill disposal is a type of modern technology that is different from conventional landfills. The screening of landfill sites is necessary to alleviate the many environmental issues, like leakage of leachate that contaminates groundwater, odor problem, and air pollution. Sanitary conditions require aerobic, anaerobic, and semi-aerobic landfill. Normally, the site developed for anaerobic landfill should be very simple and relatively inexpensive. The biogas produced from landfills can be reused, and it is therefore widely exploited. However, aerobic and semi-aerobic landfills have rigorous site requirements; thus, the price is greater than for anaerobic landfill, but mineralization of organic matter is rapid and considerably reduces the time duration of waste maturation, and it is also a constructive influence on quality (Bogner et al. 2008). Hence, there are many technique flaws, but landfill technology is still effective for large scale waste disposal in big cities. Landfill types are shown in Table 6.1.

First, it is difficult to control the contamination produced by this method of waste disposal. For solid waste landfills, leachate precipitation can cause air and water pollution. Landfill are very hard to maintain during heavy rain and the rainy season, so leachate is unavoidable. Meanwhile, characteristics of landfill leachate are complex, so it is difficult to modify and dispose of it. Second, landfills require a lot of land, and the land surrounding landfills must also be considered. Third, landfills require a long-term commitment. They need to be strictly managed during operation and after closure. Most domestic landfill tend to establish landfill gases evacuation to streamlined the transfer of landfill gases, which the main component is methane so as to obstruct the deposition of landfill gases, prevent outburst and the interactive landfill and odorous gases remedy system.

6.2.2 Composting Operations

Composting is the ecofriendly mineralization and conversion of organic matter into a stable end product call humus by the action of microorganisms under aerobic or anaerobic conditions. Through their own enzymatic activities, microorganisms oxidize a portion of organic matter into simple inorganic substances; At the same time, energy is released and utilized by the group of microorganisms, and the other group of the organism is synthesized into new cytoplasm, which promotes the multiplication and reproduction of

TABLE 6.1

Landfill Types and Its Characteristics and Application

Landfill Type	Characteristics and Application
Anaerobic landfill	Simple operation, low construction costs, recyclable methane gas, widely used
Aerobic and semi-aerobic landfill	Decomposition speed fast, garbage stabilization time is short, but the process requirements are more complex, high cost, still in the research stage

microbes and reproduces more organisms. (Sánchez et al. 2015). Sometimes, anaerobic microorganisms have played a significant role in absolute mineralization and the conversion of organic matter into stable end product. Several of developing and developed nations have applied to mix composting or anaerobic digested residues with other kinds of biodegradable waste such as food waste from kitchen or restaurant, garden or yard organic waste and agricultural waste mixed. Both technologies are best practices for separating and managing the waste fractions.

During composting, organic waste is aerobically decomposed and converted into CO_2, water vapor, and a humified end product called compost; however, a considerable amount of nutrients is retained in the residual compost. With the rapid advancement of industrial technology, composting has become an environmentally feasible method for organic waste recycling. The kinds of organic waste that can used for composting are listed in Table 6.2. The volume and water content of waste are reduced and pathogens are destroyed via composting. Microorganisms degrade highly toxic compounds into less toxic forms, depending on the temperature and soil/waste correction ratios (Blanca et al. (2005); Bo et al. (2003); Cayuela et al. (2006); Osaka /Shiga (2010); Saer et al. (2013); Topp, and Hanson (1991); Wang et al. (2014); Wentao et al. (2015); Zhanyun et al. (2014), Sanchez-Garcia et al. 2015 Antizar-Ladislao et al. 2005; Maszenan et al. 2011).

The organic contaminants rapidly mineralized in the first phase of composting produce heat through microbial metabolism, which facilitates the destruction of the pathogenic microbes and weed seeds. During the thermophilic phase, however, the maximum amount of mineralization and microbial biomass are found. In addition, the composting of the compost material with noncompost waste also increases composting efficiency as previously adapted microbes enhance the degradation of contaminants during composting (Hwang et al. 2001; Maszenan et al. 2011). In addition, Maszenan et al. (2011) have reviewed and confirm that composting is a better remedy for organic solid waste management.

TABLE 6.2

Waste Types That Can be Used for Composting

Waste Type	Main Source	Focus
Urban waste	Sewage treatment sludge and urban organic waste	Heavy metals and insects
Industrial waste	Cellulose waste, high concentration of organic wastewater, fermentation industry residues (bacteria and waste materials)	
Livestock waste	Livestock and poultry manure	Stench and pathogen
Forest agricultural waste	Plant straw	
Aquatic product waste	Seaweed, fish, shrimp, crab, and other's processing waste	

6.2.3 Anaerobic Degradation

With the steady improvements in the ability of classifying waste, the quantity of generated organic waste is also increasing, which is very suitable for anaerobic digestion. Thus, the conversion of organic waste into useful end products by anaerobic treatment is also growing. Organic waste anaerobic digestion accounts for 25% of the total organic waste in Europe (Bo et al. 2003). At present, the most widely used anaerobic digestion in China is in rural area for biogas production. However, in China, except for a small amount of wastewater sludge for anaerobic treatment, the application of anaerobic digestion is quite little used to the municipal waste. Anaerobic technology has the following advantages for these biological treatment methods: Clean energy (methane) can be obtained after anaerobic digestion; the final product of digestion can be used as a high-quality organic fertilizer and soil conditioner; compared with aerobic processes, anaerobic digestion requires less oxygen and less power consumption and is therefore less expensive. Anaerobic digestion reduces GHG emissions (Ahring 1994).

Anaerobic fermentation of domestic waste is the process by which microorganisms decompose organic matter in anaerobic conditions. First, the complex organics are converted to volatile acids by rapid value-added and pH-sensitive acid bacteria through hydrolysis and fermentation. The propionic and valeric acids are oxidized to acetic acid, molecular hydrogen, and carbon dioxide by acetic acid bacteria; the methane bacteria then convert these substances into methane. Normally, the anaerobic fermentation procedure can be split into four major phases: hydrolysis, acidification, acetic acid and hydrogen production stage (acid decline stage), and the methanation stage. The biggest difference between aerobic and anaerobic is that aerobic degradation does not produce methane gas and is almost all converted to carbon dioxide gas, but AD generates approximately 60% CH_4 and 40% CO_2 (Lebrato et al. 1995).

6.2.4 Fermentation of Solid Waste

Many microorganisms play an important role in environmental pollution purification. Due to their own physiological characteristics, microorganisms can adapt to changes in the environment through variation and heredity and other biological processes. Microorganisms can use a variety of pollutants, especially organic pollutants, as sources of nutrients, through absorption, metabolism, and a series of reactions, and they can transform the pollutants into stable and harmless inorganic matter. Scientists use and exploit this function of microorganisms to deal with environmental pollution. Bioreactors can be split into four types: anaerobic, aerobic, quasi-aerobic treatment, and mixed fermentation (Chen et al. 2001).

By utilizing the metabolism of aerobic microorganisms under aerobic conditions, aerobic fermentation changes the complex organic matter of solid waste into carbon dioxide and water. The important point is to ensure adequate oxygen supply, stable temperature, and enough water. The actual project is to inject air or oxygen into the landfill so that microorganisms in the state of aerobic metabolism. Aerobic digestion significantly increases the rate of waste degradation and stability, greatly reducing the production of methane, the amount of toxic organic matter in the leachate, and the quantity of leachate (Rao et al. 2000; Zhou et al. 2004). Anaerobic fermentation is thought process, where use the anaerobic or facultative microbial metabolism under anaerobic conditions to dispose the organic waste. The main degradation products are methane and carbon dioxide, etc., but generally need to ensure that the temperature, anaerobic or low dissolved oxygen concentration maintained through the process.

The quasi-aerobic treatment process relies on the fermentation heat generated by the decomposition of garbage to cause the internal and external temperature difference to make the air flow naturally through the landfill and promote the decomposition and stability of garbage. Quasi-aerobic fermentation has the following advantages: (1) It does not require forced ventilation, saving energy; (2) generated leachate will be collected rapidly, reducing the pollution of groundwater; and (3) relative to anaerobic treatment, garbage stabilizes faster, and the production of GHGs such as CH_4, H_2S, and others is reduced.

Mixed fermentation treatments are both aerobic and anaerobic biological treatment methods. The main advantage of this method is that it combines the simple operation of the anaerobic method with the high efficiency of the aerobic method. It increases the degradation of volatile organic acids and harmful air pollutants, which its main feature is the active rate of mineralization. (Wang et al. 2013).

6.2.5 Land Treatment Units

The technology of land remediation offers an environmentally sound alternative for the management of industrial and municipal wastes, wastewater, and sludge. The instruction and guidance make contribution in site selection, waste depiction, treatment-manifestation examination; at the same time, unit design, management, and termination as well as other propositions of land- operation are beneficial for engineer and governance of land remedy units. Land treatment is defined in the Resource Conservation and Recovery Act (RCRA) of 1976 as one of the land demolition privileges for handling hazardous waste. Land treatment depends on intervention, mineralization, and immobilization of hazardous waste components within the prescribed remedy zone to confirm conservation of groundwater, surface water, and air pollution. Land remediation is the confinement of solid, liquid, and semisolid organic wastes to the upper layers, normally the top 6–12 m^2 area of a soil system, so that they can be handled safely. The objectives of land remediation are: (1) to mitigate the movement of hazardous components to off-site land and surface waters and to the atmosphere, and (2) to furnish the wastes less or nonhazardous through management.

Treatment combines the mechanism of decomposition, alteration, and immobilization of hazardous components in a soil/waste composite.

6.3 Estimating Greenhouse Gas Emissions from Various Biogenic Treatments

Landfill is the main method of solid waste disposal in China and in most countries of the world. Controlled and uncontrolled landfilling of untreated waste has become a primary source of GHGs (Huang et al. 2015). The CH_4 emissions from landfills showed that the organic waste sector took the main responsibility for GHG emissions, which emitted approximately 700 Mt CO_2-e (estimate for 2009). (Bogner et al. 2008). The organic matter of solid waste is trapped in landfills and mineralized directly by anaerobic digestion, during this process, the gases which are released from landfills are more than 90% CH_4 and CO_2, and the rest are trace gases, including H_2S, NH_3 and other gases. (Wei et al. 2009). In terms of methane alone, as an important GHG, its 100-year global warming prospects value is 25 higher that of CO_2, the rate of subscription to the greenhouse effect is about 22% (Ma et al. 2014). The quantity of CH_4 emission in landfill gas can culminate 55% to 60%, the periodical global emissions are approximately 40 Tg, the total figure of CH_4 generation from global landfill treatments is the world's third anthropogenic methane source, balancing for 13% of global methane emissions, the greenhouse impact is corresponding to 840 million tons of CO_2 (Eggleston et al. 2006). In addition to CH_4, CO_2 is the main component of landfill gas, its turnover in landfill gas can arrive at 40%–45%, CO_2 with landfill gas and CH_4 together into the atmosphere, because of its essential, it is not part of the total GHGs emissions, but it is an considerable basis of landfill gas (Ma et al. 2014). With the increasing amounts of municipal solid waste (MSW) in landfills, the fraction of methane from landfills in GHGs emitted in China would increase from 3.83% (in 2000) to 7.19% (in 2020). Therefore, its contribution to global warming can't be ignored. The GHG emission mitigation and resource exploitation of landfill CH_4 are considerable tasks for MSW management. The physical properties of each component of landfill gas are shown in Table 6.3.

As a cost-effective engineering, which can recycle organic wastes, make them harmless and convert them into stable end products, thus composting is widely acceptable for the remediation of livestock manure and sludge. After composting, organic manure and sludge can be converted to harmless and stable mature compost that could be beneficial for plant growth. The end product generated by composting can be used as excellent organic fertilizer to enhance soil fertility.

The organic aerobic composting process is shown in Figure 6.1. During the composting process, substantial quantities of CH_4 and N_2O are generated

TABLE 6.3

Landfill Gas Composition of the Physical Properties

Project	Methane	Carbon Dioxide	Hydrogen	Hydrogen Sulfide	Carbon Monoxide	Nitrogen
Relative specific gravity (air = 1)	0.555	1.520	0.069	1.190	0.967	0.967
Flammability	Combustible			Combustible	Combustible	Combustible
Volume range for explosive mixed with air (%)	5–15			4–75.6	4.3–45.5	12.5–74
Smell	No	No		Yes	Slight	No
Toxicity	No	No		Yes	Yes	No

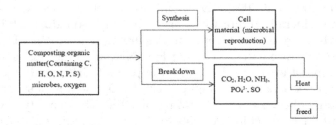

FIGURE 6.1
Organic aerobic composting process.

due to the existence of anoxic and anaerobic modification in some areas of composting where the oxygen is insufficient. The percentage of N_2O generated is approximately 0.02%–9.9% of the total nitrogen, the quantity of CH_4 is 0.1%–12.6% of the sign of total carbon, and the NH_3 release during composting can also be partially converted into N_2O in the atmosphere, consequence to allusive greenhouse impact (Huang et al. 2014). N_2O and CH_4 are GHGs that contribute significantly to the greenhouse effect. The 100-year greenhouse effect is 25 and 298 times of CO_2. Consequently, the rapid mineralization of organic matter escort by high concentrations of NH_3 volatilization which is also the main inference for nitrogen loss during the composting; however, many researchers have confirmed that ammonia volatilization during composting lead for approximately 20%–60% of the total nitrogen loss from the compost, that could not only causing air pollution but also causing nutrient loss.

Compared with other biological treatment methods, anaerobic fermentation has some advantages in reducing GHG emissions. The final gases produced by anaerobic fermentation, including methane, can be collected as an energy source. Therefore, anaerobic fermentation can reduce GHG emissions. Under anaerobic conditions, anaerobic fermentation can produce 60%–75% methane-containing biogas, which can provide humans with green bioenergy (for electricity generation, daily life, and car power). The residue of fermentation can be used to make organic fertilizer, which can be widely used in gardens, on lawns, and so on. Therefore, anaerobic digestion is an effective way to achieve the harmless and resource-efficient organic solid (Liu et al. 2017). In China, the annual production of agricultural solid waste in the billions of tons; if one-third of this amount could be used in anaerobic fermentation, it could produce 100 billion cubic meters of biogas every year, equivalent to 230 million tons of standard coal, which is about 10% of China's current annual coal production.

6.4 Mitigation of Greenhouse Gas Emissions during Biological Processing of Solid Waste

Landfills are the third largest source of methane gas after wetlands and paddy fields; therefore, it is of great significance to study the emission reduction of methane in landfills. The techniques for reducing GHG emissions can be split into two kinds, namely, natural and artificial emission mitigation. The soil contains a large amount of methane-oxidizing bacteria, and methane-oxidizing bacteria can consume both atmospheric CH_4 and methane from the soil. Methane-oxidizing bacteria grown in landfill cover can also oxidize the CH_4 produced by the waste. This kind of oxidation mainly occurs in the top-soil, and the oxidation of CH_4 by the covering soil greatly reduces CH_4 release (Ma et al. 2014). Data show that the oxidation rate of methane-oxidizing bacteria is as much as $45g/(m^2.d)$, and the amount of oxidation can reach 50% of production. Based on the above viewpoints, strengthening the operation and management of sanitary landfills, standard construction and preventing the garbage dumping as well as making full use of methane in the soil oxidation can reduce the CH_4 emission from landfills. The impact of waste management practices on GHGs and the range of values estimated for various management options are presented in Table 6.4 (United Nations Environment Programme 2010).

The main measures for artificial emission abatement are as follows: First, reduce the external displacement of garbage (mainly organic wastes) produced by the industrial production process, which can be achieved through technologies of the current promotion of cleaner production (industry) and clean vegetables into cities and other advanced technologies. Second, make full use of organic wastes to produce biogas. The technology of producing biogas from wastes has made great progress both in domestic and foreign countries, which provides the significant technical support. Third, make full use of CH_4 collected from landfills to generate electricity, also can use landfill gases for boiler heating or grid operation. During active period of landfill gases, the content of CH_4 in landfill gases is as high as 50%, which is an ideal

TABLE 6.4

Evaluation of Greenhouse Gas Generation from Waste Management Operation in Europe

Waste Management Activity	Upstream Emissions (kg CO_2-e/ton input waste)	Direct Emissions (kg CO_2-e/ton input waste)	Downstream Emissions (kg CO_2-e/ton input waste)
Open composting systems	0.2 to 20	3 to 242	145 to 19
Enclosed composting systems	1 to 60	5 to 81	145 to 19
Anaerobic digestion	3 to 46	20 to 76	−414 to −49
Landfill with flared LFG	2 to 12	−71 to 150	−5 to −140
Landfill with LFG capture and utilization	2 to 16	−71 to 150	−5 to −140
Low organic waste landfill	2 to 10	−50 to −13	0

Negative values reveal GHG savings and positive values indicate GHG emissions.

renewable energy. The use of landfill gases to generate electricity is kind of an efficient and internationally accepted greenhouse gas emission reduction operation. Using the upgrade effective measures to increase the CH_4 concentration of landfill gases to 95%, while removing dust and acid gases can make excellent performance of the pipeline gases, which can be as a substitute to urban gas products, thereby controlling the release of methane in landfills.

Some small landfills have been closed for many years because of the insufficient production of landfill gases, so they are not acceptable in economic way. Since the global warming potential (GWP) of CH_4 is 21 times that of CO_2, converting CH_4 into CO_2 by flare combustion can significantly reduce the GHG emissions of landfill gas. When landfill gas is passed through the surface of the landfill, under the action of methane-oxidizing microorganisms, CH_4 can be oxidized to CO_2, and the CH_4 oxidation rate can achieve 12%–60%. The design and installation of landfill bio-covering layer can enhance the activity of methane oxidizing microorganisms, thereby accelerating the oxidation rate of CH_4, and it is expected to realize the mitigation of landfill gases which is not suitable in the later stage of landfill operation where is mainly low concentrations of CH_4 landfill gases. (Czepiel et al. 2003).

Bioreactor landfill is the third-generation landfill technology recommended by the United States Environmental Protection Agency (USEPA). It uses both aerobic and anaerobic conditions in landfills. By controlling the temperature and humidity conditions of landfills, the stability of landfills can be accelerated, the gas production rate and CH_4 concentration of landfill gas can be increased and also the economic value of landfill gases utilization can be improved, meanwhile it can create convenient conditions for the reduction of landfill GHG emissions. The greenhouse gas emissions from composting can be reduced through attempts of various interaction, such as the formulation of composting mass and the operational process.

The influence of the characteristics of waste mixture on compost is critical to identify the quantity and nature of emissions obtained from the process. High moisture and bulk density are responsible for higher GHG generation. Excessive water subtracts from the free air space (FAS) and produces anaerobic sites that can form methane (Tamura and Osada 2006). The correct FAS level ensures proper composting of the compost material in humanmade and natural inflatable systems and prevents the growth of anaerobic bacteria (Ruggieri et al. 2009). The biochemical properties, particularly the C:N ratio, of the compost material also play a considerable role in gas emissions. However, the bioavailability of these nutrients determines the kinetics as well as carbon and nitrogen losses during the composting. (Cayuela et al. 2006). Hence, the C:N ratio appraisal should depend on the decomposable content. The organic wastes are mixed with additives to co-compost in a balanced initial mixture C:N ratio and porosity, which can enhance the biodegradability as well as minimize the GHG emissions from subsequent composting mechanism. Meanwhile, it also depends on the different industrial systems in which the composting process was performed. Generally, there are two types: open and closed composting systems. The closed system can draw off the exhaust gases from the reactor compared to the open system. Closed composting systems include implanted reactors, such as rotary and compost tunnels, as well as sealed piles (with textile covers), or composters in enclosed buildings and equipped with gas treatment systems. Plants with gas operation schemes have adequate positive effects on the environment because GHGs are not directly emitted into the atmosphere (Colón et al. 2012). In this sense, based on the conclusions of published work, the first technical proposal to minimize GHG emissions would be focus on gas processing systems. Another important process parameter is the process temperature. Higher temperatures increase the volatilization of volatile compounds, resulting in higher gas emissions.

We can study the open system as a static pile, steering pile, and inflatable open air to better understand the gas emissions dynamics associated with ventilation strategies, that is, airflow and pile rotation (Pagans et al. 2006). Different authors emphasize the importance of airflow in compost gas emissions. It is believed that high flow increases the utilization of oxygen, avoids the formation of anaerobic bags and the subsequent methane generation, and reduces gas emissions. On the other hand, the high airflow contributes to its volatilization in the compost, and it is reported that the increase in ventilation reduces methane emissions but elevates NH_3 and significant amount of N_2O emissions. The composting process is enhanced by formulation of a homogenous mixture of organic waste and bulking agents. It also provides oxygenation and its alteration for solid components (Pagans et al. 2006).

From the perspective of biodegradation, it is recommended to improve processing of composting. However, pile rotation has proven to have a negative impact on climate emissions, including GHG emissions. Rotate itself to release the entrained gas inside the pile. Ahn reported that CO_2, CH_4, and N_2O flux increased after rotation due to an elevation in the gas dispersion rate, which in turn occurred due to an increase in porosity after rotation. If the oxygen uptake and biomass temperature of the pile are under control, it is recommended that they should be avoid stacking in the first stage. (Puyuelo et al. 2011). In the second stage, when the oxygen content in the pile increases, the formation of methane is oxidized into carbon dioxide. These authors recommend the steering plan that reduces CH_4 emissions and maximizes CH_4 oxidation in the piles. The authors also noticed considerably greater emissions in the steering system than in the inflatable system. When considering CH_4 and N_2O as carbon dioxide equivalents, the non-inflated system provides greater emissions and is simulated by the steering system, the steering system expands through natural action, and finally the strain expansion system, which exhibits the lowest process emissions. (Artola et al. 2009). However, as the authors pointed out, the energy consumption of ventilation contributes to the overall carbon dioxide nonbiological source emissions when approaching the obstacle from the overall impact appraisal. Practical activities can improve the composting efficacy to reduce GHG emissions beyond the mineralization process itself.

From the preceding discussion, it can be seen that the key to effective composting with negligible gas emissions is to arrange the material in piles of appropriate size and consistency to facilitate uniform oxygen diffusion. In noninflated systems, this will enhance natural convection. In inflatable systems, it is a considerable task to modulate the aeration to ensure aerobic perforation without providing air. High airflow over oxygen demand can be shown to be reasonable in order to avoid increased emissions from high temperatures. A new innovative controller has been proposed for aerosolizing the measured oxygen uptake rate on-line. In addition to the physical properties of the raw materials and the mixture to be composted, suitable water content and a harmonious decomposable C:N ratio are needed. Whether the composting system is open or closed, it is necessary to optimize the functional supply, which means to adjust the power or fuel consumption so as to mitigate the environmental impact of the entire composting process. Finally, if possible, it is recommended to use a gas treatment (through biological filtration or other technology) to finalize the solution to reduce the gaseous emissions into the atmosphere (Nikiema et al. 2007).

Anaerobic digestion is the process of microbial catabolism in which organic matter decomposes and then produces biogas in the absence of nitrate, sulfate, and oxygen. Organic waste anaerobic fermentation biogas need to go through the basic classification, broken treatment, and then add a certain amount of water, control the appropriate temperature and maintain anaerobic conditions, through the fermentation of bacteria, hydrogen production of acetic acid bacteria, consumption of hydrogen production of acetic acid bacteria, and other metabolic effects of fermented microorganisms, through different ways of decomposition, and finally produce methane gases

(Killilea et al. 2000). Organic waste in the complex organic matter contained in the role of bacteria hydrolase; the formation of the corresponding complex organic matter, such as monosaccharides, fatty acids, amino acids; and then in the hydrolysis of bacteria intracellular enzyme, the decomposition of acetic acid, acid, butyric acid, lactic acid, ethanol, and CO_2 and H_2, and then in the hydrogen production of acetic acid or acetic acid production of acetic acid bacteria under the action of the above products into acetic acid, and then through the hydrogen production of methane or acetic acid production of methane. Under the action of bacteria, decomposition of acetic acid produces methane. The anaerobic fermentation process is illustrated in Figure 6.2.

Anaerobic digestion of MSW is not very difficult, but it requires that the waste is sorted or pretreated (the amount of moisture does not have to be limited but the MSW cannot contain woody garbage). High-temperature digestion and suitable agitation are two other measures. Different carbon and nitrogen ratios have a significant effect on the production of biogas from anaerobic fermentation of microorganisms. Wang et al. (2014) reported that when the content of N was high, high concentrations of ammonia nitrogen inhibited the activity of methanogens and inhibited the anaerobic fermentation of methane. In the anaerobic fermentation process, when the nitrogen content is very low, generation of CH_4 and the nitrogen content is reciprocal to the ammonia content increased to 2000 mg/L or more, as the fermentation broth acid-base balance is overwhelmed. The catalytic activity is reduced, and the methane production is rapidly reduced (Mata Alvarez et al. 2000). Therefore, the control of C:N at the appropriate level for the production of biogas is very important.

An appropriate increase in the temperature of the organic waste fermentation broth is conducive to the production of biogas from anaerobic fermentation. Anaerobic fermentation of organic waste is carried out at about 30°C. Properly increasing the temperature increases the production of methanogenic bacteria. The degradation of the organic matter in the waste is more thorough. When the temperature is lower than 10°C, the catalytic enzyme contained in the methanogenic bacteria cannot play its catalytic function and cannot produce methane. When the temperature is in the range of 10°C–35°C, the methane production capacity of methanogenic bacteria increases as the temperature increases in that range. When the

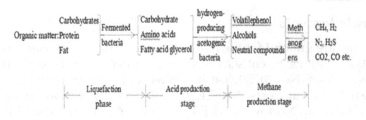

FIGURE 6.2
Three stages of anaerobic fermentation.

temperature reaches 35°C, the methane production is the maximum; when the temperature is above 35°C, the growth of methanogenic bacteria is inhibited. The impact of distinct pH values on anaerobic biogas generation is also considerable. Since methanogenic bacteria exhibit better growth under weakly alkaline conditions, the optimum pH range is 6.8–7.5. Thus, the pH value must be kept within the range of 6.8–7.5. If the pH is below this level, Sulfate-reducing bacteria (SRB) will compete and grow; thus, the amount of CO_2 will increase significantly, a large amount of water-soluble organic matter and H_2S and other S-containing compounds will be produced, and the increase in sulfide content will in turn inhibit the growth of methanogens. The pH of the traditional regulated fermentation broth can be adjusted by adding lime or adjusting the C:N ratio.

6.5 Conclusion

Solid waste has become one of the greatest threats to the health of the planet and human health, and countries around the world must find efficient and effective ways to treat solid waste.

Organic solid waste is a recyclable waste resource. Biological treatment of the organic matter in solid waste can reduce the gaseous emissions of toxic and hazardous substances but also play a considerable role in the conservation of the environment. Enhanced source recovery of the recyclable material can be a boon for organic farming and energy generation.

Biological treatment of solid waste, including composting, can be an eco-friendly alternative for environmental safety and the protection of human health.

Biological treatment methods can minimize GHG emissions. Thus, changing or updating existing facilities can improve their performance. GHG emission mitigation technologies can also help in this area.

Research and develop the technology by the optimization of process to get a balance between processing speed, processing breadth, processing effectiveness, and resource reuse.

Acknowledgments

The authors are grateful for the financial support from a Research Fund for International Young Scientists from National Natural Science Foundation of China (Grant No. 31750110469), Chins and The Introduction of talent research start-up costs (No. Z101021803), College of Natural Resources and Environment, Northwest A&F University, Yangling, Shaanxi Province 712100, PR China. We are also grateful to our all laboratory colleagues and research staff members for their constructive advice and help.

References

Ahring, B.K. 1994. Status on science and application of thermophilic anaerobic digestion. *Water Science Technology*, 30(12), 241–249.

Antizar-Ladislao, B., Lopez-Real, J., and Beck, A.J. 2005. In-vessel composting-bioremediation of aged coal tar soil: effect of temperature and soil/green waste amendment ratio. *Environment International*, 31(2), 173–178.

Artola, A., Barrena, R., Font, X., Gabriel, D., and Gea, T. 2009. Composting from a sustainable point of view: Respirometric indices as a key parameter. *Dynamic Soil Dynamic Plant*, 41(7), 71.

Bo, Z., Lili, Z., and Jianbo, X. 2003. Current situation and development on anaerobic digestion for municipal solid wastes. *China Biogas*, 21(4), 17–21.

Bogner, J., Pipatti, R., Hashinmoto, S., Diaz, C., and Mareckova, K. 2008. Mitigation of global GHGs emissions from waste: Conclusions and strategies from the Intergovernmental Panel on Climate Change (IPCC) Fourth Assessment Report. *Waste Management & Research the Journal of the International Solid Wastes & Public Cleansing Association Iswa*, 26(1), 11–32.

Cayuela, M.L., Sánchez-Monedero, M.A., and Roig, A. 2006. Evaluation of two different aeration systems for composting two-phase olive mill wastes. *Process Biochemistry*, 41(3), 616–623.

Chen, Q.J., Liu, H.B., and Hu, Y.Y. 2001. Application development of anaerobic digestion technologies of solid organic wastes. *China Biogas*, 19(4), 11–15.

Colón, J., Cadena, E., Pognani, M., Barrena, R., and Sánchez, A. 2012. Determination of the energy and environmental burdens associated to the biological treatment of source-separated municipal solid wastes. *Energy and Environment Science*, 5(2), 5731–5741.

Czepiel, P.M., Shorter, J.H., Mosher, B., Allwine, E., and Mcmanus, J.B. 2003. The influence of atmospheric pressure on landfill methane emissions. *Waste Management*, 23(7), 593–598.

Dong, J., Zhao, Y.S., Bing, L.I., Chen, Y.J., and Zhang, W.J. 2003. Bioreactor landfill methods and its feasibility. *Sichuan Environment*, 22(1), 8–12.

Eggleston, S., Buendia, L., Miwa, K., Ngara, T., and Tanabe, K. 2006. IPCC guidelines for national greenhouse gas inventories Volume 1: General guidance and reporting. City, Japan: Institute for Global Environmental Strategies.

Huang, J., Zheng, Y., Wu, X., and Zhao, H. 2015. Advanced progress of greenhouse gases and VOCs from municipal solid waste landfill. *Environmental Engineering*, 70–74.

Hwang, E.Y., Wan, N., and Park, J.S. 2001. Recycling of remediated soil for effective composting of diesel-contaminated soil. *Compost Science & Utilization*, 9(2), 143–148.

Killilea, J.E., Colleran, E., and Scahill, C. 2000. Establishing procedure for design, operation and maintenance of sewage sludge anaerobic treatment plants. *Water Science & Technology*, 41(3), 305–312.

Lebrato, J., Perez-Rodriguez, J.L., and Maqueda, C. 1995. Domestic solid waste and sewage improvement by anaerobic digestion: A stirred digester. *Resources Conservation and Recycling*, 13(13), 83–88.

Li, W., Gao, Q., Wang, L., Ma, Z., Liu, J., Li, C., and Zhang, Y. 2015. Emission characteristics of greenhouse gases from municipal solid waste treatment in China. *Research of Environmental Sciences*, 28(7), 1031–1038.

Liu, Z.M., Zhu, J.H., and Cheng, J.Y. 2017. Thermophilic anaerobic digestion of organic wastes for biogas production: a review. *Journal of Biology*, 34(1), 58–64.

Ma, Z.Y., Li, H.L., Yue, B., Gao, Q.X., and Dong, L. 2014. Study on emission characteristics and correlation of GHGs CH_4 and CO_2 in MSW landfill cover layer. *Journal of Environmental Engineering Technology*, 4(5), 399–405.

Maszenan, A.M., Liu, Y., and Ng, W.J. 2011. Bioremediation of wastewaters with recalcitrant organic compounds and metals by aerobic granules. *Biotechnology Advances*, 29(1), 111–123.

Mata Alvarez, J., Mace, S., and Llabres, P. 2000. Anaerobic digestion of organic solid wastes: An over-view of research achievements and perspectives. *Bioresource Technology*, 74(1), 3–16.

Nikiema, J., Brzezinski, R., and Heitz, M. 2007. Elimination of methane generated from landfills by biofiltration: A review. *Reviews in Environment Science and Biotechnology*, 6(4), 261–284.

Osaka/Shiga. 2010. Waste and Climate Change: Global trends and strategy framework. City, ST: United Nations Environmental Programme Division of Technology, Industry and Economics International Environmental Technology Centre.

Pagans, E., Font, X., and Sánchez, A. 2006. Emission of volatile organic compounds from composting of different solid wastes: abatement by biofiltration. *Journal of Hazardous Materials*, 131(1), 179–186.

Puyuelo, B., Ponsá, S., Gea, T., and Sánchez, A. 2011. Determining C/N ratios for typical organic wastes using biodegradable fractions. *Chemosphere*, 85(4), 653–659.

Rao, M.S., Singh, S.P., Singh, A.K., and Sodha, M.S. 2000. Bioenergy conversion studies of the organic fraction of MSW: Assessment of ultimate bioenergy production potential of municipal garbage. *Applied Energy*, 66, 75–87.

Ruggieri, L., Gea, T., Artola, A., and Sánchez, A. 2009. Air filled porosity measurements by air pycnometry in the composting process: A review and a correlation analysis. *Bioresource Technology*, 100(10), 2655–2666.

Saer, A., Lansing, S., and Davitt, N. 2013. Life cycle assessment of a food waste composting system: Environmental impact hotpots. *Journal of Cleaner Production*, 52, 234–244.

Topp, E., and Hanson, R.S. 1991. Metabolism of radioactively important trace gases by methane-oxidizing bacteria. In: Rogers JE, Whitmann W (Eds.), *Microbial Production and Consumption of Greenhouse Gases: Methane, Nitrogen Oxides and Halometanes*. Washington, DC: American Society for Microbiology, pp. 71–90.

United Nations Environment Programme. 2010. *Waste and Climate Change: Global Trends and Strategy Framework*. United Nations Environmental Programme, Osaka/Shiga.

Wang, J., Hu, Z., Xu, X., Jiang, X., Zheng, B., Liu, X., Pan X., and Kardol, P. 2014. Emissions of ammonia and greenhouse gases during combined pre-composting and vermin-composting of duck manure. *Waste Management*, 34, 1546–1552.

Wang, J., Shitao, Z., Zhiqiang, X., and Zhang, L. 2013. Summary of hydrogen-production via fermentation on solid waste. *Chinese Agricultural Science Bulletin*, 29(9), 139–148.

Wei, N., Li, X.-C., Wang, Y., and Zhi-Meng, G.U. 2009. Resources quantity and utilization prospect of methane in municipal solid waste landfills. *Rock & Soil Mechanics*, 30(6), 1687–1692.

Yang, L.L., and Yang, A. 2013. The emission reduction of greenhouse gas in sustainable development. *Guangzhou Chemical Industry*, 41(9), 46–47.

Zhou, F.C., Xian, X.F., and Xu, L.J. 2004. A review of research achievements and application on anaerobic treatment of organic solid wastes. *Journal of China Coal Society*, 29(4), 439–442.

7

Bio-Drying of Solid Waste

Rucha V. Moharir, Rena, Pratibha Gautam, and Sunil Kumar

CONTENTS

7.1 Introduction

Biodegradable waste is bio-dried to remove the moisture from the waste stream; as a result, its weight is reduced (Choi et al., 2001). Bio-drying processes are achieved by the drying rates that are amplified by using biological heat and by adding required aeration. The biologically heated major portion, available naturally by means of aerobic degradation of organic matter, is used to evaporate bound and surface water that is combined with mixed sludge. This generated heat assists in reducing the biomass moisture content without any additional fossil fuels and with negligible electricity consumption (Navaee-Ardeh et al., 2006). It takes a minimum of 7–8 days to dry the waste by this method (Sugni et al., 2005). This allows reduced cost for disposal if the landfill cost is on a per tonne basis. For refuse-derived fuels, bio-drying can be used as production process. Bio-drying generally does not affect the biodegradability of the waste, and for this reason it is not stabilized. Disruption of bio-dried waste takes place in landfills, which in addition generates harmful landfill gases and is responsible for climate change. Commercial mechanical biological treatment (MBT) included the bio-drying of plants bio-drying, but it is still a topic for concern in individual research and development (Velis et al., 2009).

7.2 Waste Management Scenario

MBT plants include mechanical as well as biological parts. The mechanical portion does the work of separating the recyclable elements from the waste stream. The biological portion includes the processes like anaerobic digestion, composting, and bio-drying (Figure 7.1). In waste management, these aerobic processes are mainly categorized as: biostabilization, composting, and bio-drying (Christensen, 2011). When biological activity takes place, some amount of heat is released because microbes are in an active condition while utilizing food for their survival and simultaneously reduces their food into smaller fractions for easy uptake (Zawadzka et al., 2010).

7.2.1 Bio-Drying

Biological treatments are always preferable over other treatments due to their attributes like complete decomposition of organic matter, not generating harmful by-products, and being eco-friendly. The products generated after drying the waste biologically are extremely useful, for example, refuse derived fuels that can be utilized in so many applications, some of which require their use.

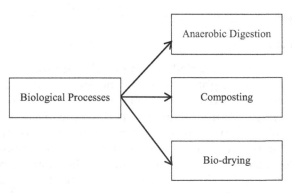

FIGURE 7.1
Categorization of biological processes.

7.2.1.1 Bio-Drying of MBT

Organic matter decomposition is a main concern in the biological treatment for waste management. Processes like composting and biostabilization have the capacity to convert the waste matter (organic matter) into valuable humus-like material that can be used as manure in plant cultivation (Adani et al., 2002). The outputs of these processes generally differ because the treatment is different for every process. For example, the final product generated from the stabilization process is stable in nature and is not biologically active, so it can be landfilled without any adverse impact on the environment (Adani et al., 2004; Araujo de Morais et al., 2008; Długosz, 2013; Grzesik and Malinowski, 2016; Myszograj et al., 2014; Scaglia et al., 2013; Suchowska-Kisielewicz, 2011; Suchowska-Kisielewicz et al., 2012, 2013, 2014; Tambone et al., 2011). As refuse-derived fuel (RDF) input is a variant of calorific value, the bio-drying reactor only degrades the easily degradable biological matter, which can be decomposed easily and has high calorific value. Velis described bio-drying variation in aerobic decomposition (Velis et al., 2009). Jewell first used the term *bio-drying*. The term *bio-drying* means the biological conversion of organic matter that is biologically decomposable and dries in the reactor when it is in the process of bioconversion activity for waste treatment. The optimum temperature required for bio-drying is as high as 46°C; for the processes like composting and stabilization, the temperature can be raised up to 71°C (Adani et al., 2002; Tambone et al., 2011).

7.2.2 Need of Automation

The main reasons for the automation of the processes are as follows:

- Workers health: It is necessary to avoid or to minimize the contact between waste and workers.
- Public health: The need for controlling emissions requires good facilities.
- Process efficiency: Efficiency increases output and thus reduces retention time.
- Quality of reliable output: Process control automation is especially useful for biological treatments where sorting of some factions is required.

Material handlers process the waste fed in. The fed waste is then moved with the help of wheel loaders, which are part of the biological treatment, and the waste is then moved with a conveyor belt, which is a part of mechanical process. Screens, magnetic metal separators, wind sifters, eddy current metal separators, and so on, be used for sorting the waste appropriately, and all are mechanical processes. Sometimes manual picking can is done to control the quality and output fraction enhancement. The factors of the biological treatment, for example, moisture, aeration, and temperature control, are computerized. Representation of the process is shown in Figure 7.2.

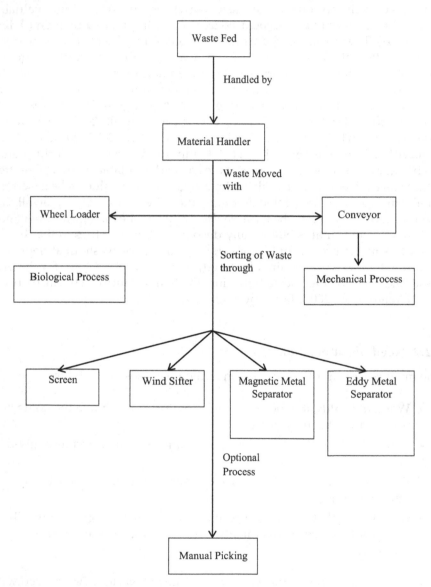

FIGURE 7.2
Schematic of the automation process.

7.3 Bio-Drying of Solid Waste

The main objective of the bio-drying process is to obtain RDF of high quality that can replace any prime fuels in co-incineration plants, industrial boilers, etc. Bio-drying of waste matter not only generates secondary fuel like RDF production during operation but also separates recyclable materials like ferrous and nonferrous metals, high-quality plastics, and hazardous waste like car batteries. Composting can be easily done if required per the treatment line designed conditions.

7.3.1 Operating Process of Bio-Drying of Solid Waste

The waste is collected in waste collection containers by using waste collection methods. The collected waste is transported by transport trucks to a transfer station. Then waste is properly discharged according to its type. The initial step is to segregate the waste appropriately and remove large fractions of waste such as batteries, and this is done by spreading the waste with the help of a wheel loader to remove any hazardous components before processing the waste further. These fractions are removed for disposal. To reduce the particle size up to 150 mm, the waste is then loaded onto shredders. This step is important for size reduction of waste and shaping the waste appropriately. There is no need for any bag-opening device as shredding is already processed. This shredded portion of waste is then passed through the magnetic separator belt for removing ferrous metals. The metals recovered from this process are stored in a separate container and then recycled. The intensive rotting is done on the waste that was fed to the drying tunnels using wheel loaders. This process normally takes 2 weeks. The tunnels are operated so that bio-drying and composting can be achieved simultaneously. The organics are partially degraded during this process. To remove any fine dirt, the final product is again passed through a 20 mm filter membrane. This bio-dried fine portion of the waste is further processed via composting or sometimes directly used in various applications, depending on the use and materials characteristic. The portion of the waste that is retained on the 20 mm filter membrane is again passed through a magnetic separator for separating ferrous metals. Then an eddy current separator extracts nonferrous metals. After this procedure is completed, all metals are further processed for recycling. The fraction that remains after this procedure is then passed through a three fraction separating wind sifter that separates the particles by their weight as light, medium, and heavy. The heavy-weight particles are mostly minerals and can be used in landfills or for construction purposes. Separate processing is required for light and medium fractions of waste, which can be done through optical sorting for recovering high-quality plastics for recycling or any other useful material recycling. The option of optical sorting is an additional facility to be provided for the quality control unit; it depends completely on the buyers and on factors like reducing pollution levels, reduction in cost, etc. High-quality plastics are generally recovered and further processed through recycling. The overall process is shown in Figure 7.3.

7.3.2 Reactor Dynamics

Reactors are basically a combination of engineered, biochemical, and physical processes. Fabricating any reactor, container is required that is either enclosed from the top or open, depending on the purpose for which it is to be used. Biochemical process includes decomposition of biologically degraded organic matter; on the other hand, physical process includes the factors that can affect the reactor on physical approach. Though bio-drying reactors are generally operated on exactly the same conditions as in composting, there is a slight difference between the two processes. Composting is generally used for decomposition of organic matter and is widely accepted in the waste management system. In this process, microorganisms play a vital role for degrading the organic matters under thermophilic conditions (Bertoldi et al., 1996; Epstein, 1997; Haug, 1993; Insam, and Bertoldi, 2007). During this biodegradation process, microorganisms utilize substrate through complex biochemical reactions for satisfying their metabolic and growth needs, which gradually leads to mineralization of organic substances (Richard, 2004). The parameters affecting composting are shown in Figure 7.4. These parameters have been discussed elsewhere (Diaz and Savage, 2007; Haug, 1993; Richard, 2004; Schulze, 1961).

7.4 Bio-Drying and Size Sorting of Solid Waste with High Water Content for Improving Energy

Many developing countries have waste with a high-moisture content due to the high proportion of food waste. On average, the amount of food waste is more than 60% (Cheng et al., 2007; He et al., 2005; Munnich et al., 2006; Norbu et al., 2005; Zhang et al., 2008). Municipal Solid Waste (MSW) having high heterogeneity along with enormous moisture content leads to difficulties in

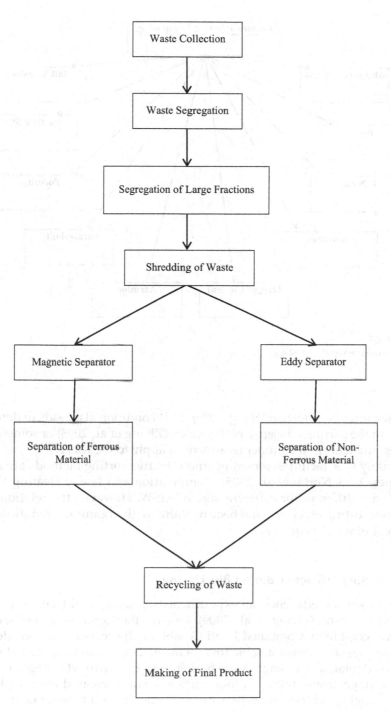

FIGURE 7.3
Overview of the bio-drying process.

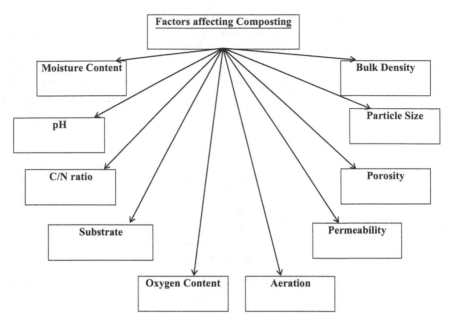

FIGURE 7.4
Factors influencing composting.

self-sustained combustion (Haug, 1993). This condition also leads to deterioration in the sorting efficiency of the waste (Zhang et al., 2008) or sometimes causes problems in the proper treatment to apply to the waste.

In many RDF facilities, screening and effective sorting methods are used (Juniper, 2005; Norbu et al., 2005). Composition and Lower Heating Value (LHV) are different for different size of MSW. However, the relationship between sorting efficiency and heating value with organic degradation and removal of water is unclear.

7.4.1 Sorting Efficiency during Bio-Drying

To check sorting efficiency, an experiment was set up and trials were performed in China (Zhang et al., 2009). During the experiment one aerobic and two combined (Combined 1 and Combined 2) process were considered. The samples underwent aerobic treatment. At 2-day intervals, the fed samples were rotated. The stages were broken into the hydrolytic stage and the aerobic stage. It was found sorting efficiency was increased after applying the bio-drying process, when previously it was difficult to separate the raw material. To determine the efficiency of bio-drying, an equation was proposed by Norbu et al. (2005):

$$SE = P_{<60} / W * 100\%$$

where:
SE = sorting efficiency
$P_{<60}$ = amount of underflow
W = total waste

The enhancement in sorting efficiency is shown in Table 7.1.

Both the size of the granule and the adhesion property determine sorting efficiency. A negative correlation was observed between the sorting efficiency and the content of MSW (correlation co-efficient, $r = -0.85$) during the experiment, where a positive correlation was also seen between degradation of organic and sorting efficiency. With respect to organic degradation, the sorting efficiency was reduced in the following order: aerobic > Combined 2 > Combined 1. However, at the end of the bio-drying, the sorting efficiency was decreased in the following order: Combined 2 > aerobic > Combined 1.

7.4.1.1 Effect of Particle Size on Drying

Two sets of the samples were analyzed at a different particle size to examine the effect of particle size. The samples were marked as G9 and G2 (Nzioka et al., 2016). The names of the samples and the sizes of the particles are listed in Table 7.2.

The temperature supplied to the different samples was in the range of 60°C, 80°C and 100°C. The results of the experiment are shown in Table 7.3.

For the sample whose particle size was >20 mm, the low drying rate coefficient, and mass transfer coefficient were linked to the low specific

TABLE 7.1

Sorting Efficiency of the Samples

Name of the Sample	Initial (%)	Final (%)
Combined 1	62	34
Combined 2	71	34
Aerobic	68	34

TABLE 7.2

Names of the Samples and the Sizes of the Particles

Name of the Sample	Size of the Particle
G9	>20 mm
G2	<20 mm

Source: Nzioka A.M. et al. *Thermochemical Foundations of Chemical Engineering,* 50, 414–421, 2016.

TABLE 7.3

Temperature and Effect of the Particle Size

Name of the Sample	Temperature	Effect of Particle Size		
		Water Content	K, L/S	Kd, Kg/(m²s)
G9	60	57	0.142	0.116
G10	60	46	0.16	0.168
G9	80	57	0.144	0.18
G10	80	61	0.166	0.227
G9	100	48	0.162	0.194
G10	100	60	0.187	0.318

Source: Nzioka, A.M. et al., *Thermochemical Foundations of Chemical Engineering,* 50, 414–421, 2016.

surface area, which ultimately affected the moisture diffusivity. The value of K and Kd was not much significant at low temperature. A similar tendency was seen with the increase in temperature at 80°C and 100°C: for particle size <20 mm, easier penetration of the drying agent could be seen due to the porous structure. Porosity also increased the contact between the surface and the drying agent, which ultimately increased the saturation pressure on the surface and the drying rate. The difference in the size of the particle also played an important role in creating a porous structure.

Bio-drying and size sorting of municipal solid waste with high water content for improving energy recovery are concerns that need to be checked. To determine the variation in the content of moisture, 5 kg of the waste sample was screened with an opening size of 25, 60, and 45 mm. The diameter of the bottom pan was kept at 500 mm (Shao et al., 2010). It has been divided into four parts according to the particle size. The names of the samples and screen sizes are shown in Table 7.4, and the instrument used for analysis is depicted in Table 7.5.

TABLE 7.4

Names of the Samples and Screen Sizes

Name of the Sample	Screen Size
Sample A (>60)	60 mm
Sample B (45–60)	45 mm
Sample C (25–45)	25 mm
Sample D (0–25)	Retained on the bottom of the Pan.

Source: Shao, L.M. et al., *Waste Management*, 30, 1165–1170, 2010.

TABLE 7.5

Instrument Used for Monitoring and Analysis of the Sample

Parameters	Instrument Used
Temperature	WMY-01 C (Huachen Co., China)
LHV	Calorimeter (6100, PARR, USA)
Inorganic chlorine	10L/KG followed by $AgNO_3$
Sulphur	An organic halide analyzer (AQF-100/ICS-1500 systems, Dionex, USA)
Metal concentration (cd, Cr, Pb, Zn, Cu, Mn, and Ni)	ICP (ICP-Optima 2001DV, Perkin-Elmer, USA)
Hg	Atomic fluorescence spectrometer /9AFS-XGY1012)

Source: Shao, L.M. et al., *Waste Management*, 30, 1165–1170, 2010.

1. **Effect on the Wet weight**

   The result shows that, after bio-drying, the wet weight of the different sample was decreased in the following order (Shao et al., 2010):

   - Sample A (>60) reduced by 54.3%
   - Sample B (45–60) reduced by 60.3%
   - Sample C (25–45) reduced by 79.7%
   - Sample D (0–25) reduced by 72.7%

2. **Effect on the fraction of particle**

   The fraction of the particle after bio-drying was ultimately reduced as follows (Shao et al., 2010):

   - Sample A (>60) reduced by 31.7%
   - Sample B (45–60) reduced by 14.8%
   - Sample C (25–45) reduced by 14.6%
   - Sample D (0–25) reduced by 38.9%

3. **Effect on the food residue**

   Sample D had the highest percentage of the food material remaining after bio-drying, i.e., 85.3% of the waste (0–25), but a gradual decrease, i.e., 27%, was seen in sample C. Also no food residues were found in Sample A (>60) and Sample B (45–60) (Shao et al., 2010).

4. **Effect on the amount of plastic and paper**

Sample B (45–60) and sample A (>60) after bio drying consisted only of paper and plastic. However, sample C was predominantly paper, having a percentage of 62.4% (Shao et al., 2010).

5. **Effect on the heavy metal concentration**

The concentration of metal was increased by 60%. However, the concentration of Cu was not changed. Among all the samples analyzed, sample D (0–25) had the highest concentration of metal (Cr, Pb, Cu, Ni, Hg, Zn). Ni did not vary much in concentration (Shao et al., 2010).

6. **Effect on the concentration of chlorine**

After applying bio-drying technique, the amount of chlorine in the final product was enhanced by 27%, 73.6%, and 92.2%. Due to the dominant amount of plastic in sample A (>60), higher concentration of chlorine was found in that sample. Other samples had less concentration of chlorine (Shao et al., 2010).

7. **Effect on the concentration of sulphur**

The concentration of sulphur was enhanced quite a bit in the entire sample of waste. The concentration of sulphur was raised by 40% in sample A (>60), 75% for sample B (45–60), 75% for sample C (25–45), and 70.8% for sample D (0–25) (Shao et al., 2010) (Figure 7.5).

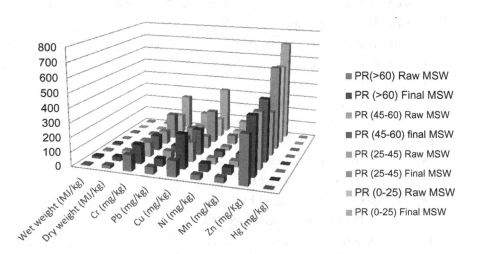

FIGURE 7.5
The effect of size sorting and bio-drying on different parameters of MSW.

7.5 Kinetic Model for Solid Waste Bio-Drying

Bio-drying is meant to remove excess water from high-water content bio-wastes using the heat generated during aerobic degradation of waste along with forced aeration (Zhao et al., 2011). It is important to determine the biodegradation behavior of wastes in this context. This behavioral study is important for an optimized design regarding the bio-drying process wherein processing time, aeration flow rate (AFR), reactor size, and initial moisture content (MC) can be considered as the main driving factors. Kinetic modeling helps to understand the transport process in a bio-drying reactor as it involves a number of variables such as air velocity in each reactor compartment, inlet air temperature, relative humidity, the bulking agent ratio, inlet feed dry solids content, residence time etc., which ultimately helps in process design and optimization.

7.5.1 Inductive Modeling Approach

Any kinetic model should be able to predict the rate of the process after considering all variables and different operating conditions. In the inductive modeling approach, the modeling process starts with data; for waste composting and bio-drying, the starting point of any inductive kinetic model is the degradation of organic matter as this supplies the free energy needed to drive the process (Godden and Penninckx, 1987). The other factors affecting the process rate are environmental factors like temperature, biomass, moisture, oxygen, particle size, etc. But the inductive approach has several limitations:

- It gives a good description of observed kinetic dependencies, but the data-oriented approach does not yield a comprehensive kinetic model, i.e., a model that embraces all major environmental factors including waste.
- The heterogeneity of the waste calls for numerous replications.
- Some of the factors like biomass, particle size, etc., are important to be defined, but they cannot be measured as no techniques are available for quantitative measurement in organic waste matrix (Mitchell and Lonsane, 1992).

7.5.2 Deductive Modeling Approach

In deductive modeling, both theory and measurements are exploited for information that may be fruitful in understanding the kinetics of degradation of organic waste. However, less work is done under this approach due to the complexities arising from the increase in the number of (unknown) initial values and the number of parameters of the constitutive relationships. This increase generally occurs due to incorporation of more (state) variables. Deductive modeling tacitly assumes that there exists a well-established quantitative theory of the phenomenon of interest. Either of these strategies tends to obscure the problem. To overcome this issue, a modified deductive approach is explored that aims at reducing the model so that it has only identifiable parameters left. By solving the model for the output, an analytical model can be described that measures the output variables.

7.5.3 Kinetic Modeling for Bio-Drying

Mason (2006) included the degradation models proposed by many researchers and kinetic equations in his study. Basically, three types of models have been used in previous studies for modeling the bioenergy production (Finger et al., 1976; Haug, 1993; Kaiser, 1996; Paredes et al., 2002):

- A model that includes first-order substrate degradation kinetics
- A model that includes Monod-type expressions
- A model that uses empirical substrate degradation equations

Due to the complexity of the compost matrix, the Monod degradation model is widely used in biodegradable process to simulate aerobic degradation process (Zavala et al., 2004) and to predict the kinetics of heat produced in composting process. But very little research is done on the application of the Monod degradation model for bio-drying systems.

Huilinir et al. (2017) studied the kinetics of bio-drying of sewage sludge based on the deductive approach. In this theoretical approach based on models reported by Pommier et al. (2008) and Pujol et al. (2011), the microbial population (X_H) was explicitly accounted for through Monod-type kinetics. In this kinetic model, Volatile Solids (VS) were divided into the following four fractions:

1. Rapidly biodegradable (Xrb)
2. Slowly biodegradable (Xsb)
3. Nonbiodegradable (or inert, XI)
4. Heterotrophic biomass (X_H)

Heterotrophic biomass consumed soluble substrate (Ss), which was hydrolyzed from Xrb and Xsb. It was an exothermic reaction, and rapidly biodegradable new biomass was generated, taking biomass decay in to account.

For the entire kinetic process, the effect of MC and temperature were also taken into account per the following equations:

$$f\,T = 1.066^{T-20} - 1.21^{\,T-60}$$

$$f\,Mc = 1/\ (e^{7.0622-17.684 f_{MC}} + 1)$$

Several kinetic models have been used to predict the behavior of VS solid degradation under different AFR (Villegas and Huiinir, 2014). The functioning of these models was developed only at different AFR and did not take into account other important factors such as MC, which is important as its high-value limits oxygen transport and obstructs microbial activity, thus retarding bio-drying. Even free air space (FAS) is affected by initial MC. The movement of air through the matrix and its quantity can be determined using FAS. The biodegradation behavior of waste is an important factor for the optimal design of a bio-drying process. A kinetic model that predicts VS biodegradation is useful to determine the degree of biodegradation. Table 7.6 shows the various kinetic models applied for this study.

In most of the kinetic modeling, a 3^2 factorial design was used to test the kinetic models. The factors involved in this design were MC and AFR. The VS biodegradation at different initial MV and high airflow rates was described by incorporating the measurement of MC, T, and FAS done during the assays.

TABLE 7.6

Kinetics Models Used

Sr. No.	Kinetic Model	References
1	$k_T = k_0 * (1.066^{T-20} - 1.21^{\,T-60}) * 1/ (e^{7.0622-17.684 f_{MC}} + 1) * 1/(e^{3.4545-23.675\,FAS} + 1)$	Haug (1993)
2	$k_T = k_{23} * a^{\,T-23}$	Haug (1993)
3	$k_T = a * e^{\,(b*T+c*\,Mc/T)}$	Kulcu and Yalzid (2004)
4	$k_T = a/\ (Mc * (T - b)) * e^{\,(c*T + d*\,Mc/T)}$	Kulcu and Yalzid (2004)
5	$k_T = e^{\,AFR-a} * k_{23} * b^{T-23}$	Kulcu and Yalzid (2004)
6	$k_T = a *(b^{\,c*T + d*\,Mc/T})/\ e^{\,f.\,FAS}$	Villegas and Huiinir (2014)
7	$k_T = a * (T* Mc + b* FAS) * e^{c*\,Mc/T + FAS*\,Mc}$	Villegas and Huiinir (2014)

k_T = Reaction rate constant (d^{-1}).
T = Process temperature (T°C).
Mc = Materials daily moisture content (% by weight).
f_{Mc} = Materials daily moisture content measured in fraction.
AFR = Air flow rate measured in L / min Kg$_{TS}$.
k_{23} = Kinetic constant at 23°C.
FAS = free air space in %, a, b, c, and d = constants.

7.6 Conclusion

Bio-drying is a process of removing moisture content from collected solid waste matter. It can be achieved manually as well mechanically. Different processes involved in bio-drying of waste material were explained in this chapter. A case study was briefly discussed to develop a better understanding of the application. Different modeling kinetics of solid waste were discussed along with their impacts and assumptions. This approach would be helpful in suggesting better pathways for applying such techniques. Pathways, processes, stages, and process kinetics of bio-drying were elaborated in this chapter.

References

Adani F., Badio D., Calcaterra E., Genevini P., 2002. The influence of biomass temperature on biostabilization-biodrying of municipal solid waste. *Bioresource Technology.* 83, 173–179.

Adani F., Tambone F., Gotti A., 2004. Biostabilization of municipal solid waste. *Waste Management.* 24, 775–783.

Araujo de Morais J., Ducom G., Achour F., Rouez M., Bayard R., 2008. Mass balance to assess the efficiency of a mechanical-biological treatment. *Waste Management.* 28, 1791–1800.

Bertoldi M., Sequi P., Lemmes B., Papi T., 1996. *The Science of Composting: Part 1.* Chapman & Hall, London, UK.

Cheng H.F., Zhang Y.G., Meng A.H., Li Q.H., 2007. Municipal solid waste fueled power generation in china: A case study of waste-to-energy in Changchun city. *Environmental Science and Technology.* 41, 7509–7515.

Choi H.L., Richard T.L., Ahn H.K., 2001. Composting high moisture materials: Biodrying poultry manure in a sequentially fed reactor. *Compost Science and Utilization.* 9 (4), 303–311.

Christensen T.H., 2011. *Solid Waste Technology & Management, Publication A.* John Wiley & Sons, 2, 515–685.

Diaz L.F., Savage G.M., 2007. Factors that affect the process. In Diaz, L.F. et al. (Eds.), *Compost Science and Technology.* Elsevier, 49–65.

Długosz J., 2013. Selected methods of treatment of leachate from municipal landfills–for review. *Archives of Waste Management and Environmental Protection.* 15 (2), 59–68.

Epstein E., 1997. *The Science of Composting.* Lancaster, PA: CRC Press, Technomic Publishing.

Finger S.M., Hatch R.T., Regan T.M., 1976. Aerobic microbial growth in semi-solid matrices: Heat and mass transfer limitations. *Biotechnology and Bioengineering.* 18, 1193–1218.

Godden B., Penninckx M., 1987. Biochemistry of manure composting: Lignin biotransformation and humification. In *Compost: Production Quality Use.* Applied Sciences Publishers, 238–244.

Grzesik K., Malinowski M., 2016. Life cycle assessment of refuse-derived fuel production from mixed municipal waste, Energy Sources, Part A. *Recovery, Utilization, and Environmental Effects.* 38 (21), 3150–3157.

Haug R.T., 1993. *The Practical Handbook of Compost Engineering.* Boca Raton, FL: Leis Publishers.

He P.J., Shao L.M., Qu X., Li G.J., Lee D.J., 2005. Effects of feed solutions on refuse hydrolysis and landfill leachate characteristics. *Chemosphere.* 59, 837–844.

Huilinir C., Perez J., Olivares D., 2017. A new model of batch biodrying of sewage sludge, Part 1: Model development and simulations. *Drying Technology.* 35 (6), 651–665.

Insam H., Bertoldi M., 2007. Microbiology of the composting process. In Diaz L.F. et al. (Eds.), *Compost Science and Technology.* Elsevier, pp. 25–48.

Juniper, 2005. *Mechanical–Biological-Treatment: A Guide for Decision Makers Processes, Policies and Markets.* CD-ROM, v1, March 2005, Juniper Consultancy Services, UK.

Kaiser J., 1996. Modelling composting as a microbial ecosystem: A simulation approach. *Ecological Modelling.* 91, 25–37.

Kulcu R., Yaldiz O., 2004. Determination of aeration rate and kinetics of composting some agricultural wastes. *Bioresource Technology.* 93 (1), 49–57.

Mason I.G., 2006. Mathematical modelling of the composting process: A review. *Waste Management.* 26, 3–21.

Mitchell D.A., Lonsane B.K., 1992. Definition, characteristics and potential. In Doelle HW, Mitchell DA, and Rolz CE (Eds.), *Solid Substrate Cultivation.* London, UK: Elsevier Applied Science.

Münnich K., Mahler C.F., Fricke K., 2006. Pilot project of mechanical-biological treatment of waste in Brazil. *Waste Management.* 26, 150–157.

Myszograj S., Kozłowska K., Krochmal A., 2014. Evaluation of biological activity of cellulose pulp by means of the static respiration index (At4). *Civil and Environmental Engineering Reports.* 14 (3), 49–62.

Navaee-Ardeh S., Bertrand F., Stuart P.R., 2006. Emerging biodrying technology for the drying of pulp and paper mixed sludges. *Drying Technology.* 24 (7), 863–878.

Norbu T., Visanathan C., Basnayake B., 2005. Pretreatment of municipal solid waste prior to landfilling. *Waste Management.* 25, 997–1003.

Nzioka A.M., Kim M.G., Hwang H.U., Yan C.Z., Ved V.E., Meshalkin V.P., Kim Y.J., 2016. Experimental investigation on the drying of loosely-packed and heterogenous municipal solid waste. *Thermochemical Foundations of Chemical Engineering.* 50, 414–421.

Paredes C., Bernal M., Cegarra J., Roig A., 2002. Bio-degradation of olive mill wastewater sludge by its co-composting with agricultural wastes. *Bioresource Technology.* 85, 1–8.

Pommier S., Chenu D., Quintard M., Lefebvre X., 2008. Modelling of moisture-dependent aerobic degradation of solid waste. *Waste Management.* 28 (7), 1188–1200.

Pujol A., Debenest G., Pommier S., Quintard M., Chenu D., 2011. Modeling composting processes with local equilibrium and local non-equilibrium approaches for water exchange terms. *Drying Technology.* 29 (16), 1941–1953.

Richard T.L., 2004. Fundamental parameters of aerobic solid-state bioconversion processes. In Lens, P. et al. (Eds.), *Resource Recovery and Reuse in Organic Solid Waste Management,* London, England: IWA Publishing, pp. 262–277.

Scaglia B., Salati S., di Gregorio A., Carrera A., Tambone F., Adani F., 2013. Short mechanical biological treatment of municipal solid waste allows landfill impact reduction saving waste energy content. *Bioresource Technology*. 143, 131–138.

Schulze K.L., 1961. Aerobic decomposition of organic waste materials. Final report. Project RG-4180 (C5R4), Washington, DC: National Institute of Health.

Shao L.M., Ma Z.H., Zhang H., Zhang D.Q., He P.J., 2010. Bio-drying and size sorting of municipal solid waste with high water content for improving energy recovery. *Waste Management*. 30, 1165–1170.

Suchowska-Kisielewicz M., 2011. The methods to intensify waste decomposition. *Civil and Environmental Engineering Reports*. 59–73.

Suchowska-Kisielewicz M., Jedrczak A., Myszograj S., 2012. Kinetic constants of decomposition of the municipal solid waste prior to and after mechanical-biological processing. *Archives of Environmental Protection*. 38, 71–86.

Suchowska-Kisielewicz M., Jedrczak A., Sadecka Z., 2014. Evaluation of biodegradability of waste before and after aerobic treatment. *Civil and Environmental Engineering Reports*. 13 (2), 121–132.

Suchowska-Kisielewicz M., Jedrczak A., Sadecka Z., Myszograj S., 2013. Effect of aerobic pretreatment of waste on the rate of anaerobic treatment processes. *Journal of Material Cycles and Waste Management*. 15 (2), 138–145.

Sugni M., Calcaterra E., Adani F., 2005. Biostabilization-biodrying of municipal solid waste by inverting air-flow. *Bioresource Technology*. 96 (12), 1331–1337.

Tambone F., Scagalia B., Scotti S., Adani F., 2011. Effects of biodrying process on municipal solid waste properties. *Bioresource Technology*. 102, 7443–7450.

Velis C.A., Longhurst P.J., Drew G.H., Smith R., Pollard S.J., 2009. Biodrying for mechanical-biological treatment of wastes: A review of process science and engineering. *Bioresource Technology*. 100 (11), 2747–2761.

Villegas M., Huilinir C., 2014. Biodrying of sewage sludge: Kinetics of volatile solids degradation under different initial moisture contents and air-flow rates. *Bioresource Technology*. 174, 33–41.

Zavala M.A.L., Funamizu N., Takakuwa T., 2004. Modeling of aerobic biodegradation of feces using sawdust as a matrix. *Water Research*. 38, 1327–1339.

Zawadzka, A., Krzystek, L., Stolarek, P., Ledakowicz, S., 2010. Biodrying of organic fraction of municipal solid wastes. *Drying Technology*. 28(10), 1220–1226. doi:10.1080/07373937.2010.483034.

Zhang D.Q., He P.J., Jin T.F., Shao L.M., 2008. Bio-drying of municipal solid waste with high water content by aeration procedures regulation and inoculation. *Bioresource Technology*. 99, 8796–8802.

Zhang D.Q., He P.J., Shao L.M., 2009. Sorting efficiency and combustion properties of municipal solid waste during bio-drying. *Waste Management*. 29, 2816–2823.

Zhao L., Gu W.M., He P.J., Shao L.M., 2011. Biodegradation potential of bulking agents used in sludge bio-drying and their contribution to bio-generated heat. *Water Research*. 45 (6), 2322–2330.

8

Recent Trends in Bioprocessing of Antibiotic Residues and Their Resistant Genes in Solid Waste

Shashi Arya, Rena, Digambar Chavan, and Sunil Kumar

CONTENTS

8.1 Introduction

Accelerating population growth and the accompanying food crisis have challenged food security around the world. The practice of using a variety of drugs, as shown in Table 8.1, as feed additives, for the curing and prevention of infections and diseases for enhanced agricultural growth, and to increase food animal production has had a dramatic impact since the early 1950s (Cromwell, 2001). The application of antibiotics for the treatment of diseased animals enhances animal well-being. The rate and fate of antibiotics in the environment, including surface water, groundwater, and soils, have received major attention from researchers around the world (Kummerer, 2001; Xiao et al., 2008). Due to some application failures and/or irrational and irresponsible use of antibiotic growth promoters, antimicrobial resistance and antibiotic residues have been increasing. Hence, antibiotics and their transformation products are considered as emerging chronic pollutants and have serious negative impacts on ecosystems (Boxall et al., 2003; Martinez, 2008).

These residues have been found in food from animal sources like meat, milk, and eggs and have threatened the health of consumers (Tufa, 2016). The consumption of contaminated meat, milk or any source of food from an animal is potentially carcinogenic due to toxic antibiotic residues and their allergic potential risks for public health (Tufa, 2016). China is the world's biggest maker and exporter of antibiotics. It makes 90% of the world's ingredients for manufacturing antibiotics. According to the Interagency Task Force on Antimicrobial Resistance, the use of antibiotics has resulted in drug resistance that is accelerating the global menace. The main cause of antibiotic resistance is incorrect use by humans, exploitation in intensive agriculture, and pollution from manufacturing. This chapter reviews the comprehensive advantages and disadvantages of antibiotic use in animals such as livestock

TABLE 8.1

Application of Drugs for Treatment of Food Animals

Drugs	Use	References
Topical antiseptics, bactericides, fungicides	Used to treat and cure affected skin, hoofs cuts and abrasions	National Research Council (US)
"Ionophores"	Used to ensure more favorable and energy efficient substrates to feed and defense against some parasites	Committee on Drug Use in Food Animals.
"Steroid and Peptide"	Used as growth promoters and enhanced production rates	Washington (DC): 1999
Antiparastite drugs	Protect against parasites attack on food animals	
Antibiotics	To control "overt, obscure and occult" diseases and enhanced growth production	

and poultry and agricultural crop production, and discusses the spread of antibiotics through various pathways, the global impact, and bioprocessing treatment.

8.2 Antibiotics and Their Fate

Antibiotics were known as the "lifeline bullet" when they were introduced in the twentieth century to kill and prevent microbial growth such as bacteria, fungi, and protozoans. These antibiotics helped to enhance life expectancy, cure diseases, and promote vaccination programs. Every year a million ton of antibiotic is produced ("the value is based on the average mass ratio of antibiotic products to residue by-products, the water contents more than 80% of 1:60") (Zhang et al., 2015).

Pasteur and Joubert found that one type of microorganism could prevent the growth of other microbes. Due to lack of information and knowledge, Pasteur and Joubert were unaware of the mechanism of antibiotics, which was produced by one bacterium to suppress or kill the growth of other bacteria. The microbes emerged as a boon in the field of medicine to cure and prevent several diseases.

8.2.1 Pharmacodynamics of Antibiotic Use

The prime role of antibiotics is to treat infections and eradicate diseases at the earliest stages. In the beginning, the antibiotics intend to bind with the microorganism to destroy the biochemical reaction of the bacterium since it is vital to the life of the microorganism (Capitano and Nightingale, 2001). However, the dose and response of antibiotics play an important role in the pharmacodynamics of the antibiotic. Therefore, the antibiotic is allowed to inhabit and act for a sufficient period of time at active microbial sites (Phillips et al., 2004b).

The correlation between the antibiotic dose-response relationship that exists at the active sites or specific target sites, "termed the area under the concentration-time curve, i.e., $Cp \times time = AUC$, is an important factor to the life and death of the bacterium" (Grant and Nicolau, 1999; Nightingale et al., 2001). In general terms, pharmacodynamics is basically an "indexing of the total drug exposure in the active sites to measure the microbiological activity of the dose against the organism" (Phillips et al., 2004b).

8.2.1.1 Benefits

The global population has increased tremendously, from 1.5 billion (Roser and Ospina, 2017) to 7.6 billion (US Census Bureau, 2018), in just 100 years, and 155 million children are estimated to be malnourished (WHO, 2016).

To meet the basic needs, natural resources are being exploited at an extreme level. The demand for food increases exponentially day by day, especially in developing countries (Krehbiel, 2013). The introduction of antibiotics for use in animals has been debated in veterinary medicine. However, such use has prevented, controlled, and treated various deadly diseases in animals since the 1940s (Forman and Burch, 1947). A few advantages of antibiotics are discussed in the following subsections.

8.2.1.1.1 Control of Animal Disease

The control and prevention of infectious diseases are of basic importance for food animals. The emergence of antibiotics to treat, control, and prevent microbial infections led to a drastic improvement in animal production.

8.2.1.1.2 Protection of Public Health

The widespread occurrence of infectious diseases in animals such as rudderpost, anthrax, and tick fever is common in many countries (Phillips et al., 2004b; Dabbir, 2017). Animals that are infected are considered to be hazardous because the infection might be transferred by contact with humans (Kimman et al., 2013). Infected animals can pose a severe health threat to humans health and to other healthy animals through microorganisms such as "Campylobacter spp., Salmonella spp., *E. coli* O157, Vibrio parahaemolyticus and Aeromonas hydrophila" (Mellata, 2013). The emergence of antimicrobial drugs (see Table 8.2) protects human health and provides reliable access to a sufficient quantity of food by controlling and minimizing diseases in animals and agriculture crop production and disrupting the spread of pathogens from diseased animals to humans (Hao et al., 2014).

TABLE 8.2

List of Significant Antibiotics Used in the Protection of Human Health

Antibiotic	Impact	References
Virginiamycin	Decrease the contamination of clostridium perfringes, "campylobacter spp." And other pathogens borned in animal carcasses as food consumption.	Hurd et al. (2005)
Salinomycin	Reduced infection of type C "Clostridium in sows and weaning piglets by 43%"	Nagraja and Taylor (1987)
Neomycin	Reduced the no. of E. coli O157:H7 in animal faeces	Doyle and Erickson (2006)
"Gentamycin"	Reduces the no. of bacterial count in poultry eggs and meat and other food animals.	
Oxytetracycline hydrochloride or norfloxacin	Reduced the occurrence of A. hydrophila in water by 46.88%–66.24% thereby reduced the risk of bacterial infection to humans	Hao et al. (2014)
Florfenicol	100% efficiency for the treatment of A. hydrophila of piaractus mesopotamicus	Carraschi et al. (2012)

TABLE 8.3

Drugs used for Enhancement of Animal Production

Antibiotic	Effect	Reference
Bambermycin, Neomycin,[a] Penicilin,[a] Roxarsone	Increased weight in chickens, beef cattle, turkeys; faster growth of pigs	Veterinary Feed Directive (VFD) ahdc.vet.cornell.edu, Taylor and Gordon (1955)
Lasalocid, Laidlomycin, Lincomycin	Increased weight gain in beef cattle	
Monensin, Salinomycin, Virginiamycin	Increased weight gain in beef cattle and proficient milk production in dairy cows	Hao et al. (2014)
Bacitracin	Promotes egg production in chicken	He et al. (2011) and Zhou et al. (2011)
Carbadox, Tylosin*	Increased weight gain in chickens and swine	Partanen et al. (2007)
Chlortetracycline, Doxycycline, Sulfonamides	Enhanced growth promotion in calves, pigs, and chicken	Hao et al. (2014)

[a] Illegal for this use in the United States.

8.2.1.1.3 Enhancement of Animal Production

As shown in Table 8.3, the enhanced growth of animal production due to antibiotics was described in mid-1940s. Within approximately five years, the use of antibiotics became a common practice.

8.2.1.2 Risks

The use of antibiotics in animals and crop production continues to be a mammoth task (Hao et al., 2014). Inappropriate antibiotic application in animals producing food causes a potential threat to human health as pathogenic-resistant organisms inserted in these animals tend to intrude in the food supply and could be widely disseminated in food products (Landers, 2012). For example, commensal bacteria found in livestock are often present in fresh meat products that act as reservoirs for resistant genes that could potentially be shifted to humans in the form of pathogenic organisms. According to State of Worlds Antibiotics (2015), around 65,000 T to 100,000 T of antibiotics produced annually are used in animal agriculture (Figure 8.1).

8.2.1.2.1 Public Health Threat

The application of antibiotics in both human healthcare and the growth of food animals leads to the resistant pathogens, a rising and serious public health threat (Manyi-Loh et al., 2018). According to the World Health Organization (WHO) (2018), antibiotic resistance has become a global crisis and can affect anyone, of any age, in any country. Also, the U.S. Centers for Disease Control and Prevention (CDC) revealed that around 23,000 Americans die annually as a direct result of drug-resistant infections (Antibiotic Resistance Threats in the US, Report, 2013).

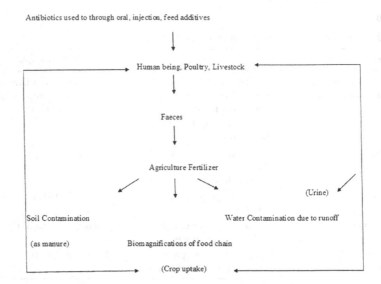

FIGURE 8.1
Flow of antibiotics and its impact.

Food animals exposed to continuous antibiotics promote the gradual evolution of bacteria that most often develops resistance against the antibiotic (Marshall and Levy, 2011). Foodborne pathogens such as salmonella and campylobacter can develop resistance and inadvertently cause a public health threat because of human consumption of contaminated food animal (DuPont, 2007).

Methicillin-resistant staphylococcus aureus (MRSA) emerged in livestock and spread to people working in and living near farms. Simultaneously, the use of growth promoter antibiotics has led to the contamination of flocks and food products by antibiotic-resistant pathogens, including campylobacter, salmonella, enterococcus, and escherichia coli, in turn leading to increased risks of human infections and other resistant pathogens. A few routes of transmission of antibiotic resistance and a few examples of countries with antibiotic concentration in the environment are shown in Table 8.4. The following is a list of possible transmission sources:

1. Water sanitation systems: Due to negligence and inefficient removal of antibiotics from wastewater treatment facilities before releasing into the environment.
2. Application of manure to fields with cultivated land.
3. Biomagnification through the food chain.
4. Poor infrastructure at harbor: The distribution of contaminated flesh of animals and fishes is also the cause of transmission.

TABLE 8.4

Few Examples of Transmissions of Antibiotics into the Environment

Country	Pharmaceutical Drugs	Concentration	Year	Reference
China	Oxytetracycline	Effluent: 1065 mg L^{-1}	1988	Qiting and Xiheng (1988)
			2008	Li et al. (2008)
	Penicillin G & its metabolites	Effluent: 44 mg L^{-1} Surfacewater 11.6 mg L^{-1}		
	Diclofenac	Surfacewater: 27 µg L^{-1}		Lin et al. (2008)
India	Fluoroquinolone	Effluent: Ciprofloxacin 31 mg L^{-1}	2007	Larson et al. (2007)
		Groundwater: Ciprofloxacin: 770 mg L^{-1}	2014	Rutgersson et al. (2014)
		Soil: Ciprofloxacin: 7.2 µg L^{-1} organic matter		
Denmark	Sulfonamide antibiotics and intermediates metabolites	Groundwater: Sulfaguanidine 1.6 mg L^{-1}	1995	Holm et al. (1995)
Norway	Bacitracin	Effluent: up to 250 kg per discharge	2005	Norwegian Environment Agency (2005)
Korea	Lincomycin	Effluent: 43.9 mg L^{-1}	2011	Sim et al. (2011)
Pakistan	Many other anitiotics	Surfacewater: Sulfamethoxazole 49 µg L^{-1}	2013	Khan et al. (2013)

5. Wildlife, insects, and pests (potential carrier of antimicrobial resistance)

6. Tourism, migration, and food imports

7. Healthcare facilities, spread between patients and healthcare staff, through contaminated tools and instruments

8.3 Antibiotic Resistance

Antibiotic resistance occurs when bacteria change their response to the particular drug or medicine. Antibiotic-resistant bacteria may harm animals and humans, and infections caused by these bacteria are very hard to treat because of their resistant to drugs (Berkner et al., 2014). Infectious diseases such as pneumonia, gonorrhea, and tuberculosis are becoming harder to treat because of overuse of antibiotics. Antibiotic resistance is growing to dangerously high levels in all parts of the world. New resistance mechanisms of bacteria are emerging and spreading globally, threatening our ability to treat common infectious diseases. Antibiotic-resistant bacteria are posing a major threat to humans and animals and to food security. Overuse of antibiotics in animals and humans accelerates the process of resistant in bacteria. An approach to prevent the misuse of antibiotics is urgently needed to deal with the major threat of antibiotic-resistant bacteria (http://www.who.int/mediacentre/factsheets/antibiotic-resistance/en/).

Use of antibiotics is not limited to the treatment of human illness. Antibiotics have also used in livestock and poultry for many decades to prevent and cure infectious diseases and to increase the production of animal products (Stokstad and Jukes, 1949; Page and Gautier, 2012).

A recent study of the U.S. Centers for Disease Control and Prevention (CCD) concerning the threat of bacterial pathogens refractory to treatment with antibiotic therapies contains a blunt warning: simply using antibiotics creates resistance (CDC, 2013a, 2013b). Antibiotic resistance is a well-known issue for microbiologists, epidemiologists, and physicians and for associated public health officials focused on the problem of antibiotic-resistant infectious diseases (Limoges and Jassim 2014). Bacteria resistant to antibiotics pose a challenge to the field of biomedical science because of their toxicity to kill the antibiotics and their ability to disrupt the physical structure of microbes (Landecker, 2015). It has been observed that antibiotic-resistant bacteria present today for any particular disease are new; genetic modifications, structural disruptions, and cultural and ecological changes accelerated the growth of antibiotic-resistant bacteria. Bacteria resistant to antibiotics have different traits and plasmids, interrelationships, temporalities, and distribution abilities than bacteria of the past. Still, it is not clear for the biomedical sciences that bacteria are only responsible for antibiotic resistance (Landecker, 2015). The major problem is the growth of antibiotic-resistant bacteria in hospitals, communities, and environment associated with their overuse for disease control. Antibiotic resistance due to the predominant role of anthropogenic activities and human interference cannot be disputed (Davies and Davies, 2010). Antibiotic resistance genes (ARGs) are increasing the numbers of toxins, creating a potential worldwide threat to human and animal health. Intensive livestock-producing farms are an important source for the increased environmental load of ARGs (Yong-Guan Zhu et al., 2012). The presence of ARGs in the surrounding environment has been caused by misuse of antibiotics in humans and animals (Knapp et al., 2010). It has been observed in the last 35 years that use of antibiotics has a direct relationship with the rise in associated resistant genes in humans as well as direct transfer of such antibiotic resistant genes from animals to humans (Levy, 1978; Smillie et al., 2011; Forsberg et al., 2012; Price et al., 2012). Uses of genome sequencing for many recent studies on antibiotics have confirmed animal-to-human transfers of ARGs (Harrison et al., 2013).

ARGs are identified as environmental pollutants; and the required steps must be taken to continue the efficacy of antibiotics (Zhu et al., 2012). See Table 8.5. The World Organization for Animal Health combined with WHO

TABLE 8.5

Occurrence of Antibiotics Resistance in Selected Organism

Organism	Occurrence	Comments
Salmonella typhimurium	Meat and meat products, unpasteurized milk Fish and shellfish Poultry, sheep and pigs Hamburgers	Multidrug resistant definitive type 104(DT 104) emerged in cattle in 1998 in England and Wales Products exported to US from Canada between 1986 to 1989 In Pacific Northwest 4% of human isolates were DT 104 in 1989 raised to 43% in 1994 1985 outbreak effected 1000 people in California with consumption of humburgers
Enterobacter and *Campylobacter*	Poultry, human and sewage plants receiving effluent from poultry abattoirs	Resistant to fluoroquinolones
Escherichia Coli	Animal feed (birds that ate animal feed or drank water) Vegetable and fruits Meat products and intestinal microflora of the pig farmers and their families as well as patients with urinary tract infections and general public municipality	Subtherapeutic use of antibiotics (penicillin, sulfethazine, chlortertracycline, oxytetacycline and neomycin) Due to use of nourseothricin) as a porcine growth promoting agent in the former East Germany
Enterococci	Isolated from sewage treatment plants, manure samples from pigs and poultry farms	Resistant to vancomycin Isolated from sewage treatment plants in Britain and small towns in Germany, Norway, and Netherlands

and US Food and Drug Administration (FDA) revised guidelines for the use of veterinary antibiotics in more than 100 developing countries (Gilbert, 2012). Antibiotics in food animal production does have major benefits, including improved food production, better animal health, and reduction in food-borne pathogens. However, this use, specifically for higher production, has increased the spread of antibiotic-resistant bacteria that are significant for humans (Mathew et al., 2007). Overuse of antibiotics can result in considerable pools of resistance genes among bacteria, which include human pathogens, and the risk posed to humans by antibiotic-resistant bacteria from farms and livestock has not been clearly defined (Mathew et al., 2007). Subtherapeutic antibiotics are extensively used in livestock, commercial farms, fish, and poultry farms for rapid production and as compensation for unsanitary farm environments (Emanuele, 2010). Today, many antibiotic drugs are significant for the human health, such as erythromycin, tetracycline, and penicillin; these drugs have been widely used for subtherapeutic purposes in livestock production (Mellon et al., 2001; Page and Gautier, 2012). In the United States, the use of antibiotics for subtherapeutic purposes in livestock was increased by 50% from 1985 to 2001 (Gerber et al., 2007), and it has been observed that around 80% of antimicrobials are used to feed livestock (US Congress, 2011). Increased use of antibiotics as a subtherapeutic purpose in poultry industry was observed to increase from 907,185 kg to 4,762,720 kg from 1980 to 2001 which was a 307% increase for an individual bird (Mellon et al., 2001). Overuse of such antibiotics in livestock and poultry industry has reduced the efficiency of antibiotics for treating major infectious diseases, a trend that has only accelerated in recent years (CDC, 2013a).

Approximately 75% of all antibiotics given to animals are not fully digested and eventually pass through the animal's body and enter the environment (Chee-Sanford et al., 2009), where they can encounter new bacteria and create additional resistant strains (Horrigan et al., 2002). With huge quantities of manure routinely sprayed onto fields surrounding confined animal feeding operations, antibiotic-resistant bacteria can leach into surface and groundwater, contaminating drinking wells and endangering the health of people living nearby (Clemans et al., 2011). Bacteria can also be spread by animals, birds, and insects that come in contact with animal waste (Graham et al., 2009; Page and Gautier, 2012). A considerable amount of pressure is being exerted on the natural microbial environment, including beneficial bacteria, and human and animal nutrition and immunity, by the antibiotics provided to humans, animals, and plants, as well as the spraying of antibiotics on fruit trees. The result is dangerous superbugs (Phillips et al., 2004b; Yan and Polk, 2004; O'Hara and Shanahan, 2006; Buffie et al., 2011).

8.3.1 Antibiotic Residue

The major sources of antibiotic residue are the pharmaceutical industries, hospitals, and medical shops; a very small percentage of antibiotic waste is also generated through common house waste. Antibiotic residue enters the environment also through patient excretion following the use of antiobitics to treat medical conditions. The antibiotic residue generally contains high moisture and strains of antibiotics that are very challenging to utilize or treat. Thus, antibiotics became victims of their own success.

One of the reports published by Science Daily (2018) mentions that the U.S. FDA approved more than 30 million pounds of antibiotics to be used in food-producing livestock in 2016. This amount is just a fraction of the total amount of antibiotics used all over the world in humans and animals.

The environment itself is a natural sink for resistance determinants. It also includes mobile genetic elements that operate as vectors. Even low concentrations of antibiotic are sufficient to provide a selective advantage for resistant over nonresistant microorganisms. Specific gene clusters encode the required protein to synthesize the code for self-protection, and these resistance-conferring proteins modify the antibiotic strain. The ARG has a phylogenic history going back years; the mutagen causes the resistive gene in particular genotype. If an antibiotic disappears from one zone, this does not mean it is eradicated because it could have been transferred to another zone.

8.3.2 Spread of Antibiotics through Human Beings

Antibiotics have a complex life cycle. When a patient takes an antibiotic, its cycle does not end with the patient. These pharmaceutical antibiotics are excreted, and the amount of antibiotic compounds excreted in urine and feces ranges from 10% to more than 90%; the exact amount varies according to the route of application and the type. Some antibiotic drugs, for example, beta-lactams, tetracyclines (fluoro), quinolones, phenols, and trimethoprim, are excreted in an amount that exceeds 50% of the administered dose (Berker et al., 2014). The antibiotics and their metabolites then ultimately go into the sewage system or wastewater treatment plant. After undergoing different steps like mechanical, biological, and chemical processes, the antibiotics present in the sewage are not fully treated fully; antibiotic residues still live within the system. As reported by Adriano Joss and his colleagues at the Swiss Federal Institute of Aquatic Science and Technology (2006), while conducting batch experiments on simulating activated sludge, the removal of antibiotics is either very slow or remains intact in the sludge itself. Several other researchers also suggested that antibiotics like fluoroquinolones are stored in the sewage sludge. When this sludge is introduced in the field as a fertilizer or filler material, the antibiotics accumulate in the environment. The pathways of antibiotic are spread via a matrix of biofilms, Waste Water Treatment Plant (WWTP)-treated effluent, sewage sludge, pharmaceutical production sites, and soil that is repeatedly fertilized with manure.

Antibiotic hot spots can increase the development of antibiotic resistance in two ways:

1. Direct transmission: Transmission of antibiotic-resistant microorganisms via the environment matrix.
2. Indirect transmission: A large number of microorganisms are exposed to antibiotic concentrations under favorable conditions.

8.3.3 Spread of Antibiotics through Cattle Feedstock, and Other Livestock

Ten million pounds of antibiotics are purchased every year for use in cows, pigs, poultry, and other livestock. When the manure of these animals contains hazardous compounds and it is reused in the field as fertilizer, the hazardous compounds, that is, the antibiotics leach into the soil, which ultimately results in land pollution. Diana Aga and Henry M. Woodburn (Buffalo College of Arts and Science, 2018) reported that the traces of antibiotic-resistant bacteria are not removed through systems like advanced anaerobic digestion and reverse osmosis filtration. An level of antibiotic residue containing both the traces of drugs or molecules that form due to the breaking up of larger molecules are left behind. The residue or the solid waste contains a higher concentration of antibiotics than unprocessed manure. Aga and Woodburn also found that none of the treatment systems are designed to process the antibiotic waste. Even the advanced anaerobic digestion (AD) used to reduce odor emissions and to produce biogas is not helpful in the removal of antibiotic compounds.

To test the effectiveness of both the technologies, two experiments were carried out. During the experiment, the solid matter of cow manure was extracted before introducing the elite waste management techniques. The leftover manure was subjected to both the treatment process that is, AD and RO. AD-enhance microorganisms to pasteurize and to break down the manure molecules and to produce gas. On the other hand, the slurry was passed through a series of RO steps to purify the water. The results show that antibiotic residues in liquid manure were reduced, but the number of antibiotics in solid matter were not reduced. Another major concern was that most of the solid matter of cows was extracted earlier, which means a huge amount of solid matter would have gone untreated. This experiment was also done to check a well-known class of antibiotics called tetracycline, and it was found that the liquid part of the sludge transferred to the solid part, resulting in a higher level of tetracycline antibiotics than the original raw manure. This experiment was done in RO to check the removal of synthetic antimicrobial known as ionosphere. These ionospheres promote growth in dairy cows and are used to cure coccidiosis parasitic diseases in cows. The results showed that the level of drug was lower in the liquid portion even after the treatment, and a high level of "ionosphere was found in the solid matter".

8.4 Global Perspective of Antibiotic Waste

The global scenario of antibiotic waste is a matter of concern. An online report (Third Pole, 2015) mentions that drugs in countries like India and China release their antibiotic waste directly into the environment, without any treatment (Zhua et al., 2013). This pollutes both the surrounding soil and the water. The antibiotic concentration is quite high near the industries that manufacture pharmaceuticals. Universities like Rice, Nankai, and Tianjin conducted research in North China and found that, for every waste treatment plant, four to five antibiotic bacteria are released into the environment. The pollution related to pharmaceuticals is very dangerous, but when antibiotics are involved, the problem is increased many times over. The antibiotic develops antimicrobial resistance wherever it is present: in soil, streams, rivers, and lakes. This antibacterial resistance can

TABLE 8.6

Antibiotics Banned in Few Countries

Antibiotics Banned	Year	Country or Region
Avoparin, Glycopeptide	1997	European Union
Virginiamycin	1998	Denmark
Antimicrobial growth promoters		
Bacitracin, Spiramycin, Tylosin	1999	European Union
and Virginiamycin	2006	
Antibiotic growth promoters		
Antibiotic growth promoters		Mexico and Taiwan

cross national and international boundaries. Humans and animals are carriers of these strains. These antibiotic-resistant microorganisms, also known as superbugs, cross the physical barrier with ease. Pharmaceutical production in the United States accounts for 39% of world production (Lim, 2012). Hospitals and other care centers discharge approximately 250 million pounds of unused antibiotic or pharmaceutical waste. Seeing the potential danger of antibiotic waste, US drug makers (Third Pole, 2015) raised a campaign to reduce the environmental impact of antibiotics manufacturing discharge by 2020 and banned the use of some drugs and antibiotics, as shown in Table 8.6.

8.4.1 Regulation of Hazardous Waste Pharmaceuticals around the World

- Antibiotics come under the category of hazardous waste. To reduce the environmental impact of pharmaceuticals, the U.S. Environmental Protection Agency (EPA) identified and classified pharmaceuticals as hazardous waste to ensure safe disposal.
- The Swann Committee in the United Kingdom recommended a ban on subtherapeutic use as animal feed to reduce or mitigate the effects of antimicrobial resistance in human beings.
- Approximately 31 chemicals were identified by the EPA in 1980 that match the Resource Conservation and Recovery Act (RCRA) hazardous waste category.
- The FDA classified an average of 30 new drugs each year since 1996.
- In 1997, WHO published a report and later released several reports linking the use of antibiotics in food animals to antibiotic resistance in humans and emphasizing the link between the spread of antibiotics resistance to the use of antibiotics in animal production.
- An amendment was proposed by U.S. EPA in 2008 to address pharmaceutical waste. The revised proposal then made available to public comment period in 2004. The revised proposal then made

available to the public in June 2014. The proposed standard for the hazardous waste pharmaceutical rule was published in the "Federal Region" on September 25, 2015.

- Several countries have initiated surveillance programs for antimicrobial resistance. A few countries, such as Denmark, Finland, France, Italy, Japan, Norway, the United States, and Sweden, are monitoring all scenarios of antibiotic-resistant bacterial infections in food (Epps and Blaney, 2016).

- Sweden is the first country to mandate and implement the Swann recommendations and eliminate the antimicrobial use.

- Various countries have banned the application of various antibiotics after several impacts on human health and the environment.

8.5 Bioprocessing of Antibiotic-Producing Industrial Solid Waste

According to the updated and revised rule proposed by U.S. EPA (2015), antibiotic residues are considered a hazardous waste. Dumping antibiotics in the open is prohibited. One cannot use residue containing an antibiotic as an animal feedstuff, fertilizer, or landfill. In recent years, many efforts were made to treat the antibiotic residue, but without success. Different treatment methods come with different drawbacks. For example, when pyrolysis was done to treat the antibiotic residue the distilled oil, carbon residue, and volatile gases were hard to return to the system (Liet et al., 2012). Activated carbon to treat the antibiotics is very costly; it also has a low strength constant and precise surface area of the product. The antibiotics sector is in dire need of a technology that can treat antibiotic waste. Many research studies are being conducted to develop a technology that is environmentally, economically, and technically viable. Bioprocessing, the process that uses complete living cells or their components (e.g., bacteria, enzymes, chloroplasts) to obtain desired products, emerged as the most eco-friendly, sustainable, and economically cost-effective method for processing antibiotic residue.

With the global menace of antibiotic resistance, several analytical techniques have been investigated to determine the levels of antibiotic residue in food animals. In particular their are two best known techniques is being used in present are confirmatory and Screening (see Figure 8.2). Screening methods are generally considered more viable due to theirs easy operation, cost-effectiveness, and good selectivity. Biosensors used in this approach make a practical solution in the large-scale detection of antibiotic residue in animals, including at farms and slaughterhouses.

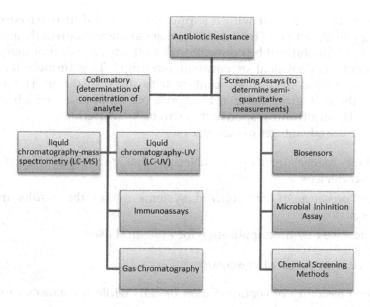

FIGURE 8.2
Steps for bioprocessing of antibiotics resistance in the environment.

One technology gaining popularity is hydrothermal technology, which uses hot water as a solvent. This technology is more efficient than other technologies as it uses water vapor to treat the high moisture content of antibiotic residue. The biodegradable property of the waste is increased to produce biogas. This technique increases not only the biodegradability of the waste for biogas (Bougrier et al., 2006) but also lessens the volume and consumes less energy (Neyens and Baeyers, 2003, Meng et al., 2012).

The complex process of intercellular dewatering can also be achieved. This technology cracks the cell walls and cell membrane to drain out biochemical contents and intercellular water. The antibiotic waste is very similar in supracolloidal structure (as the aggregate consists of mycelia and EPS) and is comprised mainly of saccharide and protein (Neyens et al., 2004; Comte et al., 2006). The dewatering phenomenon reduces the size of organic particulates in sewage sludge containing antibiotics, and they become vulnerable to microbes to undergo anaerobic digestion.

8.5.1 Methods of Detection of Antibiotic Residue

8.5.1.1 Biosensors

Biosensors are used on a large scale for the detection of antibiotic residue in food animals using a semiquantitative approach. They are a very practical solution at farms and slaughterhouses (Mungroo and Neethirajan, 2014). A biosensor includes two key elements: (1) a transuding device and

(2) a recognition element with a supporting material. It is based on the working principle that "The desired biological material (usually a specific enzyme) is immobilized by conventional methods (physical or membrane entrapment, non-covalent or covalent binding)". This immobilized biological material is in intimate contact with the transducer. The analyte binds to the biological material to form a bound analyte, which in turn produces the electronic response that can be measured.

Here are the advantages of biosensors:

- Ability to measure nonpolar molecules that are not receptive to most other devices.
- The presence of immobilized systems makes the results more specific.
- Extensive real-time application for industrial use.

Here is a disadvantage of biosensors:

- The possibility of contamination of cells while substances diffuse through the membrane.

8.5.1.2 Liquid Chromatography and Mass Specrometry

The introduction of the mass spectrometry (MS) based method for identification and quantification of antibiotics in the environment is much easier and cost effective. Coupling liquid chromatography (LC) with MS has brought more accuracy in determining the traces of antibiotic residue in the samples.

8.5.2 Methods of Bioprocessing of Antibiotic Residue

8.5.2.1 Biosorption

Biosorption plays a primary role in the removal of toxicants during biological treatment processes. The biosorption mechanism of antibiotics and hormones from the aqueous phase to sludge flocs/soils mainly occurs via absorption and adsorption. The hydrophobic interaction between aliphatic and aromatic groups with lipid molecules of the sludge cell membranes of microorganisms leads to absorption of antibiotics, while adsorption occurs through electrostatic interaction of a positively charged compound to negatively charged microbes and sludges (Li and Zhang, 2010; Luo et al., 2014).

8.5.2.2 Biodegradation

Biodegradation is one of the most important mechanisms for removing antibiotics and hormones from food animal wastes (Chen et al., 2012). The study conducted by Zheng et al. (2017) demonstrated that >60% of 11 veterinary antibiotics in swine wastewater were removed by biodegradation, while only 24% were absorbed by sludge. A study conducted at the American Institute of Chemical Engineers Environment Program found a novel antibiotic(tylosin)-degrading strain, namely, Citrobacter amalonacticus, isolated from soil deposited by pharmaceutical solid waste (PSW). The strain is capable of degrading almost 100% of tylosin in PSW with an initial concentration of 50 mg/L after 72 hours of incubation (Ma et al., 2014). Co-metabolic biodegradation plays a crucial role in the removal of antibiotics and hormones during biological treatment processes due to its low concentration to act as a direct growth substrate (Majewsky, 2014).

8.5.2.3 Anaerobic Digestion Process

The removal of antibiotics such as ampicillin, florfenicol, sulfamethoxydiazine, tylosin, and trimethoprim from swine and cattle manure through the anaerobic digestion process has been explored. However, the process of digestion studied so far at the mesophilic temperature indicated increased removal efficiency of the antibiotics.

8.5.2.4 Membrane Bioreactor Process

Compared to other conventional treatment processes, membrane bioreactors (MBRs) are more efficient for the processing and treatment of antibiotic residue in waste samples. An MBR process is a combination of adsorption, biodegradation, and membrane separation processes that facilitate a high quality of effluent with very low amounts of total suspended solids (TSSs).

8.5.2.5 Activated Sludge Process

Activated sludge process (ASP) for water treatment was established in the last century as a biological treatment process. It has been playing a major role for control of the conventional water pollutants, like total solids, nutrients (N/P), COD, BOD, etc. Bacterial diversity is high in active suspended biomass compared to any soil sample and human excreta. ASP in wastewater treatment was identified as a key hot spot for the release of ARGs into surface and subsurface water bodies. ARG dissemination through active sludge has risks to humans and livestock.

ARGs generated in ASP may travel to the bulk water of the aeration tank and sedimentation unit, and may be released with the final effluent. The ARGs remaining in the final effluent are released into the surrounding water bodies, and sludge containing resistant bacteria is spread into the soil by sludge disposal or land application. The resistant bacteria generated in active sludge could pollute the soil and water bodies when sludge is used as a soil fertilizer/conditioner, which is the current practice in many countries. The effluent from wastewater treatment plants generally carries bacteria of 105 to 106 cells/liter, which contributes continuously to the resistance pool in the surface water body and even in soil if treated water is used for irrigation. Various major ARGs like β-lactam (25), tetracyclines (26), have been found in the ASP unit of wastewater treatment plants (http://resistancecontrol.info/2016/amr-in-food-water-and-the-environment/antibiotics-and-resistance-genes-in-wastewater-treatment-plants/). Considering the number of wastewater treatment plants with ASPs worldwide, the importance of the dissemination of ARGs cannot be ignored. However, studies carried out on the development of ARGs in the ASP of the wastewater treatment plants are very few, so it has not been entirely overlooked. However, a precise and comprehensive knowledge of ARG profiles in wastewater treatment plants is critical for understanding the dissemination of ARGs in water bodies.

8.6 Conclusion

The magnitude of antibiotic consumption in various fields such as human health, food animals, poultry, etc., has revolutionized the tremendous growth in the number and types of infections. However, the high rate of consumption of antibiotics in human and veterinary medicine means that most of the microorganisms and pathogens have developed resistance mechanisms. The antibiotic residue has become a growing worldwide concern due to its presence in the environment. The unmetabolized (generally 55%–80%) compounds of antibiotics are disposed of openly. Most important, the major sources of contamination of antimicrobial agents are through its veterinary use because the contamination occurs in soil, surface water runoff after application of manure, fodder, etc. Therefore, bioprocessing of antibiotics has become a necessary and important aspect to combat the spread of the use of antibiotics and their subsequent residue in the environment.

References

Antibiotic Resistance Fact Sheet. (2018). World Health Organization. http://www.who.int/mediacentre/factsheets/antibiotic-resistance/en/.

Antibiotic Resistance Threats in the United States, Report. (2013). U.S. Department of Health and Human Services Centres for Disease Control and Prevention, pp. 1–114. https://www.cdc.gov/drugresistance/pdf/ar-threats-2013-508.pdf.

Berkner S, Konradi S, and Schonfeld J. (2014). Antibiotic resistance and the environment there and back again. *EMBO Reports* 15(7):740–744.

Bougrier C, Delgenes JP, and Carrère H. (2006). Combination of thermal treatments and anaerobic digestion to reduce sewage sludge quantity and improve biogas yield. *Process Saf Environ Prot* 84:280–284.

Boxall ABA, Rudd MA, Brooks BW, Caldwell DJ, Choi K, and Hickmann S. (2012). Pharmaceuticals and personal care products in the environment: What are the big questions? *Environ Health Perspect* 120:1221–1229.

Buffalo College of Arts and Science. (2018). http://www.buffalo.edu/news/releases/2018/04/022.html.

Buffie C, Jarchum I, Equinda M, and Lipuma L. (2011). Profound alterations of intestinal microbiata following a single dose of clindamycin results in sustained susceptibility to Clostridium difficile-induced colitis. *Infect Immun* 80:62–73.

Capitano B and Nightingale CH. (2001). Optimizing antimicrobial therapy through use of pharmacokinetic/pharmacodynamic principles. *Med Infect Dis* 21:1–8.

Carraschi SP, da Cruz C, Machado Neto JG, Ignácio NF, Barbuio R, and Machado MR. (2012). Histopathological biomarkers in pacu (*Piaractus mesopotamicus*) infected with *Aeromonas hydrophila* and treated with antibiotics. *Ecotoxicol Environ Saf* 83:115–120. doi:10.1016/j.ecoenv.2012.06.016.

Centers for Disease Control and Prevention (CDC). (2013a). *Antibiotic Resistance Threats in the United States, 2013*. Atlanta, GA: US Department of Health and Human Services. https://www.cdc.gov/drugresistance/threat-report-2013/index.html.

Centers for Disease Control and Prevention (CDC). (2013b). Vital signs: Carbapenem-resistant Enterobacteriaceae. *Morb Mortal Wkly Rep* 62(9):165–170.

Chee-Sanford JC, Mackie RI, Koike S, Krapac IG, Lin YF, Yannarell AC, Maxwell S, and Aminov RI. (2009). Fate and transport of antibiotic residues and antibiotic resistance genes following land application of manure waste. *J Environ Qual* 38:1086–1108.

Chen Y, Zhang H, Luo Y, and Song J. (2012). Occurrence and assessment of veterinary antibiotics in swine manures: A case study in east China. *Chin Sci Bull* 57:606–614.

Clemans D, Francoeur S, Liggit P, and West B. (2011). Antibiotic resistance, gene transfer, and water quality patterns observed in waterways near CAFO farms and wastewater treatment facilities. *Water Air Soil Pollut* 217(1–4):473–489.

Comte S, Guibaud G, and Baudu M. (2006). Relations between extraction protocols for activated sludge extracellular polymeric substances (EPS) and EPS complexation properties Part I. Comparison of the efficiency of eight EPS extraction methods. *Enzyme Microb Technol* 38:237–245.

Cromwell GL. (2001). Antimicrobial and promicrobial agents. In: Lewis A and Southern L (eds.) *Swine Nutrition*, 2nd ed. CRC Press, Boca Raton, FL, pp. 401– 426.

Dabbir BR. (2017). Prevention of anthrax epidemic in sheep and goats with Anthracinum 200. *Indian J Res Homoeopathy* 11(4):244–248.

Dapcevich M. 2018. Treated farm waste contains traces of antibiotics, possibly contributing to spread of resistance. The lighter side of science. http://www.iflscience.com/environment/treated-farm-waste-contains-traces-ofantibiotics-possibly-contributing-to-spread-of-resistance. Accessed July 9, 2018.

Davies J and Davies D. (2010). Origins and evolution of antibiotic resistance. *Microbiol Mol Biol Rev* 74(3):417–433. doi:10.1128/MMBR.00016-10.

Doyle MP and Erickson MC. (2006). Reducing the carriage of foodborne pathogens in livestock and poultry. *Poul Sci* 85(6):960–973.

DuPont HL. (2007). The growing threat of foodborne bacterial enteropathogens of animal origin. *Clin Infect Dis* 45(10):1353–1361.

Emanuele P. (2010). Antibiotic resistance. *AAOHN J* 58(9):363–365. doi:10.3928/08910162-20100826-03.

Epps AV and Blaney L. (2016). Antibiotic residues in animal waste: Occurrence and degradation in conventional agricultural waste management practices. *Curr Pollution Rep* 2:135–155.

Forman CR and Burch JE. (1947). Use of sodium sulphonamides as single injection specific treatment in foot rot. *J Am Vet Med Assoc* 111:208–214.

Forsberg KJ, Reyes A, Wang B, Selleck EM, Sommer MO, and Dantas G. (2012). The shared antibiotic resistome of soil bacteria and human pathogens. *Science* 337(6098):1107–1111.

Gerber P, Opio C, and Steinfeld H. (2007). Poultry production and the environment—A review. FAO Publishing. http://www.fao.org/ag/againfo/home/events/bangkok2007/docs/part2/2_2.pdf.

Gilbert N. (2012). Rules tighten on use of antibiotics on farms. *Nature* 481(7380):125.

Global Health Observatory (GHO). Data. World Health Organization. http://www.who.int/gho/child-malnutrition/en/.

Graham JP, Boland JJ, and Silbergeld E. (2007). Growth promoting antibiotics in food animal production: An economic analysis. *Public Health Rep* 122(1):79–87.

Grant EM and Nicolau DP. (1999). Pharmacodynamic considerations in the selection of antibiotics for respiratory tract infections. *Antibio Clinic* 3, Suppl. 1, 21–8.

Hao H, Cheng G, Iqbal Z, Ai X, Hussain HI, Huanh L, Dai M, Wang Y, Liu Z, and Yuan Z. (2014). Benefits and risks of antimicrobial use in food-producing animals. *Fron Mircrob* 5:1–11.

Harrison EM, Paterson GK, Holden MT, et al. (2013). Whole genome sequencing identifies zoonotic transmission of MRSA isolates with the novel mecA homologue mecC. *EMBO Mol Med* 5:509–515.

He S, Zhou Z, Meng K, Zhao H, Yao B, and Ringo E. (2011). Effects of dietary antibiotic growth promoter and *Saccharomyces cerevisiae* fermentation product on production, intestinal bacterial community, and nonspecific immunity of hybrid tilapia (*Oreochromis niloticus* female × *Oreochromis aureus* male). *J Anim Sci* 89:84–92. doi:10.2527/jas.2010–3032.

Holm JV, Rugge K, Bjerg PL, and Christensen TH. (1995). Occurrence and distribution of pharmaceutical organic compounds in the groundwater downgradient of a landfill (Grindsted, Denmark). *Environ Sci Tech* 29(5):1415–1420.

Horrigan L, Lawrence RS, and Walker P. (2002). How sustainable agriculture can address the environmental and human health harms of industrial agriculture. *Environ Health Perspect* 110(5):445–456.

Hurd HS, Gailey JK, McKean JD, and Griffith RW. (2005). Variable abattoir conditions affect *Salmonella enterica* prevalence and meat quality in swine and pork. *Foodborne Pathog Dis* 2:77–81. doi:10.1089/fpd.2005.2.77.

http://www.thethirdpole.net/en/2015/12/22/antibiotic-waste-is-on-the-rise-in-china. Accessed July 9, 2018.

https://www.epa.gov/hwgenerators/proposed-rule-management-standards-hazardous-waste-pharmaceuticals. Accessed July 8, 2018.

https://www.ncbi.nlm.nih.gov/pmc/articles/PMC5390938/-Antibiotic Resistance and the Biology of History, Monitoring Editors: Lisa Blackman, Hannah Landecker.

https://www.sciencedaily.com/releases/2018/04/180412154509.htm. Accessed July 7, 2018.

Joss A, Zabczynski S, Göbel A, Hoffmann B, Löffler D, McArdell CS, Ternes TA, Thomsen A, and Siegrist H. (2006). Biological degradation of pharmaceuticals in municipal wastewater treatment: Proposing a classification scheme. *Water Res* 40:1686–1696.

Khan GA, Berglund B, Khan KM, Lindgren PE, and Fick J. (2013). Occurrence and abundance of antibiotics and resistance genes in rivers, canals and near drug formulation facilities-A study in Pakistan. *PLoS ONE* 8(6):e6271. doi:10.1371/journal.pone.0062712.

Kimman T, Hoek M, and Jong MCM de. (2013). Assessing and controlling helth risks from animal husbandry. *NJAS-Wagenin J Life Sci* 66:7–14.

Knapp CW, Dolfing J, Ehlert PAI, and Graham DW. (2010). Evidence of increasing antibiotic resistance gene abundances in archived soils since 1940. *Environ Sci Technol* 44(2):580–587.

Krehbiel C. (2013). The role of new technologies in global food security: Improving animal production efficiency and minimizing impacts. *Anim Front* 3:4–7.

Kummerer K. (2001). Drugs in the environment: emission of drugs, diagnostic aids and disinfectants into wastewater by hospitals in relation to other sources – a review. *Chemosphere* 45:957–969.

Landeckar H. (2016). Antibiotic resistance and the biology of History. *Body Soc* 22(44):19–52.

Landers TF. (2012). *A Review of Antibiotic use in Food Animals: Perspective, Policy and Potential.* New York: Centre for Interdisciplinary Research to Reduce Antibiotic Resistance, Columbia University School of Nursing. 127:5–21.

Larsson DGJ, de Pedro C, and Paxeus N. (2007). Effluent from drug manufactures contains extremely high levels of pharmaceuticals. *J Hazard Mater* 148:751–755.

Levy SB. (1978). Emergence of antibiotic-resistant bacteria in the intestinal flora of farm inhabitants. *J Infect Dis* 137(5):689–690.

Li B, and Zhang T. (2010). Biodegradation and adsorption of antibiotics in the activated sludge process. *Environ Sci Technol* 44(9):3468–3473.

Li YX, Li W, Zhang XL, and Yang M. (2008). Simultaneous determination of fourteen veterinary antibiotics in animal feces by solid phase extraction and high performance liquid chromatography. *Chin J Analyt Chem* 40:213–217.

Li Z, Zuo J, Tian B, Yang J, Yu X, Chen P, and Zhao Y. (2012). Thermal—Alkaline pretreatment on the decomposition of the streptomycin bacterial residue. *Biotechnol Biotechnol Equip* 26:2971–2975.

Limoges RG and Jassim SA. (2014). Natural solution to antibiotic resistance: Bacteriophages 'The Living Drugs'. *World J Mircobiol Biotechnol* doi:10.1007/s11274-014-1655-7.

Marshall BM and Levy SB. (2011). Food animals and antimicrobials: Impacts on human health. *Clini Micro Review* 24:718–733.

Mathew AG, Cissell R, and Liamthong S. (2007). Antibiotic resistance in bacteria associated with food animals: A United States perspective of livestock production. *Foodborne Pathog Dis* 4(2). doi:10.1089/fpd.2006.0066.

Mellon M, Benbrook C, and Benbrook KL. (2001). Hogging it: Estimates of antimicrobial abuse in livestock. Publishing UCS Web. http:// www.ucsusa.org/assets/documents/food_and_agriculture/hog_ front.pdf. Accessed March 9, 2018.

Meng D, Jiang Z, Kunio Y, and Mu H. (2012). The effect of operation parameters on thehydrothermal drying treatment. *Renew Energy* 42, 90–94.

National Research Council (US) Committee on Drug use in food animals. (1999). *The use of Drugs in Food Animals: Benefits and Risks.* Washington, DC: National Academies Press.

Neyens E and Baeyens J. (2003). A review of thermal sludge pre-treatment processes to improvedewaterability. *J Hazard Mater B* 98:51–67.

Neyens E, Baeyens J, Dewil R, and De heyder B. (2004). Advanced sludge treatmentaffects extracellular polymeric substances to improve activated sludgedewatering. *J Hazard Mater B* 106:83–92.

Nightingale CH, Murakawa T, and Ambrose PG., Eds. (2001). *Antimicrobial Pharmacodynamics in Theory and Clinical Practice.* New York: Marcel Dekker.

Page SW and Gautier P. (2012). Use of antimicrobial agents in livestock. *Rev Sci Tech Off Int Epiz* 31(1):145–188.

Phillips I, Casewell M, Cox T, De Groot B, Friis C, Jones R, Nightingale C, Preston R, and Waddell J. (2004a). Antibiotics use in animals. *J Antimicrob Chemother* 53:885.

Phillips I, Casewell M, Cox T, De Groot B, Friis C, Jones R, Nightingale C, Preston R, and Waddell J. (2004b). Does the use of antibiotics in food animals pose a risk to human health? A critical review of published data. *J Antimicrob Chemother* 53:28–52.

Price LB, Stegger M, Hasman H, et al. (2012). Staphylococcus aureus CC398: Host adaptation and emergence of methicillin resistance in livestock. *MBio* 3(1):00305–00311.

Roser M and Ospina EO. (2017). World Population Growth. Our World in Data. https://ourworldindata.org/world-population-growth.

Smillie CS, Smith MB, Friedman J, Cordero OX, David LA, and Alm EJ. (2011). Ecology drives a global network of gene exchange connecting the human microbiome. *Nature* 480(7376):241–244.

Stokstad ELR and Jukes TH. (1949). Further observations on the animal protein factor. *Proc Soc Biol Exp Med* 73:523–528.

Tufa TB. (2016). Veterinary drug residues in food-animal products: Its risk factors and potential effects on public health. *J Veterinar Sci Tech* 7:1–7.

US Congress. (2011). H.R.965 (112th): Preservation of antibiotics for medical treatment act of 2011. 112th Congress, 2011–2013. Publishing Govtrack Web. https://www.govtrack.us/congress/ bills/112/hr965/text. Accessed March 12, 2018.

Wallace J, Garner E, Pruden A, and Aga SD. (2018). Occurrence and transformation of veterinary antibiotics and antibiotic resistance genes in dairy manure treated by advanced anaerobic digestion and conventional treatment methods. *Environ Pollut.* 236:764. doi:10.1016/j.envpol.2018.02.024.

Waxman HA, Corr B. (2016). Waste from pharmaceutical plant in India and china promotes antibiotic assistant superbugs. https://www.statnews.com/2016/10/14/ superbugs-antibiotic-resistance-india-china. Accessed July 9, 2018.

World Population Growth. (2018). United Nations Population Division. Department of Economic and Social Affairs. http://www.worldometers.info/ world-population/.

Zhang G, Li C, Ma D, Zhang Z, and Xu G. (2015). Anaerobic digestion of antibiotic residue in combination with hydrothermal pretreatment for biogas. *Bio resource Technology* 192:257–265.

Zhua YG, Johnsonc TA, Su JQ, Qiaob M, Guo GX, Stedtfeld RD, Hashshamc SA, and Tiedje JM. (2013). Diverse and abundant antibiotics resistance genes in Chinese.

9

Bio-Processing of Mining Solid Waste and Resource Recovery

Amjad Ali, Fazli Wahid, and Di Guo

CONTENTS

9.1 Introduction

Rapid industrialization and sustainable economic growth require the exploration of natural resources to feed and clothe the large population around the world. Mining provides raw materials to different industries. Mining also creates waste rocks, tailing and overburden. The dust and wastewater released from mining sites is flushed into water channels (canals, rivers, etc.) and become part of the aquatic resources. Water scarcity in arid regions has forced farming communities to use the mine wastewater for irrigation purposes. However, mine wastewater is laden with harmful wastes like heavy metal (HMs), acidic substances, and minerals (Jamali et al., 2008;

Ma et al., 2015). The wastewater poses threats to the marine environment and to soil, and the dust travels long distances to harm ecosystems and humans. Mining operations can disrupt soil structure and alter soil biological properties and vegetative composition due to the removal of the topsoil layer. Mining of a target mineral creates several times more waste, which contains HMs. The HMs present in these wastes is mobilized by chemical or biological leaching and reaches water sources to find their way into the food chain. The dust released from the mining sites reduces air quality in the areas far away from the main site (Alvarez et al., 2017).

The contaminated soil, mine tailings, overburden, and rock piles from mining operations are deprived of flora due to extreme acidity and HM toxicity (Kelly et al., 2014). Mining activities have considerably increased the HM contamination in the food chain and the surrounding environment. Mining activities stress the safety of our environment and increase human health hazards (Spokas et al., 2009; World Health Organization [WHO], 2017). Proper reclamation of mine-contaminated soils after mining operations is needed to reduce human health hazards and enhance vegetation (Fellet et al., 2014; Pazferreiro et al., 2014). The conventional remediation technologies for HM removal include chemical precipitation, adsorption, ion exchange, and membrane separation. These methods are costly and generate residues with no economic value (Oliveira et al., 2017).

HM recovery from contaminated sites is very laborious when conventional methods are employed. These conventional methods have variable efficiencies and depend on the nature of soil contaminants, soil basic characteristics, and the geography of the location (Marques et al., 2009). Similarly, the in-situ stabilization method of applying stabilizing agents (biosolids, sawdust, wood ash, biochar, fly ash, etc.) aims to decrease the HMs labile pool, thus reducing the bioavailability and toxicity of HMs. Adsorption by organic materials can provide a nutrient supply; promote water retention and organic matter (Forestier et al., 2017; Hattab et al., 2014); and enhance metal adsorption through surface charge, organic metal complexes, and precipitates formation (Ahmad et al., 2014; Komárek et al., 2013). The organic amendments immobilize/adsorb HMs (Gadepalle et al., 2007). However, the possible release of HMs via the decomposition of organic matter can be a potential drawback of this process (Herwijnen et al., 2007). Organic material like biochar has a huge potential to stabilize HMs in the polluted sites (Lomaglio et al., 2016; Pazferreiro et al., 2014).

The remediation and rehabilitation of mine sites and hazardous waste can be attained by utilizing the phytostabilization, biotechnological, and organic amendments. These can be reliable and cost-effective rehabilitation strategies to promote the revegetation process, lower the transfer of pollutants (that is, HMs), and restore ecological indicators (Fellet et al., 2011). These goals can't be achieved without soil amendments. The addition of biochar and metals resistant microbial consortium to polluted soil/leftover rocks raise soil acidity, H_2O holding capacity, and fertility; lower the mobility of available pollutants and promote revegetation and soil enzymatic activities. (Anawar et al., 2015; Fellet et al., 2014; Kelly et al., 2014; Reverchon et al., 2015). Hence, innovative remediation strategies must be developed to reduce the impact of mining on mine landscapes to ensure a healthy ecosystem (Moreno-Barriga et al., 2017a).

9.2 Impacts of Mining

Metalliferous mining results in vastly degraded ecosystems, reduces vegetation cover and faunal diversity, changes soil quality/structure, changes the morphology of the area, and creates large areas of derelict land resulting from various mineral extraction and on-site dumping of residues (Zawadzki et al., 2016). The possible effects of the metalliferous mine residues include low acidity, high concentrations of HMs/metalloids, and extremely low soil organic matter content. These results hinder plant root growth in the soil and microbial activities (Martínez-Pagán et al., 2011). Mining results in acid mine drainage, which is a source of water contamination (Johnson and Hallberg, 2005).

9.2.1 Impacts on the Environment

Global mining activities have a negative impact on the environment due to the deposition of waste materials. The hazardous wastes commonly include acid-generating sulfide minerals, and HMs (Zn, Pb, As, Cd, Cu, Sn, Hg, Mn, Fe, Al, Sb, Tl, U, Ag, Th, and W) (Modabberi et al., 2013). The soil and water erosion, constant dispersal, leaching into the soil profile, and atmospheric transport of tailings via wind currents may distribute HMs in the aquatic and terrestrial systems (Anawar et al., 2011). The disposal (burial, pedogenesis) of mine waste and the formation of oxyhydroxide minerals (Mn, Fe, and Al) can control HMs fractionations and acid mine drainage on residential and agricultural lands as well as in marine ecosystems. Therefore, mining practices cause huge environmental impacts with water and soil pollution and biodiversity extinction (Anawar, 2015).

9.2.2 Impact on Human Health

The extension of agricultural activities and urbanization has helped the conversion of a number of mining sites worldwide into residential areas and farmlands, posing a human health risk (Anawar, 2015). The HMs released from the mining sites pose threats to humans, especially in the local proximity. The impact of HMs on humans is classified in Table 9.1.

TABLE 9.1

List of Heavy Metals, Their Risks to Human Health, and Permissible Limits

HMs	Effects	Permissible Limit (mg/L)	References
As	Bronchitis, dermatitis, poisoning, respiratory disorders.	0.02	Abdelhafez et al. (2014)
Cd	Renal dysfunction, blood pressure, bronchitis, Itai-Itai disease, pulmonary adenocarcinomas, osteoporosis, prostatic lesions, kidney dysfunction.	0.06	Sharma and Archana (2016)
Pb	Encephalopathy in children, liver, kidney.	0.10	Abdelhafez et al. (2014)
Mn	Damage to central nervous system.	0.26	Simate and Ndlovu (2014)
Hg	Spontaneous abortion, tremors, leading to tremors, shyness, memory loss. Anxiety, depression, drowsiness, fatigue, insomnia, and ulcer.	0.01	Alina et al. (2012)
Zn	Harm nervous membrane.	15.0	Simate and Ndlovu (2014)
Cr	Damage nervous system, fatigue, irritability.	0.05	Gill et al. (2015)
Cu	Damage Liver, stomach, hypochromic anemia, leukopenia, and osteoporosis in human.	0.10	Goswami and Das (2016)
Ni	Cause dermatitis (nickel itch), carcinogenic in nature and can lead to lungs, throat, nose and stomach cancer, immunotoxic, genotoxic, and hepatotoxic.	—	Ali et al. (2013)

9.2.3 Impacts on Plants

HMs not only severely affect human health but also pose threats to vegetation and ecosystems. The low pH and organic matter in the soil in and around mines impede plant propagation and microbial activities. The various impacts of HMs on plants are classified in Table 9.2.

9.2.4 Socioeconomic Impacts of Mining

Mining not only has adverse effects on humans and plants but also on society. Local people usually do not reap the benefits from mining, and the revenues generated are used in other sectors by the government. The non-provision of jobs to the local people is also a common problem in the mining areas. Mines are usually away from developed and urban areas. Child labor is also common in such mining areas. There is a possibility of the onset of armed conflict and the dependence of a country or a nation on one or two

TABLE 9.2

Effects of Specific HMs on Plant Health

HMs	Effects	References
Cd	Reduces germination, reduces lipid content, cause leaf chlorosis, inhibits auxin homeostasis.	Rizwan et al. (2016)
Pb	Reduces chlorophyll production, increases superoxide dismutase, stunted growth, root cessation and even leaf senescence.	Shu et al. (2012)
Ni	Reduces germination, protein production, chlorophylls content and enzymes.	Alam et al. (2007)
Hg	Reduce photosynthesis, H_2O uptake, lower antioxidant enzymatic activities, accumulates phenolic compound and proline.	Simate and Ndlovu (2014)
Zn	Lower Ni toxicity and improve plant growth.	Simate and Ndlovu (2014)
Cr	Reduces enzyme activity, damage membrane and root overproduction of ROS and lead to chlorosis.	Adrees et al. (2015)
Cu	Inhibition of photosynthesis, plant reproduction, impose changes in respiratory process, inactivate enzymatic activities.	Singh et al. (2007)
As	Reduce seed germination, leaf area, tiller and dry matter production, reduction in fruit yield, stunted growth, leaf chlorosis and plant wilting.	Chibuike and Obiora (2014)

natural resources that can be easily exploited, such as diamonds, gold, and oil. Radioactive substances emitted from uranium mines has devastating effects on the environment. The huge amount of fresh water used by mining companies and the waste discharged into clean water reservoirs are the key causes of water pollution, human disease, and economic problems. The natural biodiversity and the habitat are usually dead after the initiation of mining activities.

9.3 Remediation Methods for Mine Wastes

Mining has been in existence throughout human history, but rehabilitation activities in mining sites have only been initiated in the past few decades. To protect the environment, farmland, and fresh H_2O resources as well as to ensure human health, various remediation practices are in use. Remediation/rehabilitation schemes are required for the restoration of ecological indicators of mining sites and to reduce human health hazards. Remediation of mining sites is a very complex process due to the presence of a variety of HMs and soil conditions, which influence the remediation process (Ma et al., 2015). Some of the remediation schemes are low cost and environmentally sustainable, but most are expensive and not suitable to adopt in many cases. Some of the more widely acceptable, cheaper, and most recent techniques are mentioned below. Techniques for HM removal from aqueous solutions are classified as chemical, physical, and biological. The conventional methods are filtration, chemical precipitation, ion exchange, membrane technologies, electrochemical treatment, adsorption, and evaporation. Chemical precipitation and electrochemical treatment are not effective at high concentrations (1–100 mg L^{-1}) in solution and produce sludge as a secondary pollutant. Similarly, some techniques, like ionic exchange and membrane technologies, are costly for treating HM-contaminated wastewater (Wang and Chen, 2009).

9.3.1 Removal and Recovery of HMs by Biosorbents from Mining Solid Waste

The term *biosorption* is defined as "the removal of metal or metalloid species, compounds and particulates from solution by biological material" (Gadd, 1993). In the biosorption process, both living and dead biomasses and their products (polysaccharides) are utilized for HM retention. This method can be employed in treating mine wastewater. HM pollution is a common environmental problem associated with mining. Mining and smelting discharge different HMs into the environment, cause serious environmental pollution, and threaten human health and the natural ecosystem (Wang and Chen, 2006). Recently, biotechnology has been given more attention as a potential method for removing metals. Biosorption uses natural sorbents, like fungi, bacteria, algae, and yeast, etc. These microorganisms sequester HMs and reduce the content of HMs in aqueous solution (ranging from ppm to ppb), and it is employed to treat a high volume of mine wastewater (Wang and Chen, 2006). The function of different biosorbents depends on the chemical formation of the microbial cells (Volesky et al., 1993). Mechanisms responsible for biosorption of metallic ions in wastewater may use a single process or a combination of ion exchange, adsorption, coordination, electrostatic interaction, complexation, chelation, and precipitation (Vijayaraghavan and Yun, 2008; Wang and Chen, 2006).

Different materials that have been explored as biosorbents for the elimination of HMs are shown in Table 9.3. These sorbents are categorized as bacteria, fungi, yeast, algae, industrial wastes, agricultural wastes, and polysaccharide materials (Vijayaraghavan and Yun, 2008). Bacteria, fungi, yeast, and algae has been well studied, interpreted, and confirmed to bind HMs to various extents. Abundant cellulosic materials have the potential to biosorb HMs (Wang and Chen, 2009).

TABLE 9.3

Biosorption of Precious Metals by Different Biosorbents

Metals	Biosorbent	Q_{max} (mg/g)	Sorption Condition	References
Au(I)	Fungal (*Penicillium chrysogenum*)	1.42	pH 2	Niu and Volesky (1999)
	Hen eggshell membrane	147	pH 3	Ishikawa et al. (2002)
	PEI-modified bacterial biosorbent fiber	421.1	pH 5.5	Park et al. (2012)
Au(III)	*Cladosporium cladosporioides* Strain 1	94.2	pH 4.0	Pethkar et al. (2001)
	Streptomyces erythraeus	6	pH 4.0	Savvaidis (1998)
	Sulfur derivative of chitosan	630.29	pH 3.2	Arrascue et al. (2003)
Pt(IV)	*Desulfovibrio desulfuricans*	62.5	pH 2.0	Vargas et al. (2010)
	Thiourea derivative of chitosan	386.9	pH 2.0	Guibal et al. (2015)
	Poly (allylamine hydrochloride)-modified *E. coli*	348.8	pH 2.5	Mao et al. (2010)
Pd(II)	EN-lignin	22.7	0.5 M HCl	Parajuli et al. (2008)
	Bayberry tannin immobilized collagen fiber membrane	33.4	1 M HCl	Ma et al. (2006)
	Racomitrium lanuginosum	37.2	pH 5	Sari et al. (2009)
Ru	Decarboxylated PEI-modified *C. glutamicum* biomass	47.1	Ru-containing acetic acid wastewater	Song et al. (2013)
	PEI-modified *C. glutamicum* biosorbent fiber	110.5	Ru-containing acetic acid wastewater	Kwak et al. (2013)
Zn	Hybrid pectin-based biosorbents	17.7–25.4	pH 4–7	Jakóbik-Kolon et al. (2017)
Cd	Chicken eggshells	1.57	Acid mine drainage	Zhang et al. (2017)
Ni	*Lysinibacillus* sp. BA2	238.04	pH 6, bauxite mine	Prithviraj et al. (2014)
Pb	*Penicillium* sp. MRF-1	0.357, 0.014	pH 4	Velmurugan et al. (2010)
Pb	Chicken eggshells	146.4	Acid mine drainage	Zhang et al. (2017)
Cu	Chicken eggshells	387.51	Acid mine drainage	Zhang et al. (2017)

Source: Won, S.W. et al., *Bioresour. Technol.*, 160, 203–212, 2014.

9.3.2 Stabilization and Phytomanagement of HMs by Biochar from Mining Solid Waste

Biochar is an aromatic, porous organic material produced from biomass under limited or oxygen-free conditions. Biochar possesses organic functional groups, i.e., alcohol (R-CH2-OH), carboxyl (R-COOH), carbonyl (R-C=O-R/H/OR), phenolic (aromatic-OH), and anomeric (O-C-O carbons), with negative surface charge, high surface area, and adsorption capability for HMs (Ippolito et al., 2017; Lehmann et al., 2011; Lomaglio et al., 2016). Biochar can be used on polluted mining sites for the simultaneous adsorption of HMs and the enhancement of vegetative growth. Biochar reduces the bioavailability/phytotoxicity of HMs, and improves the enzymatic activities and nutritional components (N and P) in polluted soil (Anawar et al., 2015; Lehmann et al., 2011).

9.3.2.1 Impact of Biochar on Phytostabilization in the Polluted Soil of Mining Sites

The incorporation of organic amendments improves the quality of mine soils, allows the growth of vegetation, and reduced phytotoxicity (Asensio et al., 2013; Pazferreiro et al., 2014). The goal of phytostabilization is to stabilize the site against potential migration pathways by retaining the HMs in the roots or precipitation within the root zone. This method is appropriate for treating HMs in large polluted sites, such as abandoned mines and smelters (Pignattelli et al., 2012). Mine wastes are not safe for reuse without proper clean-up. Common treatment methods for mine waste are associated with huge economic costs. Thus, phytoremediation of HMs offers a viable solution. Phytoremediation is considered a cost-effective, efficient, and ecologically sound method to remediate polluted mining soils (Ali et al., 2013, 2017a). Phytoremediation practices have become much more common in the past two decades and have been successful for in situ remediation of polluted soils, mines tailings, and overburdens. For example, water hyacinth, sorghum, and brassica have been used as phytoremediators for Ursk tailings and contaminated mining soil in Russia and China, respectively (Ali et al., 2017a; Romanova and Shuvaeva, 2016). The profitable application of phytoremediation for harvesting precious metals (Ag, U, Tl, Ni, Co, and Au), termed *phytomining*, has been widely explored in recent years (Anderson et al., 1999; Brooks et al., 1999; Harris and Bali, 2008; Lamb et al., 2001; Leblanc et al., 1999). Hyperaccumulator plant species can uptake, accumulate, and tolerate high HM contents in plant biomass (Evangelou et al., 2004). *Brassica juncea, Sorghum bicolor, Lolium perenne, Salix caprea,* and *Cucurbita pepo* produce high biomass and can be used for phytoextraction of HMs (Quartacci et al., 2006). However, nitrogen, phosphorus and potassium (NPK) fertilization is needed with biochar to improve plant biomass (Beesley et al., 2013). Numerous studies reported that organic amendments (sewage sludge, biochar, fly ash, wood chip, and manure) promote phytoextraction processes (Wei et al., 2011), but only a few studies have assessed the combined effect of phytoextraction and organic amendments in polluted mine landscapes (Novo et al., 2013a).

Biochar incorporation for rehabilitating mining sites is widely reported (Anawar et al., 2015; Fellet et al., 2014; Kelly et al., 2014; Rodríguez-Vila et al., 2015a). Biochar enhances plant growth and stabilization of mining waste (Fellet et al., 2011) by raising soil pH, Soil organic matter (SOM) content, electrical conductivity (EC), and H_2O holding potential, and lowering HM phytoavailability and toxicity (Kelly et al., 2014; Novo et al., 2013b; Rodríguez-Vila et al., 2015b). Biochar incorporation in sulfur-rich mining waste reduced the production of acid mine drainage (AMD), increased the pH/neutralization, and lowered metal concentrations from AMD (Anawar, 2015; Kim et al., 2013). The alkaline and adsorptive nature of biochar makes it a promising agent for reclamation of acid mines and HM-polluted soil (Rodríguez-Vila et al., 2015a). Biochar addition to mine-contaminated soil can simultaneously serve to improve soil fertility and mitigate the HM contamination (Hossain et al., 2015).

Metal bioavailability reduction is well documented for acidic mine drainage (Ippolito et al., 2017) and HM-polluted soil (Al-Wabel et al., 2014; Ali et al., 2017a). A number of pot trials showed that biochar significantly lowered biomagnification of HMs in vegetable (Beesley et al., 2013; Khan et al., 2015; Rodríguez-Vila et al., 2015b) and many other plants (Houben et al., 2013; Houben and Sonnet, 2015). Most of the biochar amended trials in mine spoils used phytostabilization with phytoremediation (Table 9.4). Limited data are available on HM translocation in plant species grown on biochar-amended, highly acidic mine soil (Jain et al., 2017; Rodríguez-Vila et al., 2016). Crop plants grown on mining landscapes have been reported to accumulate high concentrations of HMs and/or metalloids as a result of soil, water, and atmospheric contamination. Higher Au content was reported in *cassava tubers* in areas where Au is extracted from arsenopyrite ores, whereas Zn and Pb content come from Au mining area (Alcantara et al., 2017; Golow and Adzei, 2002).

9.3.2.2 Impact of Biochar on Microbial Activities

Polluted soils around mines have low pH and reduced microbial activities (De Boer and Kowalchuk, 2001). The liming effect of biochar amendment can significantly change the microbial activity (Bruun et al., 2011). Biochar

TABLE 9.4

Effects of Different Doses of Biochar Application on Phytostabilization of Mine Solid Waste

Biochar	Biochar Dose	Waste Type	Effects	Remediation Type	References
Biochar from orchard prune residues	0%, 1%, 5%, and 10%	Mine tailing	Proportional increase of pH, CEC, and water-holding capacity; decreased bioavailability of Cd, Pb, and Zn	Phytostabilization	Fellet et al. (2011)
Biochar from pruning residues manure	0%, 1.5%, and 3%	Mine tailing	Change in pH, EC, and CEC; reduced shoot Cd and Pb concentration	Phytostabilization	Fellet et al. (2014)
Biochar from pine wood	0%, 10%, 20%, and 30%	Mining waste rock	Increase in soil pH, OM, and NO^{-3} concentration; low bulk density and extractable Al, Cd, Cu, Pb, and Zn content in mining waste	Phytostabilization by *Spinacia oleracea, Brassica napus* and *Triticum aestivum*	Kelly et al. (2014)
BritishOak, Ash, sycamore, andBirch biochar	20%	Mine soil	Reduced pore water Cu and Pb concentration	Phytostabilization by ryegrass	Karami et al. (2011)
Technosol and biochar	20%, 40%, 80%, and 100%	Mine soil	Increased soil pH from 2.83 to 6.18 and shoot biomass from 0.74 to 2.95 g, reduced metal concentration in plant	Phytostabilization by *Brassica juncea* L.	Rodríguez-Vila et al. (2015b)
Jarrah biochar	37 and 74 t ha⁻¹	Spent mine sites	Improved soil pH, C content, C/N ratio, and biological N fixation rates; decreased soil C	Phytostabilization by *Acacia Tetragonophylla*	Reverchon et al. (2015)
Biochar mixed with compost	20%, 40%, 80%, and 100%	Cu mine settling pond soil	Increased pH, C and TN concentration in soil; decreased extractable Co, Cu, Ni and soil acidity	Phytostabilization by *Brassica juncea* L.	Rodríguez-Vila et al. (2014)

(Continued)

TABLE 9.4 (Continued)
Effects of Different Doses of Biochar Application on Phytostabilization of Mine Solid Waste

Biochar	Biochar Dose	Waste Type	Effects	Remediation Type	References
Poultry litter biochar	NA	Mine soil	Salt effects on seed germination at lowest biochar treatment (0.5% weight/weight); increased productivity of vegetation and forage yield (40%)	Phytostabilization by lettuce, rye, and birdsfoot trefoil	Mcdonald et al. (2014)
Biochar	NA	Mining sites	Enhanced activities of soil microbial C, phosphatase and dehydrogenase; lowered mobile Cd, Cu and Zn concentrations	Phytostabilization	Hanauer et al. (2012)
Pinewood biochar		Mine technosols	Increased the pH and electrical conductivity	Phytostabilization by *Salix*	Lebrun et al. (2016)
Pruning residues, fir tree pellets and manure pellets	0%, 1.5%, and 3%	Mine tailings	Changed pH, EC, CEC and availability of Cd and Pb	Phytostabilization by *Anthyllis vulneraria, Noccaea rotundifolium, Poa alpina* L.	Fellet et al. (2014)
Piptatherum miliaceum biochar	10 and 20 g C kg^{-1}	Technosols derived from pyritic tailings	Increase pH, microbial biomass and enzyme activities, and reduced methane	Phytomanagement	Moreno-Barriga et al. (2017b)
Bamboo biochar	0%, 1.0%, 2.5%, and 5.0%	Zn-Pb smelter and Au mining soil	Increase soil pH, EC, reduced Pb, Cd, Zn and Cu uptake, enhanced soil and plant enzymatic activity	Phytostabilization by *Brassica juncea* L.	Ali et al. (2017b)

Source: Anawar, H.M. et al., *Pedosphere*, 25, 654–665, 2015.

may potentially sorb soil enzymes fraction, add nutrients (N and P), and impose detrimental C and N limitations (Schomberg et al., 2012). Many studies have reported an increase in nitrification after biochar amendments in acidic soils (Deluca et al., 2009; Lehmann et al., 2011), while no change or a reduction in nitrification and N availability is also reported (Schomberg et al., 2012; Spokas et al., 2010). The increase in nitrification is due to the adsorption of phenolic compounds to the surface of char (Zackrisson et al., 1996), the improved food supply (C), protective biofilm formation by the microbes in acidic conditions, and the improved water availability and soil microbial activity after char addition (Kelly et al., 2014).

Soil enzymatic activities are used as a natural index for the growth and activity of microbes in the soil, which represent soil health, ecology, physical/chemical properties, and organic matter turnover. Soil enzymes have a close relation with nutrient transformations and are highly sensitive to natural and anthropogenic changes posed by mining activities and waste disposal (Abdelhafez et al., 2014; Lehmann et al., 2011). Biochar improves the physiochemical properties and microbial abundance (Houben and Sonnet, 2015). Biochar characteristics like exchange sites, sorption and desorption, porosity, and surface area affect the activities of enzymes, all of which could be possible reasons for the variable effects of biochar on the soil enzymatic activities in polluted soil (Pazferreiro et al., 2014).

9.3.3 Phytomining of Precious Heavy Metals

Reclaimed HMs have economic benefits through revenue generation by phytomining in the mine-polluted landscapes. The global remediation market is around 34–54 billion US$ (Evangelou and Deram, 2014). Some of the hyperaccumulators can extract as much as 20 mg kg^{-1} Au. Many companies earn profits from HM recovery and converting plant biomass to energy (Sheoran et al., 2013). Recently, a number of phytomining enterprises have been established in the United States, Latin America, Canada, Europe, Australia, Japan, and China.

Plant biomass will play a vital role in phytomining operation in future agro-farming. HM value was within 1.793–39368.59$ kg^{-1} range for Pb and Au, respectively, in March 2016. The candidate HMs for phytomining are Ni, Au, Tl, and Co due to their high market values and contents in plant biomass (Mahar et al., 2016). The high market value of some HMs extracted by phytomining can be substituted with low biomass production, and vice versa. The prices of HMs depend on the economic situation and growing capacity. Current prices of HMs can't guarantee a crop for phytomining. The HM-rich dried biomass produced at the polluted site is combusted to ash and stored until market prices of HMs increase (Brooks et al., 1999). The revenue (in US$) obtained by the grower, based on the price of the HMs, is shown in Table 9.5.

TABLE 9.5

Economic Values of Phytomining of Precious Metals Extracted by Plants

Hyperaccumulators	Metals	Biomass (kg/ha)	Metal Concentration (mg/kg)	(kg/ha)	Price $/kg March, 2016	Profit $/ha	References
Iberis intermedia	Tl	8000	4055	32.44	7.03	228.05	Brooks (1977)
Iberis intermedia	Tl	10000	4000	40	7.03	281.2	Leblanc et al. (1999)
Biscutella	Tl	4000	14000	56	7.03	393.68	Leblanc et al. (1999)
Iberis intermedia	Tl	8000	3070	24.56	7.03	172.65	Leblanc et al. (1999)
Berkheya coddii	Au	20000	10	0.2	39368.59	7873.72	Msuya et al. (2000)
Daucus carota	Au	—	3.8	0.779	39368.59	30668.13	Msuya et al. (2000)
Daucus carota (induced)	Au	—	3.8	1.45	39368.59	57084.46	Msuya et al. (2000)
Haumaniastrum robertii	Co	4000	10200	40.8	23.205	946.76	Brooks (1977)
Alyssum murale	Ni	20000	22000	440	8.62	3792.80	Li et al. (2003)
Alyssum corsicum	Ni	90000	800	72	8.62	620.64	Li et al. (2003)
Streptanthus polygaloides	Ni	10000	10000	100	8.62	862.00	Chaney et al. (1998)
Alyssum bertolonii	Ni	9000	8000	72	8.62	620.64	Robinson et al. (1997)
Berkheya coddii	Ni	22000	5500	121	8.62	1043.02	Robinson et al. (1997)
Alyssum serpyllifolium	Ni	9370	6515	61.05	8.62	526.21	Morais et al. (2015)
Alyssum serpyllifolium	Ni	8890	7037	62.55	8.62	539.26	Morais et al. (2015)
Haumaniastrum katangense	Cu	5000	8356	41.78	5.06	211.41	Brooks (1977)
Atriplex confertifolia	U	10000	100	1	63.382	63.38	Cannon (1964)
Thlaspi caerulescens	Cd	4000	3000	12	2.06	247.2	Reeves et al. (1996)
Thlaspi rotundifolium	Pb	4000	8200	32.8	1.793	58.81	Reeves and Brooks (1983)
Macadamia neurophylla	Mn	30000	55000	1650	1.91	3151.50	Jaffré (1980)
Astragalus pattersoni	Se	5000	6000	30	14.68	14.68	Cannon (1964)

9.4 Technological Gaps and Future Perspectives

Post-mining operations can have an impact on certain ecological factors, i.e., plant and soil as well as plant and microbe interaction, which play a key role in improving mining landscapes. The introduction of hyperaccumulator plant species, microbial consortium, and organic amendment for remediation will play a key role for sustainable rehabilitation in mining sites. Limited information is available on plant-microbe interactions and their molecular mechanisms in the soil of abandoned mines. Therefore, extensive and applied investigations on the identification of microbial communities in mine landscapes as well as rhizosphere and phyllosphere environments of surviving plants will be of great importance. The evaluation of bioavailable HMs in the mine landscapes, in tailings, and in water ecosystems should be the next priority so that we can learn about biotransformation of HMs using microbial communities for reestablishment of vegetation in mining sites and about the native metallophytes (i.e., perennial grasses, small shrubs, and trees) that are introduced to the mining areas and can grow well (Thavamani et al., 2017). Biosorbents have limited reuse efficiency. Biochar application to polluted soil includes locking or oversupplying of some of the beneficial nutrients like N and P to reduce crop productivity (Houben and Sonnet, 2015), increases in soil pH and EC, cementing effects, and increases in the cost of crop production (Al-Wabel et al., 2014). Biochar also increases nonbiodegradable SOM and changes in soil albedo (light reflection from earth to space). Biochar can delay crop maturity and fruit ripening, and reduce the efficiency of agrochemicals (pesticides, herbicides, and fertilizers). Biochar application can face lack of feedstock and can cause shifts in agrochemical efficiency due to sorption on biochar surface. Biochar prepared from the feedstock grown on the contaminated soil is also a major concern for environmental safety. The economics of phytomining is influenced by the HM content in soil and vegetation, biomass production per annum, and the possibility of recovery of the energy of biomass combustion and its sale (Evangelou and Deram, 2014).

9.5 Conclusion

Mining operations pose threats to plants, human health, and the natural ecosystem. The wastes can be managed by applying biological, organic, and phytomining technologies. Complete recovery of the HMs from mining wastes can't be attained. A limited proportion of the HMs can be removed from the soil to make it environmentally safe and reduce secondary pollution to the ecosystem. Biosorption, biochar amendments, and phytomining can help in resource recovery at mining sites to reduce soil, water, and air contamination. Biosorption helps in the sorption of the HM ions from mine wastewater and disposed wastes. Biochar can reduce the bioavailability and phytotoxicity of HMs, provide soil nutrients in the polluted environment, increase soil pH, and help in the fixation of HM ions in the soil clay matrix. Phytomining is used as a secondary option after phytoremediation practices for resource recovery and monetary returns in heavily polluted mine sites.

References

Abdelhafez, A. A., Li, J., and Abbas, M. H. 2014. Feasibility of biochar manufactured from organic wastes on the stabilization of heavy metals in a metal smelter contaminated soil. *Chemosphere*, 117, 66–71.

Adrees, M., Ali, S., Iqbal, M., Aslam, B. S., Siddiqi, Z., Farid, M., Ali, Q., Saeed, R., and Rizwan, M. 2015. Mannitol alleviates chromium toxicity in wheat plants in relation to growth, yield, stimulation of anti-oxidative enzymes, oxidative stress and Cr uptake in sand and soil media. *Ecotoxicology and Environmental Safety*, 122, 1–8.

Ahmad, M., Lee, S. S., Lim, J. E., Lee, S. E., Cho, J. S., Moon, D. H., Hashimoto, Y., and Ok, Y. S. 2014. Speciation and phytoavailability of lead and antimony in a small arms range soil amended with mussel shell, cow bone and biochar: EXAFS spectroscopy and chemical extractions. *Chemosphere*, 95, 433–441.

Alam, M. M., Hayat, S., Ali, B., and Ahmad, A. 2007. Effect of 28-homobrassinolide treatment on nickel toxicity in Brassica juncea. *Photosynthetica*, 45, 139–142.

Alcantara, H. J. P., Doronila, A. I. and Kolev, S. D. 2017. Phytoextraction potential of Manihot esculenta Crantz. (cassava) grown in mercury- and gold-containing biosolids and mine tailings. *Minerals Engineering*, 114, 57–63.

Ali, A., Guo, D., Mahar, A., Wang, P., Ma, F., Shen, F., Li, R., and Zhang, Z. 2017a. Phytoextraction of toxic trace elements by *Sorghum bicolor* inoculated with Streptomyces pactum (Act12) in contaminated soils. *Ecotoxicology and Environmental Safety*, 139, 202–209.

Ali, A., Guo, D., Zhang, Y., Sun, X., Jiang, S., Guo, Z., Huang, H., Liang, W., Li, R., and Zhang, Z. 2017b. Using bamboo biochar with compost for the stabilization and phytotoxicity reduction of heavy metals in mine-contaminated soils of China. *Scientific Reports*, 7, 2690.

Ali, H., Khan, E., and Sajad, M. A. 2013. Phytoremediation of heavy metals-concepts and applications. *Chemosphere*, 91, 869–881.

Alina, M., Azrina, A., Mohd Yunus, A. S., Mohd Zakiuddin, S., Mohd Izuan Effendi, H., and Muhammad Rizal, R. 2012. Heavy metals (mercury, arsenic, cadmium, plumbum) in selected marine fish and shellfish along the Straits of Malacca. *International Food Research Journal*, 19, 135–140.

Alvarez, A., Saez, J. M., Davila Costa, J. S., Colin, V. L., Fuentes, M. S., Cuozzo, S. A., Benimeli, C. S., Polti, M. A., and Amoroso, M. J. 2017. Actinobacteria: Current research and perspectives for bioremediation of pesticides and heavy metals. *Chemosphere*, 166, 41–62.

Al-Wabel, M. I., Usman, A. R. A., El-Naggar, A. H., Aly, A. A., Ibrahim, H. M., Elmaghraby, S., and Al-Omran, A. 2014. Conocarpus biochar as a soil amendment for reducing heavy metal availability and uptake by maize plants. *Saudi Journal of Biological Sciences*, 95, 503–511.

Anawar, H. M. 2015. Sustainable rehabilitation of mining waste and acid mine drainage using geochemistry, mine type, mineralogy, texture, ore extraction and climate knowledge. *Journal of Environmental Management*, 158, 111–121.

Anawar, H. M., Akter, F., Solaiman, Z. M., and Strezov, V. 2015. Biochar: An emerging panacea for remediation of soil contaminants from mining, industry and sewage wastes. *Pedosphere*, 25, 654–665.

Anawar, H. M., Canha, N., and Freitas, M. D. C. 2011. Evaluation of atmospheric particle dispersion at a contaminated mine using biomonitors. *International Journal of Environment & Health*, 5, 84–92.

Anderson, C. W. N., Brooks, R. R., Chiarucci, A., Lacoste, C. J., Leblanc, M., Robinson, B. H., Simcock, R., and Stewart, R. B. 1999. Phytomining for nickel, thallium and gold. *Journal of Geochemical Exploration*, 67, 407–415.

Arrascue, M. L., Garcia, H. M., Horna, O., and Guibal, E. 2003. Gold sorption on chitosan derivatives. *Hydrometallurgy*, 71, 191–200.

Asensio, V., Vega, F. A., Andrade, M. L., and Covelo, F. 2013. Technosols made of wastes to improve physico-chemical characteristics of a copper mine soil. *Pedosphere*, 23, 1–9.

Beesley, L., Marmiroli, M., Pagano, L., Pigoni, V., Fellet, G., Fresno, T., Vamerali, T., Bandiera, M., and Marmiroli, N. 2013. Biochar addition to an arsenic contaminated soil increases arsenic concentrations in the pore water but reduces uptake to tomato plants (Solanum lycopersicum L.). *Science of the Total Environment*, 454–455, 598.

Brooks, R. R. 1977. Copper and cobalt uptake by Haumaniastrum species. *Plant and Soil*, 48, 541–544.

Brooks, R. R., Anderson, C., Stewart, R., and Robinson, B. 1999. Phytomining: Growing a crop of a metal. *Biologist*, 46, 201–205.

Bruun, E. W., Hauggaard-Nielsen, H., Ibrahim, N., Egsgaard, H., Ambus, P., Jensen, P. A., and Dam-Johansen, K. 2011. Influence of fast pyrolysis temperature on biochar labile fraction and short-term carbon loss in a loamy soil. *Biomass & Bioenergy*, 35, 1182–1189.

Cannon, H. L. 1964. *Geochemistry of Rocks and Related Soils and Vegetation in the Yellow Cat Area, Grand County, Utah*. Washington, DC: United States Geological Survey (Bulletin No. 1176)

Chaney, R. L., Angle, J. S., Baker, A. J. M., and Li, J. M. 1998. Method for phytomining of nickel, cobalt, and other metal from soil. US Patent # 5, 711–784.

Chibuike, G. U., and Obiora, S. C. 2014. Heavy metal polluted soils: Effect on plants and bioremediation methods. *Applied & Environmental Soil Science*, 2014, 1–12.

De Boer, W., and Kowalchuk, G. A. 2001. Nitrification in acid soils: Microorganisms and mechanisms. *Soil Biology and Biochemistry*, 33, 853–866.

Deluca, T. H., MacKenzie, M. D., and Gundale, M. J. 2009. Biochar effects on soil nutrient transformations. In *Biochar for Environmental Management: Science and Technology*, ed. J. Lehmann and S. Joseph, pp. 251–270. London, UK: Earthscan.

Evangelou, M. W., Daghan, H., and Schaeffer, A. 2004. The influence of humic acids on the phytoextraction of cadmium from soil. *Chemosphere*, 57, 207–213.

Evangelou, M. W. H., and Deram, A. 2014. Phytomanagement: A realistic approach to soil remediating phytotechnologies with new challenges for plant science. *International Journal of Plant Biology & Research*, 2, 1023.

Fellet, G., Marchiol, L., Delle, V. G., and Peressotti, A. 2011. Application of biochar on mine tailings: Effects and perspectives for land reclamation. *Chemosphere*, 83, 1262–1267.

Fellet, G., Marmiroli, M., and Marchiol, L. 2014. Elements uptake by metal accumulator species grown on mine tailings amended with three types of biochar. *Science of the Total Environment*, 468–469, 598.

Forestier, L. L., Motelica-Heino, M., Coustumer, P. L. and Mench, M. 2017. Phytostabilisation of a copper contaminated topsoil aided by basic slags: Assessment of Cu mobility and phytoavailability. *Journal of Soils & Sediments*, 17, 1262–1271.

Gadd, G. M. 1993. Interactions of fungi with toxic metals. *New Phytologist*, 124, 25–60.

Gadepalle, V. P., Ouki, S. K., Herwijnen, R. V. and Hutchings, T. 2007. Immobilization of heavy metals in soil using natural and waste materials for vegetation establishment on contaminated sites. *Journal of Soil Contamination*, 16, 233–251.

Gill, R. A., Zang, L., Ali, B., Farooq, M. A., Cui, P., Yang, S., Ali, S., and Zhou, W. 2015. Chromium-induced physio-chemical and ultrastructural changes in four cultivars of Brassica napus L. *Chemosphere*, 120, 154–164.

Golow, A. A., and Adzei, E. A. 2002. Zinc in the surface soil and cassava crop in the vicinity of an alluvial goldmine at Dunkwa-on-Offin, Ghana. *Bulletin of Environmental Contamination & Toxicology*, 69, 638.

Goswami, S., and Das, S. 2016. Copper phytoremediation potential of *Calandula officinalis* L. and the role of antioxidant enzymes in metal tolerance. *Ecotoxicology and Environmental Safety*, 126, 211–218.

Guibal, E., Vincent, T., and Mendoza, R. N. 2015. Synthesis and characterization of a thiourea derivative of chitosan for platinum recovery. *Journal of Applied Polymer Science*, 75, 119–134.

Hanauer, T., Jung, S., Felix-Henningsen, P., Schnell, S., and Steffens, D. 2012. Suitability of inorganic and organic amendments for in situ immobilization of Cd, Cu, and Zn in a strongly contaminated Kastanozem of the Mashavera valley, SE Georgia. I. Effect of amendments on metal mobility and microbial activity in soil. *Journal of Plant Nutrition & Soil Science*, 175, 708–720.

Harris, A. T., and Bali, R. 2008. On the formation and extent of uptake of silver nanoparticles by live plants. *Journal of Nanoparticle Research*, 10, 691–695.

Hattab, N., Motelica-Heino, M., Bourrat, X., and Mench, M. 2014. Mobility and phytoavailability of Cu, Cr, Zn, and As in a contaminated soil at a wood preservation site after 4 years of aided phytostabilization. *Environmental Science & Pollution Research International*, 21, 10307.

Herwijnen, R. V., Laverye, T., Poole, J., Hodson, M. E., and Hutchings, T. R. 2007. The effect of organic materials on the mobility and toxicity of metals in contaminated soils. *Applied Geochemistry*, 22, 2422–2434.

Hossain, M. K., Strezov, V., and Nelson, P. F. 2015. Comparative assessment of the effect of wastewater sludge biochar on growth, yield and metal bioaccumulation of cherry tomato. *Pedosphere*, 25, 680–685.

Houben, D., Evrard, L., and Sonnet, P. 2013. Mobility, bioavailability and pH-dependent leaching of cadmium, zinc and lead in a contaminated soil amended with biochar. *Chemosphere*, 92, 1450–1457.

Houben, D., and Sonnet, P. 2015. Impact of biochar and root-induced changes on metal dynamics in the rhizosphere of *Agrostis capillaris* and *Lupinus albus*. *Chemosphere*, 139, 644–651.

Ippolito, J. A., Berry, C. M., Strawn, D. G., Novak, J. M., Levine, J., and Harley, A. 2017. Biochars reduce mine land soil bioavailable metals. *Journal of Environmental Quality*, 46, 411–419.

Ishikawa, S., Suyama, K., Arihara, K., and Itoh, M. 2002. Uptake and recovery of gold ions from electroplating wastes using eggshell membrane. *Bioresource Technology*, 81, 201–206.

Jaffré, T. 1980. *Etude écologique du peuplement végétal des sols dérivés de roches ultrabasiques en Nouvelle Calédonie*. Paris, France: Travaux et documents de ORSTOM.

Jain, S., Singh, A., Khare, P., Chanda, D., Mishra, D., Shanker, K., and Karak, T. 2017. Toxicity assessment of Bacopa monnieri L. grown in biochar amended extremely acidic coal mine spoils. *Ecological Engineering*, 1008, 211–219.

Jakóbik-Kolon, A., Bok-Badura, J., Karoń, K., Mitko, K., and Milewski, A. 2017. Hybrid pectin-based biosorbents for zinc ions removal. *Carbohydrate Polymers*, 169, 213–219.

Jamali, M. K., Kazi, T. G., Arain, M. B., Afridi, H. I., Jalbani, N., Memon, A. R., and Shah, A. 2008. Heavy metals from soil and domestic sewage sludge and their transfer to Sorghum plants. *Environmental Chemistry Letters*, 5, 209–218.

Johnson, D. B., and Hallberg, K. B. 2005. Acid mine drainage remediation options: A review. *Science of the Total Environment*, 338, 3–14.

Karami, N., Clemente, R., Morenojiménez, E., Lepp, N. W., and Beesley, L. 2011. Efficiency of green waste compost and biochar soil amendments for reducing lead and copper mobility and uptake to ryegrass. *Journal of Hazardous Materials*, 191, 41.

Kelly, C. N., Peltz, C. D., Stanton, M., Rutherford, D. W., and Rostad, C. E. 2014. Biochar application to hardrock mine tailings: Soil quality, microbial activity, and toxic element sorption. *Applied Geochemistry*, 43, 35–48.

Khan, S., Waqas, M., Ding, F., Shamshad, I., Arp, H. P. H., and Li, G. 2015. The influence of various biochars on the bioaccessibility and bioaccumulation of PAHs and potentially toxic elements to turnips (Brassica rapa L.). *Journal of Hazardous Materials*, 300, 243–253.

Kim, W. K., Shim, T., Kim, Y. S., Hyun, S., Ryu, C., Park, Y. K., and Jung, J. 2013. Characterization of cadmium removal from aqueous solution by biochar produced from a giant Miscanthus at different pyrolytic temperatures. *Bioresource Technology*, 138, 266–270.

Komárek, M., Vaněk, A., and Ettler, V. 2013. Chemical stabilization of metals and arsenic in contaminated soils using oxides–a review. *Environmental Pollution*, 172, 9–22.

Kwak, I. S., Won, S. W., Chung, Y. S., and Yun, Y.-S. 2013. Ruthenium recovery from acetic acid waste water through sorption with bacterial biosorbent fibers. *Bioresource Technology*, 30–35.

Lamb, A. E., Anderson, C. W. N., and Haverkamp, R. G. 2001. The induced accumulation of gold in the plants Brassica juncea, Berkheya coddii and chicory. Wellington, New Zealand: Massey University.

Leblanc, M., Petit, D., Deram, A., Robinson, B. H., and Brooks, R. R. 1999. The phytomining and environmental significance of hyperaccumulation of thallium by Iberis intermedia from Southern France. *Economic Geology*, 94, 109–113.

Lebrun, M., Macri, C., Miard, F., Hattab-Hambli, N., Motelica-Heino, M., Morabito, D., and Bourgerie, S. 2016. Effect of biochar amendments on As and Pb mobility and phytoavailability in contaminated mine technosols phytoremediated by Salix. *Journal of Geochemical Exploration*, 182, 149–156.

Lehmann, J., Rillig, M. C., Thies, J., Masiello, C. A., Hockaday, W. C., and Crowley, D. 2011. Biochar effects on soil biota–A review. *Soil Biology & Biochemistry*, 43, 1812–1836.

Li, Y., Chaney, R., Brewer, E., Roseberg, R., Angle, J. S., Baker, A., Reeves, R., and Nelkin, J. 2003. Development of a technology for commercial phytoextraction of nickel: Economic and technical considerations. *Plant and Soil*, 249, 107–115.

Lomaglio, T., Hattab-Hambli, N., Bret, A., Miard, F., Trupiano, D., Scippa, G. S., Motelica-Heino, M., Bourgerie, S., and Morabito, D. 2016. Effect of biochar amendments on the mobility and (bio) availability of As, Sb and Pb in a contaminated mine technosol. *Journal of Geochemical Exploration*, 182, 138–148.

Ma, H. W., Liao, X. P., Xin, L., and Bi, S. 2006. Recovery of platinum(IV) and palladium(II) by bayberry tannin immobilized collagen fiber membrane from water solution. *Journal of Membrane Science*, 278, 373–380.

Ma, S. C., Zhang, H. B., Ma, S. T., Wang, R., Wang, G. X., Shao, Y., and Li, C. X. 2015. Effects of mine wastewater irrigation on activities of soil enzymes and physiological properties, heavy metal uptake and grain yield in winter wheat. *Ecotoxicology and Environmental Safety*, 113, 483–790.

Mahar, A., Wang, P., Ali, A., Awasthi, M. K., Lahori, A. H., Wang, Q., Li, R., and Zhang, Z. 2016. Challenges and opportunities in the phytoremediation of heavy metals contaminated soils: A review. *Ecotoxicology and Environmental Safety*, 126, 111–121.

Mao, J., Lee, S. Y., Won, S. W., and Yun, Y. S. 2010. Surface modified bacterial biosorbent with poly(allylamine hydrochloride): Development using response surface methodology and use for recovery of hexachloroplatinate(IV) from aqueous solution. *Water Research*, 44, 5919–5928.

Marques, A. P. G. C., Rangel, A. O. S. S., and Castro, P. M. L. 2009. Remediation of heavy metal contaminated soils: Phytoremediation as a potentially promising clean-up technology. *Critical Reviews in Environmental Science and Technology*, 39, 622–654.

Martínez-Pagán, P., Faz, A., Acosta, J. A., Carmona, D. M., and Martínez-Martínez, S. 2011. A multidisciplinary study for mining landscape reclamation: A study case on two tailing ponds in the Region of Murcia (SE Spain). *Physics & Chemistry of the Earth*, 36, 1331–1344.

McDonald, L., Cook, J., Poudelacharya, S., and Skousen, J. 2014. Poultry litter biochar to promote reclamation of surface mine soils. *Abstract: Korea Soil Fertilizer Society Conference* 103, 138–142.

Modabberi, S., Alizadegan, A., Mirnejad, H., and Esmaeilzadeh, E. 2013. Prediction of AMD generation potential in mining waste piles, in the Sarcheshmeh porphyry copper deposit, Iran. *Environmental Monitoring & Assessment*, 185, 9077–9087.

Morais, I., Campos, J. S., Favas, P. J. C., Pratas, J., Pita, F., and Prasad, M. N. V. 2015. Nickel accumulation by Alyssum serpyllifolium subsp. lusitanicum (Brassicaceae) from serpentine soils of Bragança and Morais (Portugal) ultramafic massifs: Plant–soil relationships and prospects for phytomining. *Australian Journal of Botany*, 63, 17–30.

Moreno-Barriga, F., Díaz, V., Acosta, J. A., Muñoz, M. Á., Faz, Á., and Zornoza, R. 2017a. Organic matter dynamics, soil aggregation and microbial biomass and activity in Technosols created with metalliferous mine residues, biochar and marble waste. *Geoderma*, 301, 19–29.

Moreno-Barriga, F., Faz, Á., Acosta, J. A., Soriano-Disla, M., Martínez-Martínez, S., and Zornoza, R. 2017b. Use of Piptatherum miliaceum for the phytomanagement of biochar amended Technosols derived from pyritic tailings to enhance soil aggregation and reduce metal(loid) mobility. *Geoderma*, 307, 159–171.

Msuya, F. A., Brooks, R. R., and Anderson, C. W. N. 2000. Chemically-induced uptake of gold by root crops: Its significance for phytomining. *Gold Bulletin*, 33, 134–137.

Niu, H., and Volesky, B. 1999. Characteristics of gold biosorption from cyanide solution. *Journal of Chemical Technology & Biotechnology*, 74, 778–784.

Novo, L. A. B., Covelo, E. F., and Gonzalez, L. 2013a. Phytoremediation of amended copper mine tailings with Brassica juncea. *International Journal of Mining, Reclamation and Environment*, 27, 215–226.

Novo, L. A. B., Covelo, E. F., and Gonzalez, L. 2013b. Phytoremediation of amended copper mine tailings with Brassica juncea. *International Journal of Mining Reclamation and Environment*, 27, 215–226.

Oliveira, F. R., Patel, A. K., Jaisi, D. P., Adhikari, S., Lu, H., and Khanal, S. K. 2017. Environmental application of biochar: Current status and perspectives. *Bioresource Technology*, 246, 110–122.

Parajuli, D., Inoue, K., Kawakita, H., and Ohto, K. 2008. Recovery of precious metals using lignophenol compounds. *Minerals Engineering*, 21, 61–64.

Park, S. I., Kwak, I. S., Bae, M. A., Mao, J., Won, S. W., Han, D. H., Chung, Y. S., and Yun, Y. S. 2012. Recovery of gold as a type of porous fiber by using biosorption followed by incineration. *Bioresource Technology*, 104, 208–214.

Pazferreiro, J., Lu, H., Fu, S., Méndez, A., and Gascó, G. 2014. Use of phytoremediation and biochar to remediate heavy metal polluted soils: A review. *Solid Earth Discussions*, 5, 2155–2179.

Pethkar, A. V., Kulkarni, S. K., and Paknikar, K. M. 2001. Comparative studies on metal biosorption by two strains of Cladosporium cladosporioides. *Bioresource Technology*, 80, 211–215.

Pignattelli, S., Colzi, I., Buccianti, A., Cecchi, L., Arnetoli, M., Monnanni, R., Gabbrielli, R., and Gonnelli, C. 2012. Exploring element accumulation patterns of a metal excluder plant naturally colonizing a highly contaminated soil. *Journal of Hazardous Materials*, 227–228, 362–369.

Prithviraj, D., Deboleena, K., Neelu, N., Noor, N., Aminur, R., Balasaheb, K., and Abul, M. 2014. Biosorption of nickel by Lysinibacillus sp. BA2 native to bauxite mine. *Ecotoxicology and Environmental Safety*, 107, 260–268.

Quartacci, M. F., Argilla, A., Baker, A. J., and Navariizzo, F. 2006. Phytoextraction of metals from a multiply contaminated soil by Indian mustard. *Chemosphere*, 63, 918.

Reeves, R. D., Baker, A. J. M., and Brooks, R. R. 1996. Abnormal accumulation of trace metals by plants. *Mining Environmental Management*, 1–8.

Reeves, R. D., and Brooks, R. R. 1983. Hyperaccumulation of lead and zinc by two metallophytes from mining areas of Central Europe. *Environmental Pollution*, 31, 277–285.

Reverchon, F., Yang, H., Ho, T. Y., Yan, G., Wang, J., Xu, Z., Chen, C., and Zhang, D. 2015. A preliminary assessment of the potential of using an acacia—biochar system for spent mine site rehabilitation. *Environmental Science & Pollution Research*, 22, 2138.

Rizwan, M., Ali, S., Abbas, T., Zia-Ur-Rehman, M., Hannan, F., Keller, C., Al-Wabel, M. I., and Yong, S. O. 2016. Cadmium minimization in wheat: A critical review. *Ecotoxicology and Environmental Safety*, 130, 43–53.

Robinson, B. H., Brooks, R. R., Howes, A. W., Kirkman, J. H., and Gregg, P. E. H. 1997. The potential of the high-biomass nickel hyperaccumulator Berkheya coddii for phytoremediation and phytomining. *Journal of Geochemical Exploration*, 60, 115–126.

Rodríguez-Vila, A., Asensio, V., Forján, R., and Covelo, E. F. 2015a. Chemical fractionation of Cu, Ni, Pb and Zn in a mine soil amended with compost and biochar and vegetated with Brassica juncea L. *Journal of Geochemical Exploration*, 158, 74–81.

Rodríguez-Vila, A., Covelo, E. F., Forján, R., and Asensio, V. 2014. Phytoremediating a copper mine soil with Brassica juncea L., compost and biochar. *Environmental Science & Pollution Research International*, 21, 11293.

Rodríguez-Vila, A., Covelo, E. F., Forján, R., and Asensio, V. 2015b. Recovering a copper mine soil using organic amendments and phytomanagement with Brassica juncea L. *Journal of Environmental Management*, 147, 73–80.

Rodríguez-Vila, A., Forján, R., Guedes, R. S., and Covelo, E. F. 2016. Nutrient phytoavailability in a mine soil amended with technosol and biochar and vegetated with Brassica juncea. *Journal of Soils & Sediments*, 17, 1–9.

Romanova, T. E., and Shuvaeva, O. V. 2016. Fractionation of mercury in water hyacinth and pondweed from contaminated area of gold mine tailing. *Water Air & Soil Pollution*, 227, 1–9.

Sari, A., Mendil, D., Tuzen, M., and Soylak, M. 2009. Biosorption of palladium(II) from aqueous solution by moss (Racomitrium lanuginosum) biomass: Equilibrium, kinetic and thermodynamic studies. *Journal of Hazardous Materials*, 162, 874–879.

Savvaidis, I. 1998. Recovery of gold from thiourea solutions using microorganisms. *Biometals*, 11, 145.

Schomberg, H. H., Gaskin, J. W., Harris, K., Das, K. C., Novak, J. M., Busscher, W. J., Watts, D. W., et al. 2012. Influence of biochar on nitrogen fractions in a coastal plain soil. *Journal of Environmental Quality*, 41, 1087.

Sharma, R. K., and Archana, G. 2016. Cadmium minimization in food crops by cadmium resistant plant growth promoting rhizobacteria. *Appllied Soil Ecolology*, 107, 66–78.

Sheoran, V., Sheoran, A. S., and Poonia, P. 2013. Phytomining of gold: A review. *Journal of Geochemical Exploration*, 128, 42–50.

Shu, X., Yin, L., Zhang, Q., and Wang, W. 2012. Effect of Pb toxicity on leaf growth, antioxidant enzyme activities, and photosynthesis in cuttings and seedlings of Jatropha curcas L. *Environmental Science and Pollution Research*, 19, 893–902.

Simate, G. S., and Ndlovu, S. 2014. Acid mine drainage: Challenges and opportunities. *Journal of Environmental Chemical Engineering*, 2, 1785–1803.

Singh, D., Nath, K., and Sharma, Y. K. 2007. Response of wheat seed germination and seedling growth under copper stress. *Journal of Environmental Biology*, 28, 409–414.

Song, M. H., Won, S. W., and Yun, Y. S. 2013. Decarboxylated polyethylenimine-modified bacterial biosorbent for Ru biosorption from Ru-bearing acetic acid wastewater. *Chemical Engineering Journal*, 230, 303–307.

Spokas, K. A., Baker, J. M., and Reicosky, D. C. 2010. Ethylene: Potential key for biochar amendment impacts. *Plant & Soil*, 333, 443–452.

Spokas, K. A., Koskinen, W. C., Baker, J. M., and Reicosky, D. C. 2009. Impacts of woodchip biochar additions on greenhouse gas production and sorption/degradation of two herbicides in a Minnesota soil. *Chemosphere*, 77, 574–581.

Thavamani, P., Samkumar, R. A., Satheesh, V., Subashchandrabose, S. R., Ramadass, K., Naidu, R., Venkateswarlu, K., and Megharaj, M. 2017. Microbes from mined sites: Harnessing their potential for reclamation of derelict mine sites. *Environmental Pollution*, 230, 495.

Vargas, I. D., Macaskie, L. E., and Guibal, E. 2010. Biosorption of palladium and platinum by sulfate-reducing bacteria. *Journal of Chemical Technology & Biotechnology*, 79, 49–56.

Velmurugan, N., Hwang, G., Sathishkumar, M., Choi, T. K., Lee, K.-J., Oh, B.-T., and Lee, Y.-S. 2010. Isolation, identification, Pb(II) biosorption isotherms and kinetics of a lead adsorbing Penicillium sp. MRF-1 from South Korean mine soil. *Journal of Environmental Sciences*, 22, 1049–1056.

Vijayaraghavan, K., and Yun, Y. S. 2008. Bacterial biosorbents and biosorption. *Biotechnology Advances*, 26, 266.

Volesky, B., May, H., and Holan, Z. R. 1993. Cadmium biosorption by Saccharomyces cerevisiae. *Biotechnology & Bioengineering*, 41, 826–829.

Wang, J., and Chen, C. 2006. Biosorption of heavy metals by Saccharomyces cerevisiae: A review. *Biotechnology Advances*, 24, 427.

Wang, J., and Chen, C. 2009. Biosorbents for heavy metals removal and their future. *Biotechnology Advances*, 27, 195.

Wei, S., Zhu, J., Zhou, Q. X., and Zhan, J. 2011. Fertilizer amendment for improving the phytoextraction of cadmium by a hyperaccumulator *Rorippa globosa* (Turcz.) Thell. *Journal of Soils and Sediments*, 11, 915–922.

Won, S. W., Kotte, P., Wei, W., Lim, A., and Yun, Y. S. 2014. Biosorbents for recovery of precious metals. *Bioresource Technology*, 160, 203–212.

World Health Organization (WHO). 2017. Don't pollute my future! The impact of the environment on children's health. Geneva, License: CC BY-NC-SA 3.0 IGO.

Zackrisson, O., Nilsson, M. C., and Wardle, D. A. 1996. Key ecological function of charcoal from wildfire in the Boreal forest. *Oikos*, 77, 10–19.

Zawadzki, J., Przeździecki, K., and Miatkowski, Z. 2016. Determining the area of influence of depression cone in the vicinity of lignite mine by means of triangle method and LANDSAT TM/ETM+ satellite images. *Journal of Environmental Management*, 166, 605.

Zhang, T., Tu, Z., Lu, G., Duan, X., Yi, X., Guo, C., and Dang, Z. 2017. Removal of heavy metals from acid mine drainage using chicken eggshells in column mode. *Journal of Environmental Management*, 188, 1–8.

10

Biodegradation of Persistent Organic Pollutants

Kanchan Kumari[1,2], Ankur Khare[1,2], Siratun Montaha
S. Shaikh[1], and Pradip S. Jadhao[1,2]

1-Environmental Impact and Sustainability Division, CSIR-NEERI, Nagpur
2-Academy of Scientific and Innovative Research (AcSIR), Ghaziabad-201002, India

CONTENTS

10.1 Introduction

Persistent organic pollutants (POPs) are class of synthetic organic compounds distinguished by their ability to persist in the environment for a long time by resisting degradation through ambient environmental process. They are lipophilic and semivolatile and are thereby capable of bio-accumulating in adipose tissue, leading to their amplification in biota through the food chain and long-range transport to regions where they have never been used or produced, such as the polar regions (Artic and Antarctica) and open oceans. (Jones & Voogt, 1999). Globally, POPs can pose a potential risk of adverse effects to human health and the environment.

POPs are synthesized either intentionally or unintentionally as industrial products, agrochemicals, or unintended by-products of industrial synthesis or combustions. They include halogenated (chlorinated or brominated) aromatic or linear hydrocarbons, for example, polychlorinated biphenyls (PCBs), polybrominated diphenyl ethers (PBDEs), chlorinated paraffins, polychlorinated dibenzo-p-dioxins and -furans (PCDD/Fs), organochlorine pesticides (e.g., dichlorodiphenyltrichloroethane [DDT] and its metabolites, toxaphene, chlordane, Lindane, etc.), and many more (Jones & Voogt, 1999).

Considering the severe health effects (malignant tumors, congenital defects, immune and reproductive system diseases, more vulnerability to nervous system–related diseases) of POPs at low concentrations and its transboundary movement pattern across the globe, the international community of Stockholm Convention on May 22, 2001, called for an urgent global action to reduce and eliminate its production, emission, and trade by classifying 12 initial POPs into three categories (United Nations Environment Programme, 2011). To date, 21 chemicals have been ratified by the convention (Table 10.1). India has became the signatory to Stockholm Convention in year 2006, while CSIR-National Environmental Engineering Research Institute, Nagpur, India was endorsed as Regional Centre on POPs for Asia region in 2011 for assisting and providing technical guidance and capacity building on POPs in Asia region (Kumari et al, 2014). India has ratified a total of 19 (12 old and 7 newly listed) POPs.

The fate and behavior of POPs when they are released into the environment are governed by its physiochemical characteristics such as vapor pressure; aqueous solubility; half-lives in environmental matrices such as soil, water, and air; and the partition coefficient between air:solid or liquid and water:solid phases (Jones & Voogt, 1999).

The long-term fate of these chemicals is influenced and operated by a complex assemblage of processes, including their atmospheric dilution, i.e., transboundary movement from the source to the region where it is not used; physical occlusion or binding with soil and sediment residues; volatilization; leaching into aquifer or groundwater; redox reaction potential in the presence of electron donors and acceptors such as hydroxyl radicals; biological transfer into

TABLE 10.1

Persistent Organic Pollutants of Global Concern. Annex A: Elimination, Annex B: Restriction, Annex C: Unintentional Production

Old POPs			
Category	Chemical	Stockholm Convention Annex	Use
Pesticides	Aldrin	A	Insecticide
	Chlordane	A	Insecticide, termiticide
	DDT	B	Insecticide
	Dieldrin	A	Insecticide
	Endrin	A	Insecticide, rodenticide
	Heptachlor	A	Insecticide, termiticide
	Hexachlorobenzene	A	Fungicide
	Mirex	A	insecticide, termiticide
	Toxaphene	A	Insecticide
Industrial Chemicals	Hexachlorobenzene	A	By-product of manufacture (chlorinated solvents, pesticides), pesticides, incineration of HCB containing wastes
	Polychlorinated biphenyls	A	Industry manufacture; co-planar PCBs are a by-product of combustion
Unintended By-Products	Dioxins	C	By-product
	Furans	C	By-product

New POPs		
Conference of Meeting	Chemicals	Stockholm Convention Annex
Fourth meeting May 4 to 8, 2009	Alpha hexachlorocyclohexane (alpha HCH)	A
	Beta hexachlorocyclohexane (beta HCH)	A
	Chlordecone	A
	Hexabromobiphenyl (HBB)	A
	Lindane (gamma hexachlorocyclohexane)	A
	Octabromodiphenyl ether (c-octa BDE)	A
	Pentabromodiphenyl ether (c-penta BDE)	C
	Pentachlorobenzene	B
	Perfluorooctane sulfonate (PFOS)	A
Fifth meeting May 25 to 29, 2011	Endosulfan and its related isomers	A
Sixth meeting April 28 to May 10, 2013	Hexabromocyclododecane	A

(Continued)

TABLE 10.1 (*Continued*)

Persistent Organic Pollutants of Global Concern. Annex A: Elimination, Annex B: Restriction, Annex C: Unintentional Production

Conference of Meeting	Chemicals	Stockholm Convention Annex
Seventh meeting May 4 to 15, 2015	Hexachlorobutadiene	A
	Pentachlorophenol and its salts and esters	A
	Polychlorinated naphthalenes	A & C
Eighth meeting April 24 to May 5, 2017	Short Chain Chlorinated Paraffin	A
	Decabromodiphenyl ether (c-deca BDE)	A
	Hexachlorobutadiene (HCBD)	C

Source: Stockholm Convention on Persistent Organic Pollutants, 2017, Link-chm.pops.int/tabid/208/Default.aspx.

the food chain; and biologically mediated degradation in the water column, soil, and sediments (Voogt & Jansson, 1993; Jones & Voogt, 1999; Reid et al., 2000).

The major prerequisites for environmental biodegradation of POPs are bioavailability (accessibility to the organism) and inherently biodegradability of the compound, which sequentially is a function of the compounds' chemical structure adsorption, and solubility; environmental conditions; toxicity; ecology of the environment; and biological activities such as interaction with other available compounds, growth rates, metabolism and kinetics, acclimation effect, co-metabolic process, etc. (Gavrilescu, 2005; Reid et al., 2000).

Microbiota, including algae, bacteria, fungi, and protozoa, are the key organisms for biological degradation. Besides the chemical structure of the compound, the rate of degradation is determined by the availability of water; oxygen; and electron acceptors such as sulphate, nitrate, and iron (III). Although hydrolysis and redox reactions also contribute to the decomposition of the compound at the surface of the ground, decomposition of compounds at the subsurface of soil or in groundwater is dominated by biodegradation process (Mulligan & Yong, 2004; Zoeteman et al., 1980).

Biodegradation of POPs has gained importance in the past few years, so much so that it has found bioremedial applications for its mitigation from the environment. Bioremediation is one of the techniques that is economically viable, environment friendly, and socially acceptable. It is the only permanent solution for removal of indigenous xenobiotics, which could often lead to complete mineralization of these organic pollutants. This technique can be used effectively for lower concentration of contaminants where the clean-up by either physical or chemical methods would generally not be feasible. It is a noninvasive technique that leaves no negative impact on the ecosystem (Cornelissen et al., 1997; Jones & Voogt, 1999; Luthy et al., 1997; Pignatello et al., 1993; Sarkar et al., 2005)

To prevent risk and minimize impact on human health and the environment, technologies have emerged and have been improvised and developed to enhance solubilization of compound, and assess bioavailability, biodegradation mechanisms, and subsequent monitoring. This interdisciplinary field of remediation encompassing microbiology, biochemistry, molecular biology, analytical chemistry, and environmental and chemical engineering, each with its own individual approach, have actively contributed to the development of bioremediation progress in recent years (Sheehan, 1997).

10.2 Emerging Technologies to Analyze Natural Attenuation and Bioremediation of POPs

Natural attenuation refers to the use of biological, physical, and chemical process for transformation, reduction, containment, and/or destruction of pollutants to minimize their toxicity, concentration, mass, or mobility in the environment. The process involves biodegradation, dilution, volatilization, dispersion, biochemical stabilization and sorption. However, for attenuation of pollutants the most frequently considered primary mechanism is biodegradation. Natural attenuation is therefore also termed as intrinsic remediation, intrinsic bioremediation, or bioattenuation (Declercq et al., 2012; Mulligan & Yong, 2004; Sanchez et al., 2000; Sarkar et al., 2005; United State Environmental Protection Agency (USEPA), 1999).

The success of this process is highly dependent upon the hydrogeological conditions and microbiological niche. Benzene, toluene, ethyl benzene, and xylene (BTEX), and chlorinated hydrocarbons have been attenuated naturally; thus, the remediation of pesticides and other organic compounds through natural attenuation is also probable. Although natural attenuation is an effective remediation strategy, it requires considerable time, which may mean dispersal of the pollutant in the environment. Long-term monitoring is a prerequisite to ensure that there is no environmental impact. Synthetic chemicals are likely to exhibit resistance toward biodegradation by the indigenous microbiota. For effective management of the polluted sites within a reasonable time frame, the drawback of the natural attenuation process is overcome by combining it with engineering technologies in a process known as bioremediation (Diez, 2010; Mulligan & Yong, 2004; Rochkind-Dubinsky et al., 1987).

The term *bioremediation*, also known as either *biorestoration* or *bioreclamation*, is the mechanism for degradation of pollutants in the environment by biological methods. The metabolism of pollutants to less toxic by-products is a function of the microorganism's biochemical process under favorable environmental condition. The major strategies for accomplishment of bioremediation are bioaugmentation and biostimulation (Collin, 2004; Perelo, 2010; Sarkar et al., 2005; Scragg, 2005).

Bioaugmentation: involves introduction of appropriate cultured micro-organisms with the ability to catabolize various forms of hydrocarbon. These organisms supplement the indigenous microbial population and also increase the rate of biodegradation. Once these cultured microorganisms are introduced into the contaminated site, several inhibitory factors such as pH, presence of other toxic contaminants, redox reaction, bioavailability of contaminants, and the absence of co-substrates are likely to jeopardize the survival and activity of the microbes. Thus, the key factor for successful bioaugmentation lies in the selection of appropriate microbial strains from the source habitat (Perelo, 2010; Sarkar et al., 2005; Thompson et al., 2005).

Halogenated organic hydrocarbons such as PCDD/Fs and PCBs are likely to adhere and adsorb onto the soil and sediments. By bioaugmentation, degradation and detoxification of these halogenated pollutants by reductive dehalorespiration or dechlorination is a promising mechanism for remediation. Under reducing environmental conditions, chlorinated compounds are used as the terminal electron acceptors by dehalorespiring microorganisms; in the reductive dechlorination process, microorganisms gain energy by substituting a halogen for a hydrogen atom (Bedard et al., 2007; Bunge et al., 2003; Perelo, 2010; Scragg, 2005).

Biostimulation is the process of stimulating indigenous pollutant-degrading microorganisms by adding nutrients, and electron acceptors and donors such as nitrogen, phosphorous, carbon, or oxygen to the contaminated site to promote co-metabolism (Adams et al., 2015; Kanissery & Sims, 2011; Mrozik & Piotrowska-Seget, 2010). This process also aids in rendering the pollutant bioavailable for its degradation. Surfactants are used as agents to enhance the solubility of the pollutant by removing it from the soil (Jeong, 2005; Johnson et al., 2004; West & Harwell, 1992). Biosurfactants such as sodium taurocholate, bile salts, and synthetic surfactants (for example, sodium dodecyl sulphate and triton X-100) have proven to be effective in increasing the bioavailability of PCBs and Poly Aromatic Hydrocarbons (PAHs) (Ahrens et al., 2001; Nakajima et al., 2005; Perelo, 2010; Rouse et al., 1996; Voparil & Mayer, 2004).

10.2.1 Strategizing Bioremedial Monitoring

10.2.1.1 Monitored Natural Attenuation

The efficacy of the implemented bioremediation process is demonstrated by rigorous monitoring through various technical protocols using biological models and environmental indicators (Andreoni & Gianfreda, 2007; Megharaj et al., 2011). Ensuring the existence of ambient environmental conditions for natural attenuation to occur by continuous monitoring is called monitored natural attenuation (MNA). As defined by The U.S. Environmental Protection Agency (EPA), MNA is "reliance on natural attenuation process (within the context of a carefully controlled and monitored site clean-up approach) to achieve site-specific remediation objectives within a time frame that is reasonable compared to that offered by other more active methods" (EPA, 1999).

MNA is an in-situ bioremediation monitoring strategy relying on the process of natural attenuation to monitor degradation of pollutants by naturally occurring processes in different environmental matrices. The essential aspects of MNA are long-term monitoring and source control; the efficiency of the former lies in careful characterization and assessment of the location and frequency for data collection, which depend on the objectives of bioremediation and technology used to achieve the objectives (Mulligan & Yong, 2004; Schwartz et al., 2006). To evaluate the efficiency of bioremediation by natural attenuation "three lines of evidence," i.e., primary, secondary, and tertiary established by (National Research Council, 1993), are used:

Primary: reduction in pollutant concentration over time; to indicate occurrence of biodegradation, site specific attenuation rates may also be considered.

Secondary: assessment of chemical or geochemical indicators; microbiological activity; utilization of oxygen, sulphate, nitrate; or production of Mn (II), Fe (II) and methane; conductivity, alkalinity, pH, redox potential, and concentration of organic carbon, phosphorous, and chlorine (for chlorinated hydrocarbons).

Tertiary: Microbial ex-situ microcosm: to determine microbial activity and rate of pollutant degradation at the contaminated site.

Among the three lines of evidence, it is suggested that the secondary and tertiary lines are required if the primary line shows an insignificant trend. Regulatory agencies determine the time frame and source control for establishment of bioremediation goals. Monitoring frequency depends on the impact of the pollutant on biological receptors, plume behavior, and remedial goals. For the Office of Solid Waste Emergency Response (OSWER) policy on natural attenuation, the occurrence of natural attenuation as expectation, identification of toxic by-products, length of plume expansion, identification of hindrances in the process of natural attenuation by new releases or changes in environmental conditions, and subsequent clean-up objectives should be demonstrated. A contingency plan should also be prepared if the objectives of natural attenuation are not met (Mulligan & Yong, 2004).

The time frame required for complete remediation of a contaminated site is decided after considering the type and concentration of pollutant, the magnitude of the pollution event, and the environmental conditions persisting at the contaminated site. The technologies used for monitoring of the first two lines of evidence are pollutant- and contaminated site–specific depending on the microbial niche to a major extent (Declercq et al., 2012; Schwartz et al., 2006).

10.2.1.2 Technologies for Bioremedial Monitoring

Evaluation of the effectiveness of any bioremediation process (either bio-augmentation or biostimulation) with traditional cultivation methods to ensure the survival and metabolic adaptation of the introduced microbes is difficult, time consuming, and unreliable due to the lack of technological sensitivity and specificity. Therefore, understanding the dynamics of the microbial community in remediation systems has been considered as the ecological "black box" (Jansson et al., 2000; Manefield et al., 2002; Satoh et al., 2003). However, with recent advances in molecular technologies along with chemical analysis, the bioremedial process can be effectively assessed and will not be useful in determining the rate of degradation but also in predicting the fate of contaminants by mathematical modeling (Jørgensen et al., 2010).

10.2.1.3 Monitoring of Target Bacteria

For optimizing and monitoring biodegradation process, the detection of bacteria responsible for degradation process is essential. Specific bacterial detection methods up to the single-cell level have emerged to enumerate and monitor bacterial activity in complex microbial communities (Iwamoto & Nasu, 2001; Sayler et al., 1982; Suzuki & Giovannoni, 1996).

Fluorescence in-situ hybridization (FISH) using rRNA-targeted oligonucleotide probes allows assessment and identification of specific microbial communities in situ (Iwamoto & Nasu, 2001). The highly conserved sequence domains of intracellular 16s rRNAs that remain distributed within the variable region make the method suitable to be used in FISH for hybridization with fluorescence-labeled oligonucleotide probe (Amann et al., 1990; Amann et al., 1995; Gutell et al., 1994; Van de Peer et al., 1996). FISH involves permeabilization of the target bacterial cell, hybridization with oligonucleotide probe, and removal of the unbound probe by washing and microscopic detection of labeled cells (Glöckner et al., 1996; Maruyama & Sunamura, 2000). Epifluorescence microscopy and flow cytometry enable the identification and enumeration of hybridized cells in a short time (Amann et al., 2001; Amann et al., 1990).

The activity of hybridized bacterial cells in-situ can be assessed by using standard FISH with mono-fluorescein isothiocyanate (FITC) probe due to the large number of rRNAs produced by metabolically active cells (Hahn et al., 1992; Kenzaka et al., 1998; Manz et al., 1993). A new FISH technique, HNPP-FISH, reported by Yamaguchi et al. (1996), enhances the sensitivity of the fluorescence signals up to eightfold by using 2-hydroxy-3-naphthoic acid 2'-phenylanilide phosphate (HNPP), and Fast Red TR. The sensitivity can further be improved by using Cy3 oligonucleotide probe (Alfreider et al., 1996; Glöckner et al., 1996). Construction of phylogenetic trees by comparative sequence analysis using signature sequences unique to individual species can be used for identification (Iwamoto & Nasu, 2001).

Assessment of microbial activity to evaluate the effectiveness of the bioremediation process can be further enhanced by combining FISH technique with microelectrode measurements with high spatial resolution (de Beer et al., 1993; de Beer et al., 1997; Zhang & Bishop, 1996). This combination of techniques serves as a powerful monitoring tool by providing direct, time-dependent, and reliable indications of the fate of the pollutant concerning the in-situ microbial metabolic activity. It has been used for studying nitrification in biofilms by estimating concentration of NH_4^+, NO_3, NO_2, and O_2 and can be used effectively for bioremedial monitoring of POPs (Satoh et al., 2003).

Terminal-restriction fragment length polymorphism (T-RFLP) is a method for profiling the dynamics of microbial diversity from a homologous amplicon by measuring the polymorphism of terminal restriction fragments (T-RFs). PCR-amplified gene is analyzed for polymorphism in restriction fragments by automated sequencing. The primer used in amplification is fluorescently labeled at 5' terminus for detection and quantification of T-RFs obtained after digestion with restriction endonucleases. The length of the terminal fragments is calculated in base pairs by comparing with the DNA size standard based on the peak retention time. The fragments are then sequenced and compared with the existing database for profiling (Caffaro-Filho et al., 2007; Liesack & Dunfield, 2004; Tiquia, 2010).

To effectively apply T-RFLP for microbial community analysis, distribution of restriction sites on 16s rDNA and the association of terminal fragment size with phylogeny should be well known. This technique has found major application in monitoring the microbial flux at contaminated sites concerning the characteristics of the pollutant and biodegradation (Marsh et al., 2000; Pandey et al., 2007).

Amplified ribosomal DNA restriction analysis (ARDRA) aids in identification and classification of microbial species from the natural environment as isolates and of microbial clones. The amplified fluorescent-labeled amplicons are subjected to restriction digestion by restriction enzymes. The fragments are separated by electrophoresis and sequenced using the automated gel sequencing method. The rDNA database sequences of known bacterial isolates are used to compare the restriction pattern of the unknown bacteria. Although this technique is widely used for evaluating changes in microbial community dynamics, it is unable to quantify and assess the diversity or association among the species phylogenetically. This method has been used by several researchers for understanding the differences between heavy metal contaminated and uncontaminated soil in microbial communities. It is a potential method for assessing bacterial dynamics at sites contaminated with POPs (Chikere, 2013).

Automated rRNA intergenic spacer analysis (ARISA) estimates the diversity and composition of the microbial community by construction of a small-subunit rRNA gene clone library. This technique analyzes the length variability of intergenic spacer of rRNA gene between large (23S) and small (16S) subunits on the ribosomal RNA (rrn) operon by PCR amplification. Conserved regions of 16S and 23S genes are targeted for binding of oligonucleotide primers. The intergenic region between 16S and 23S likely to encode tRNAs exhibits heterogeneity in the nucleotide sequence as well as length base pair (bp). Both the heterogeneities are assessed to distinguish among the closely related bacterial strains. PCR-amplified products are subjected to electrophoresis followed by silver staining for visualization. The complex DNA bands obtained post-electrophoresis indicate the original assemblage of the microbial community, with each band of DNA corresponding to one organism. To improve the resolution of the technique and intensity of the bands, fluorescence-labeled primers can be used. This method has been successfully applied to assess the changes in microbiota following stress conditions such as antibiotic treatment, exposure to mercury etc. It is therefore applicable for evaluating and monitoring microorganisms involved in the biodegradation of POPs. (Fisher & Triplett, 1999; Ranjard et al., 2001).

In microbial ecology, denaturing gradient gel electrophoresis (DGGE) was introduced by Muyzer et al. (1993). It is a powerful molecular ecological technique to assess microbial diversity. It is applied to analyze PCR-amplified 16S rDNA, i.e., the 16s rRNA gene. Amplified products of the variable region(s), including 200–400 bp intervening the conserved domains, from the entire bacterial DNA community are separated by DGEE. The sequence of the variable region determines the melting behavior. Theoretically, the number of bands obtained in DGGE estimates the species richness, where each band corresponds to a single microbial species. The relative abundance of each species is further estimated by the relative intensity of each band (Strathdee & Free, 2013).

For the complex DGGE banding patterns obtained from environmental samples, a quantitative pattern analysis such as the Shannon Diversity Index can be applied for better interpretation (Eichner et al., 1999). For assessment of bacterial diversity, the Shannon Diversity Index (H) calculates the number (species richness) and relative intensity (species abundance) of band obtained in each lane of the gel (Atlas, 1984; Iwamoto et al., 2000). This technique can be applied to any gene locus exhibiting interspecies variable regions flanked by the conserved domains for primer binding (Strathdee & Free, 2013).

Biomarkers and whole-cell biosensors are alternative techniques for tracking the bacterial survival and metabolic gene activity. Biomarkers are molecules, genes, or any compounds that are organism-specific, thus conferring a distinct phenotype or genotype that can be isolated for monitoring a large scale of population and community dynamics. Promising biomarkers used for bioremedial monitoring are fluorescent, bioluminescent, and chromogenic marker genes, i.e., bacterial luciferase, *luxAB*; firefly luciferase gene, *luc*; green fluorescent protein (GFP), *gfp*; and reporter genes. Biomarker selection depends on the monitoring system, bacterial strain, specificity, and sensitivity of detection required (Andreoni & Gianfreda, 2007; Jansson et al., 2000; Lebeau, 2011).

Biomarkers are used as specific bioactive components for designing whole-cell biosensors and have the unique ability to yield the monitoring signal by interacting in accordance with the biological system of the host.

Recombinant plasmid is used as the biological component and includes a specific promoter that is sensitive to the target pollutant and the biomarker or reporter system for signal generation. The activity of the latter is directly proportional to the expression of the former (Purohit, 2003). The intensity of the light emitted as a function of the signal generation is the direct measure of the number of metabolically active viable cells in the environment. The emitted bioluminescent signal can be detected using fiber optics and photon counting modules, thereby, facilitating the continuous in-situ online monitoring. This bioremedial monitoring technique has several advantages, such as easy detection of bioluminescence with less substantial input, self-contained process lacking the need for additional exogenous chemicals, continuous real-time online profiling of the bioremediation process, and monitoring of pollutant degradation and bioavailability using microorganisms as chemical sensors (Jansson et al., 2000; Sayler & Ripp, 2000).

Lux-based biosensors have been used for bioremedial monitoring of hydrocarbon-contaminated soil by Bundy et al. (2004, 2001) and Bhattacharyya et al. (2005); for detection of BTEX in aqueous solution, Stiner and Halverson (2002) constructed a GFP-based *Pseudomonas fluorescens* strain A506 biosensor. The selected biosensor may function either as the specifically induced biosensor or constitutively expressed biosensor for measuring the presence of a specific analyte or associated polluted burden, respectively (Andreoni & Gianfreda, 2007; Bundy et al., 2002; Dawson et al., 2008).

A DNA microarray, also known as DNA chip, gene chip, or DNA microchip, consists of complimentary DNA probes of approximately 500–5000 nucleotides in length or 15 to 70 bp oligonucleotides immobilized on a solid matrix. Fluorescent-labeled target complimentary sequences are introduced into the assay system for hybridization with probes complimentary to the corresponding DNA targets. Nonspecific or unattached targets are removed by washing. Emission of fluorescent signals due to binding of complementary target sequences with probes indicates confirmation of the study. The emitted fluorescence intensity can be used for microbial population quantification because the intensity of the signal is directly proportional to the number of targets hybridized to probes (Li & Liu, 2003). Although this technique is similar to FISH, it provides simultaneous analysis of microbial population dynamics using 16s rRNA or gene expression (Chikere, 2013).

Based on the targeted gene, six categories of microarrays, i.e., functional genomic arrays (FGAs), community genome arrays, phylogenetic oligonucleotide arrays, metagenomics array, whole-genome open-reading frame arrays, and internal transcribed spacer (ITS) oligonucleotide arrays applicable to microbial ecology have been identified (Chikere, 2013; Grace Liu et al., 2011; Lebeau, 2011). The microarray techniques applied most frequently are described below:

Functional gene arrays (FGAs) measure and identify functional protein- or enzyme-coding genes involved in metabolic processes and other environmental biogeochemical process by direct estimation of mRNA (Rhee et al., 2004; Zhou, 2003). In bioremediation, this technique has been used for the detection of PCB degrading genes in soil, i.e., aromatic oxygenase genes (Denef et al., 2003).

Community genome arrays (CGAs) are used to detect and describe a community of specific microorganisms based on the whole genomic DNA.

Phylogenetic oligonucleotide arrays (POAs) are used for comparing the relatedness of microbial communities using 16s rRNA gene.

10.2.2 Bioremedial Monitoring of Pollutant Degradation

For biostimulation, conventional approaches for amendment bioremedial monitoring involve tracking metabolic products and contaminant concentration by direct sampling, and geochemical parameters such as CH_4 production or O_2 consumption. Although, monitoring of contaminated sites by direct sampling and assessment of geochemical parameters are used for analyzing change in bioremediation processes and may also provide sufficient data for fate prediction, risk assessment, and mathematical transport modeling, it does not demonstrate bioremediation of the concerned pollutant due to its indirect assessment as the changes may also result from various other physical process (reduction in pollutant concentration due to dilution) (Baldwin et al., 2008). For conclusive tracking of the bioremediation process, monitoring may be carried put with the push-pull test with passive multilevel sampler, time-elapsed electrical resistivity tomography (ERT), and stable isotope probing (SIP).

The push-pull test with passive multilevel sampler is designed to document the biodegradation of the target pollutant concurrently with the microbial activity in-situ; the latter is stimulated by introducing the test solution consisting of amendment and tracer that is introduced into the subsurface through monitoring wells. Microbial communities are assessed from collected groundwater samples over time by analyzing the concentration of pollutants and geochemical parameters. However, this test does not form the basis for assessing the spatial distribution of the important species of the microbial community, which are likely to have an impact on the efficiency of the bioremediation process. To overcome the lacunae, passive multilevel samplers (MLSs) are used in conjunction with the push-pull method. These samplers consist of a solid matrix such as Accusand (90% quartz, 9% Na-, K-, and Ca-feldspar, and 1% mica) for colonization of a microbial community. To investigate the in-situ microbial microcosms, MLSs are deployed at various multiple elevations of the study site prior to and following placement of the amendment. Characterization of the associated microbiota are carried out by real-time quantitative polymerase chain reaction (Q-PCR) (Baldwin et al., 2008; Istok et al., 1997; Lehman et al., 2001).

This combination of techniques permits evaluation of geochemical and redox conditions. It is an inexpensive approach for concurrent spatial and temporal determination of biodegradation of the target pollutant and the associated microbiota in response to biostimulation (Baldwin et al., 2008).

Time-lapse electrical resistivity tomography (ERT), coupled with radar techniques, is a proven geophysical method for conducting temporal change assessment in subsurface electrical conductivity. This technique overcomes the low-density post-amendment emplacement monitoring in the target vadose region by enhancing and complementing conventional sampling-based methods for assessment of biostimulation-based emplacement remediation. For effective management of subsurface bioremediation systems, in addition to the existing biogeochemical properties of pore fluid or sediment and redox zonation, information regarding changes in the biogeochemical characteristics due to fluid introduction or replacement by amendment are often required. Microbial mediated biogeochemical changes have also been monitored using ERT (Johnson et al., 2015; Snieder et al., 2007). Biofilm formation; microbial growth; microbial metabolic activity; and the associated by-products such as secondary minerals, biogenic gases, or mineral dissolution have been shown to change the interfacial electric property of the geologic media (Davis et al., 2006; Snieder et al., 2007).

Alterations in fluid chemistry due to biodegradation in hydrocarbon-contaminated sites resulting in higher electrical conductivity have been studied by Atekwana et al. (2005). The changes in electrical conductivity due to biodegradation can be detected using ground-penetrating radar (GPR). The increase in radar signal corresponds to the microbial degradation metabolic by-product and not to the mere formation of biofilms or presence of microbial cells (Atekwana & Slater, 2009; Cassidy, 2007; Che-Alota et al., 2009; McGlashan et al., 2012). Lane et al. (2006) evaluated the emplacement of vegetable oil emulsion by introducing it into subsurfaces for stimulating biodegradation of chlorinated biodegradation. Radar methods were used to monitor the emplacement of vegetable oil emulsion, wherein the degradation of the vegetable oil by microbial activity resulted in an increase of electromagnetic wave attenuation compared to prior background data.

Radar techniques are sensitive to electrical conductivity and dielectric permittivity. Measurable differences in pore fluid chemistry (fluid conductivity) are detected through time-lapse radar; cross-hole radar is used for monitoring of dynamics of vadose zone and tracer test. Radar is useful for effective monitoring on land, between land and borehole, or among boreholes. However, direct sampling is also needed along with geophysical monitoring to detect changes in biogeochemical conditions (Johnson et al., 2015).

Stable isotope probing (SIP) is used to distinguish and confirm biodegradation of the chlorinated and aromatic hydrocarbons from other abiotic processes such as volatilization and sorption during natural attenuation. This technique tracks the incorporated stable-isotope-labeled substrates in the phylogenetically informative biomarkers of the substrate assimilating microorganisms. The stable isotopes used are ^{13}C, ^{15}N, ^{2}H, and ^{18}O. When introduced into the environment, the stable isotopes are incorporated by the metabolically active cells into their biomarkers, which are then recovered for species identification of the microbial community responsible for biodegradation of the pollutant in the natural environment (Manefield et al., 2004; Manefield et al., 2002; Megharaj et al., 2011; Uhlik et al., 2013; Whiteley et al., 2006).

The first stable-isotope was introduced as ^{13}C-labeled acetate in the sediments, which resulted in the assimilation of ^{13}C-enriched polar-lipid-derived fatty acid (PLFA). The phylogenetic resolution is achieved by DNA-SIP and RNA-SIP, where isotope-labeled nucleic acids are recovered by density gradient centrifugation and analyzed for informative taxonomy. RNA-SIP is preferred over DNA-SIP: it is a more responsive biomarker due to its higher rate of synthesis and replication-independent labeling (Dunford & Neufeld, 2010; Manefield et al., 2002; Neufeld et al., 2007; Uhlik et al., 2013). However, investigations into the dechlorination of chlorinated hydrocarbons under reducing conditions, and the degradation of PCB in contaminated soil biofilm reactors by incorporating ^{13}C-PCB (2,2-dichlorobiphenyl) are being carried out. The assimilated isotopes probes into the PLFA and nucleic acids (DNA and RNA) can be analyzed by isotope ratio mass spectrometry (IRMS) and DGGE, respectively (Lebeau, 2011; Scow & Hicks, 2005).

SIP functions by assimilating the isotope probed substrate into the biomarkers, so its utility depends on the chemical transformation of the substrate into the assimilatory microorganisms. Nonassimilatory processes such as co-oxidation do not fall under the applicability of this technique (Andreoni & Gianfreda, 2007).

10.2.3 Mathematical Modeling

In recent years, mathematical models have gained importance in the discipline of bioremediation for predicting the fate and transport of pollutants in different environmental matrices, evaluating the implemented bioremediation process and the impacts on the biological receptors, and determining the effect of removal or addition of one or more factors into the bioremediation process. Model selection depends on a wide range of environmental conditions such as hydrogeology, heterogeneity of the surface and subsurface, site complexity, accurate information about the pollutant, and the bioremediation process. Application of a particular model to a contaminated site to determine the effectiveness of the implemented bioremediation process (natural attenuation, bioaugmentation, or biostimulation) requires analytical contaminant transport data concerning the persisting environmental condition (Mulligan & Yong, 2004).

So far various predictive models have been designed to study natural attenuation of petroleum fuel release, chlorinated solvents, and hydrocarbons by aerobic and anaerobic biodegradation. These would indicate the possibility of designing mathematical models by combining or modifying the existing models considering the characteristics of POPs in natural field conditions and their biodegradation. Some of the mathematical models (Mulligan & Yong, 2004) applicable in bioremediation are mentioned below:

BIOSCREEN is an analytical model for natural attenuation of petroleum fuel release and was developed by Groundwater Services (Houston, Texas) for Air Forces Center for Environmental Excellence. Key information regarding biodegradation like a decrease in electron acceptors (sulphate, nitrate, and dissolved oxygen) and formation of metabolic by-products (methane and Fe (II)) is considered key aspects determining feasibility of natural attenuation.

BIOCHLOR imitates the bioremediation of chlorinated solvents by including the reductive dechlorination process in sequential steps in different dechlorination rate zones. A database has been developed for assistance in predicting the degradation rate.

BIOSCREEN was developed for estimating the time duration required for complete BTEX bioremediation. Data such as hydraulic conductivity, biodegradation rate, retardation factor, and porosity gradient obtained from field monitoring were incorporated using benzene as the key compound (Suarez & Rifai, 2002).

BIOPLUME III is a two-dimensional model simulating natural attenuation by considering ion exchange, dispersion, advection, and biodegradation. The basis of the model was formed in July 1989 by the U.S. Geological Survey (USGS) and has been used for evaluating flow and transport of groundwater. For simulating biodegradation rates, Monods kinetics and first-order decay reactions were incorporated concerning aerobic as well as anaerobic electron acceptors, along with production and transport of Fe (II).

Bioredox-MT3DMS is a three-dimensional multi-species fate and transport model for predicting the efficacy of accelerated or intrinsic bioremediation of petroleum hydrocarbons and chlorinated solvents. This model works in conjunction with the redox-reaction database, MODFLOW.

The suitability of the model depends on the assumptions, site characteristics, and limitations of the bioremediation process. Incorporation of the biotic factors in a well-thought-out conceptualized model for accurate prediction of the biodegradation and pollutant concentration along with the site characteristics and available field data is key to successful bioremediation (Corapcioglu & Baehr, 1987; Mulligan & Yong, 2004).

10.3 Evolution of Catabolic Pathways and Microbial Adaptation to POPs (Industrial Chemicals and Pesticides)

Catabolic pathways are cellular processes mediated by enzymes and transport systems and modulated by regulatory and structural genes. Evolutionary acquisition of catabolic pathways elucidated in several theories is justified by the genetic framework of the organism, mutations, gene transfer process, and selective pressure that may have favored the organisms with new gene combinations. In addition to the changes in the genetic framework, the diversity of microbial species and higher growth rate have empowered faster evolution and thereby adaptation to altering, environmentally challenging conditions. The microbial genetic diversity accounting for metabolic versatility, particularly metabolism of POPs, has been called the evolutionary greenhouse (Díaz, 2004; Lovley, 2003; Timmis & Pieper, 1999).

The benzene ring, alongside glycosyl residues, is the most widely spread chemical structure in the environment. Persistence of the benzene ring in the environment is higher due to its thermodynamic stability. Therefore, most of the aromatic compounds are categorized as pollutants (Dagley, 1986). However, a microbial catabolic pathway has evolved to utilize organic pollutants as the sole source of carbon and energy (Harayama, 1992). The new mosaic gene structures evolved within the organism by gene transfer or mutation have led the microbe to develop strategies for utilizing the available pollutant to meet the energy requirement under favorable environmental conditions or selective pressure using alternative final electron acceptor (ferric ions, sulfate, and nitrate). This justifies the biodegradation of nitrobenzene, atrazine, chlorobenzene, and phenoxyalkanoic acids (Van der meer & Sentchilo, 2003).

ICElands, also known as genomic islands, are mobile conjugative and integrative DNA elements (third type of Mobile Genetic Elements (MGE)) that integrate into the host chromosome or replicon after transfer. They have been discovered to constitute aromatic pollutant catabolic genes. They carry gene modules and integrase gene-related to plasmids conjugation modules for site-specific integration (Dirk Springael & Top, 2004). Two examples are the chlorocatechol degradation (clc) element in *Pseudomonas sp.* strain B13 for the degradation of chlorocatechol and chlorobenzene, and Tn4371 in *Ralstonia oxalatica* A5 for (chloro)biphenyl (CBP) degradation (Springael et al., 2001). The clcRABD gene cluster present in clc element encodes for the modified ortho-cleavage pathway for mineralization of chlorocatechols. Similar gene clusters have been identified in different microbial species constituting intb13 gene with different catabolic information encoding CBP and naphthalene degradation. These elements have unique conserved features that bridge the gap between degradation and pathogenicity functions by constituting a genetic framework for several degradation pathways as well as pathogenicity (Top et al., 2002; Springael & Top, 2004; Van der meer & Sentchilo, 2003).

The role of MGEs in the evolution of catabolic pathways by horizontal transfer have been strongly correlated by following four major observations:

- Bacteria from distant locations constituting evolutionary related gene clusters and catabolic genes (Whyte et al., 2002).
- Non-congruency in the phylogeny of the catabolic genes with that of the corresponding host's 16sRNA genes (Mcgowan et al., 1998; Vallaeys et al., 1999).
- Association of pollutant degradation genes with mobile genetic elements (MGEs) such as transposons and plasmids (Top et al., 2002).
- Involvement of the evolutionary related gene modules and catabolic gene clusters with degradation of structurally similar xenobiotic compounds (Johnson et al., 2002; Smejkal et al., 2001; Van Der Meer et al., 1998).

Thus, it can be stated that mobilization of catabolic genes into suitable hosts from different organisms led to the formation of novel catabolic pathways that assisted in the microbial adaptation by degrading organic pollutants (Top & Springael, 2003).

The adaptation of microorganisms to the appearance of xenobiotics by expressing modified catabolic pathways is found to be a function of resisting the molecule's potential toxic effects or using it for beneficiary characteristics as an alternative nutrient and energy source. Under anaerobic environmental conditions, chlorinated organics are used as final electron acceptors for (de)halorespiration, which is the detoxification process of organohalogens (Copley, 2000; Holliger et al., 1999; Springael & Top, 2004; Top & Springael, 2003). The bacteria that carry out the dehalorespiring process belong to the genus of Gram-positive bacteria with low GC content, i.e., *Dehalobacter* and *Desulfitobacterium*, the Proteobacteria, and the genus Dehalococcoides. The latter bacterial genus utilizes vinyl chloride as final electron acceptor and can grow on PCBs, chlorobenzenes, and chlorinated dioxins (Pieper et al., 2004) (Table 10.2).

Certain species like *Bacillus, Dietzia, Burkholderai, Gordonia, Desulfovibrio, Mycobacterium, Rhodococcus*, etc., show chemoorganotrophy assists in the degradation of POPs.

Gram negative species like *Pseudomonas ptuida, P. fluorescens* have the highest potential to biodegrade POPs by assimilating the pollutants in their catabolic pathway and catalyzing the degradation of complex molecules and releasing energy. The gram positive species like *Cornrybacterium, Mycobacteruim*, and *Rhodococcus* are equally effective in the degradation of POPs and PAHs.

Certain strains showing cometabolism like *Aspergillus, Bacillus, Microbacteruim, Micrococcus, Nocardia*, etc., have the potential to degrade the chemicals without using them as a substrate (Nzila, 2013). Sakakibara

TABLE 10.2

Microbial Degradation of POPs

Name of Compound	Mechanism	Bacteria
Aldrin, Endrin, Dieldrin, PCB, Polychlorinated Dioxin, Polychlorinated Dibezofuran Chlordecone α-, β and γ hexachlorocyclohexane Tetrabromodiphenyl ether	Dioxygenation	*Pseudomonas sp., Pseudomonas vesicularis, Sphingomonas paucimobilis, Trametes hirsutus, Phanerochaete chrysosporium, Cyathus bulleri, Phanerochaete sordida Rhodococcus sp., Rhodococcus erythropolis, Pseudomonas CH07, Pleurotus ostreatus, Dehalococcoides sp., Sphingomonas wittichi, Shingomonas yanoikuyae, Panellus Stypticus, Klebsiella sp., Sphingomonas sp., Paenibacillus, Rhizobium, Pseudomonas mendocina, Phlebia lindtneri*
DDT, Heptachlor, Mirex, Toxaphene	Reductive Dechlorination	*Aerobacter, Agrobacterium, Alcaligenes, Bacillus, Clostridium, Dehalospirilum multivorans, Hydrogenomonas, Klebsiella, Pseudoxanthomonas jiangsuensis, Staphylococcus, Stenotrophomonas, Streptomyces, Xanthomonas, Xerocomus chrysenteron*
HCB	Oxidative Dechlorination	*Pseuodomonas quisquiliarum, Sphingobium lucknowense, Eupenicillium baarnense, Eupenicillium crustaceum*

Source: Chakraborty, J. & Das, S. (2016).Environ Sci Pollut Res. 23: 16883. https://doi.org/10.1007/s11356-016-6887-7.

et al. (2011) reported a novel aerobic bacteria *Pseodomonas sp.* Strain KSF27 for degradation of Dieldrin isolated from a consortium of bacteria capable of degrading various Organo Chlorines (OCs) along with dieldrin and aldrin. Zhang et al. (2013) reported a strain of *Pseudomonas stutzeri* strain BFR01. Hexachlorobenzene a recalcitrant pollutant was found to be degraded aerobically by *Nocardioides sp.* Strain PD653 reported by Matsumoto et al. (2008).

For nonhalogenated aromatic compound degradation, the aromatic compound is initially transformed to dihydroxybenzene derivatives such as protocatachuate, homoprotocatechuate, catechol, homogentisate, gentisate, or hydroquinone. A similar principle applies for the degradation of chloroaromatic compounds; however, prior to transformation into one of the central derivatives, substrates are dehalogenated (Schlömann, 1994).

Degradation of chlorobenzenes, chlorosalicylates, chlorodiphenyl ethers, chloroanilines, chlorotoluenes, chlorobiphenyls, and chlrodibezofurans have been reported to be carried out via chlorocatechol. This modified ortho-cleavage pathway of chlorocatechol is analogous to the 3-oxoadipate pathway yielding chlorosubstituted derivatives (Schlömann, 1994). Considering the similarity in the pathways, it is suggested that the evolution of the ortho-cleavage pathway for the degradation of chlorocatechol is from catechol branch of the ubiquitous 3-oxoadipate pathway.

Another possibility for the existence of the modified ortho-cleavage pathway is single origin and subsequent distribution via horizontal gene transfer. The enzymes for the degradation of chlorocatechol are induced in response to the appearance of chloroaromatic compounds and are encoded by plasmids. Acknowledging the substrate specificity, delahogenation process, and modulation of additional enzymes/genes, it may be reasonable to consider the coevolution of the pathway as an answer to anthropogenic pollution of chloroaromatic compounds (Ghosal et al., 1985; Schlömann, 1994).

Other POP catabolic genes, for example, DDT 2,3-dioxygenase, biphenyl dioxygenase, and angular dioxygenase, have been known to play a role in the degradation of organochloride pesticides, biphenyl, and dioxins or furans, respectively (Table 10.3). The strategy of activating hydrocarbons by hydroxylation in the initial phase of degradation by aerobic hydrocarbons may be ineffective with some xenobiotic compounds as they possess electron-withdrawing groups such as nitro-, halo-, or azo- substituents. The high redox potential created with these substituents leads to less susceptibility of hydrocarbons to hydroxylation by di- or mono-oxygenases, and thus they resist degradation. Hence, reductive transformation, a mechanism adopted by anaerobic bacteria where a nucleophilic or reductive attack on the xenobiotic compounds generating electron deficient π-electron systems,

TABLE 10.3

Catabolic Genes for POPs Degradation in Different Bacterial Genera

Organisms	Involved Genes	Location of Gene	Compounds
Alcaligenes sp.	*bphA, bphB, bphC, bphD*	Plasmid	PCB
Pseudomonas sp.	*bphEGFA1A2A3BCDA4R*	Genome	PCB
Alcaligenes eutrophus	*tfdA, tfdB, tfdC, tfdD, tfdE*	Plasmid and Genome	2,4-D
Bradyrhizobium sp.	*tfdF, tfdR, cad RABKC operon*		
Dehalococcoides	*rdh, cbrA*	Genome	Hexachlorobenzene
Terrabacter sp.	*dbfA1A2*	Genome	Polychlorinated Dioxins
Nocardioides sp.	*dfdABC*	Genome	Polychlorinated dibenzofuran
Sphingomonas paucimobilis	*linA1A2BCXD*	Genome	α-, β and γ hexachlorocyclohexane
Arthrobacter sp. *Rhodococcus sp.*	*esd operon, ese operon*	Genome	Endosulphane
Sphingobium chlorophenolica	*pcp A, B, C, D operon*	Genome	Pentachlorophenol (PCP)

Source: Chakraborty, J. and Das, S. *Environ. Sci. Pollut. Res.*, 23, 16883, 2016.

enhances degradation of pollutants by adding auxiliary electron donors. This process has been observed for the degradation of azo dyes, polynitroaromatic compounds, and polyhalogenated ethenes (Haderlein et al., 2000; Rieger et al., 2002).

Rapid changes and the broad range of investigations that have listed degradation pathways for persistent organic chemicals such as PAHs and pesticides have broadened the scope of research regarding catabolic pathways involved in the breakdown of these organic pollutants to simpler forms and the microbial interactions in doing so (Van der Meer et al., 1992). Thus, biochemical and regulatory studies for degradation of industrial wastes and chemicals in various isolates would provide a good perspective about those episodes that occur at initial stages of development of any new pathway (Kivisaar, 2011).

10.4 Analysis of Organic Pollutants Biotreatment in Confined Environments

Bioremediation is multi-phasic in nature and thus is an extremely complex process governed by restricted mass transfer, microbial community dynamics, and characteristics of the target pollutant. The major impediment in achieving time-stipulated efficacy of the biodegradation process is the nonuniformity of microbial spatial distribution. The overall mineralization rate of the biodegradation process is accelerated by providing direct contact between microbiota and the target pollutant in a confined environment. For this, several ex-situ bioremedial technologies are employed in slurry and solid phases such as bioreactors, land farming, biobeds, and composting (Doick & Semple, 2003; Mohan et al., 2006; Prasanna et al., 2008; Vasile Pavel & Gavrilescu, 2008; Venkata Mohan et al., 2009).

Slurry phase ex-situ bioremediation technology consists of bioreactors, also known as slurry reactors or aqueous reactors. This technique employs an engineered containment system for treating contaminated soil, sediments, sludge, or water that is excavated or pumped from the contaminated plume by subjecting it to three-phase mixing conditions, i.e., solid, liquid, and gas. This process involves formation of water slurry by mixing a predetermined quantity of water to the soil and sediments to enhance degradation of water-soluble and soil-bound pollutants (Vasile Pavel & Gavrilescu, 2008). The quantity of water to be added depends on pollutant concentration, biodegradation rate, and physiochemical properties of the soil (EPA, 2003). Slurry bioreactors (SBs), one of the most effective in situ and ex situ technologies for bioremediation of sites, are being studied for a wide range of recalcitrant pollutants and further application in large agrochemical industries that release enormous organic by-products. In spite of significant improvements in reactor operation and design optimization, the SB-based studies for characterizing microorganisms are still in their early stages of development (Robles-González et al., 2008).

In solid-phase bioremediation, the excavated contaminated soil is placed in piles and subjected to pollutant degradation by stimulating bacterial growth. A network of pipes is distributed throughout the piles to provide moisture and ventilation for microbial respiration (EPA, 2001; Vasile Pavel & Gavrilescu, 2008). Depending on the matrix employed and the optimized environmental parameters, solid-phase bioremediation can be categorized as landfarming, biocells/biobeds, and composting. However, the biodegradation rate in bioreactors has been recorded to provide higher efficiency in the entire bioremediation process due to the highly manageable, controlled, and confined environment than any other technologies employed in solid-phase bioremediation (Vasile Pavel & Gavrilescu, 2008; Vidali, 2001).

Isolates of indigenous or genetically engineered pollutant-degrading microbiota are subjected to optimized ex-situ environmental conditions, i.e., temperature, moisture, micro- and macronutrients, aeration, C:N ratio, etc., to enhance the catabolism, microbial uptake, and intrinsic activity of each cell (Beškoski et al., 2011; Johnsen et al., 2005; Singh & Ward, 2004; Venkata Mohan et al., 2009). Nutrient addition in the form of pyruvate as a carbon source has proven to stimulate microbial growth and subsequent biodegradation of PAHs (Lee et al., 2003). Spent mushroom compost (SMC) is also utilized for treating sites contaminated with organic pollutants (Trejo-Hernandez et al., 2001). In addition to providing enzymes, microorganisms, and nutrients for microbial growth, SMC was also observed to reduce toxicity of the pollutant (Agamuthu et al., 2013).

The bacterial transport system has gained importance in bioremediation due to its ability to uptake and degrade or transform pollutants; it is the major factor considered for monitoring of biotreatment in a confined environment (DeFlaun et al., 2001). Bacteria, yeast, and fungi are the predominant hydrocarbon degraders (Wentzel et al., 2007). Accessibility of the pollutants to the microbiota can be enhanced by reducing surface tension or by direct contact (Baboshin & Golovleva, 2012). The former is achieved by using biosurfactants such as phospholipids, lipoproteins, polymeric lipids, or biopolymers; for the latter, hydrophobicity of the cell surface is increased by synthesizing adhesion structures, i.e., lipopolysaccharides, hydrophobic exopolymers, proteins, and mycolic acids (Abbasnezhad et al., 2011; Das et al., 2008; Fuentes et al., 2014).

In the microorganism transport system, mechanisms to degrade hydrocarbons and generate metabolic intermediates to participate in the central metabolic pathways have evolved over time. In nutrient-limited environmental conditions, microorganisms oxidize these pollutants as substrate by adding hydroxyl groups to the main hydrocarbon compound to initiate its aerobic catabolism. The enzymes responsible for adding molecular oxygen to the substrate in the degradation pathway are oxygenases, i.e., monooxygenases or dioxygenases. Contrary to the oxidation pathway, coupling of fumarate or CO_2 with hydrocarbons activates the degradation of hydrocarbons under reducing conditions using sulfates and nitrates as terminal electron acceptor (Callaghan et al., 2012; Fuentes et al., 2014; So et al., 2003; Wentzel et al., 2007).

The genes encoding the mono- and dioxygenases are used as a molecular marker for monitoring biodegradation of hydrocarbons, assessment of phylogenetic distribution, and divergence in gene sequence, and they become highly enriched when subjected to hydrocarbons (Iwai et al., 2011; Wang et al., 2010). Enzymes for biodegradation of PCBs and biphenyls are synthesized in *Burkholderia*, *Achromobacter*, *Pseudomonas*, *Sphingomonas*, and *Rhodococcus* strains encoded by *bph* genes (Chain et al., 2006; Fuentes et al., 2014; Pieper & Seeger, 2008; Seeger et al., 1997).

The biodegradation of POPs by the organotrophic bacteria can take place in both oxic and anoxic conditions, resulting in different breakdown products. (Figure 10.1). Another pathway mostly studied for aerobic biodegradation is of lindane by *Shingomonas paucimobilis UT26*, where lindane yields pentachlorocyclohexane (PCCH) after being subjected to dehydrochlorination by the microbe. Degradation of Lindane by *Pseudomonas sp.* under oxic conditions produces tetrachlorocyclohexane (TCCH) along with PCCH (Tu, 1976). The novel enzymes for complete mineralization of lindane involved in the degradation pathway identified in *S. paucimobilis* are encoded by *linA*, *linB*, *linC*, *linD*, *linE*, *linR*, and *linX*; maleylacetate and b-ketoadipate are formed as common metabolites during the mineralization process of aromatic compounds (Nagata et al., 1999). Similar lindane degradation pathways have been identified in some fungal and bacterial species that form trichlorobenzenes (TCBs) and dichlorophenols (DCPs) as the metabolites (Francis et al., 1975; Nagasawa et al., 1993; Phillips et al., 2005; Yule et al., 1967).

In addition to lindane, white-rot fungi have been observed to degrade other classes of pesticides such as atrazine, diuron, terbuthylazine, metalaxyl, DDT, gamma-hexachlorocyclohexane (g-HCH), dieldrin, aldrin, heptachlor, chlordane, and mirex (Kennedy et al., 1990; Mougin et al., 1997; Singh & Kuhad, 1999). The ability of fungi to penetrate the contaminated soil or sediments by forming fungal hyphae and synthesizing extracellular enzymes for pollutant degradation has several advantages over bacteria (Aprill & Sims, 1990; Garon et al., 2004; Mohan et al., 2006). Lignin-degrading fungal species *Phanerochaete chrysosporium* and *Trametes hirsutus* have been shown to degrade lindane by producing extracellular lignin-degrading peroxidases with broad substrate specificity (Bumpus et al., 1985; Kennedy et al., 1990; Mougin et al., 1997; Phillips et al., 2005; Singh & Kuhad, 1999).

The operons dispersed on the bacterial genome and the subsequent catabolic pathways representing the regulatory networks as a prerequisite for interaction among the species, growth of the microbiota, and biodegradation are coordinated in response to the presence of substrate, i.e., pollutant (Gottesman, 1984; Heinaru et al., 2005; Mohan et al., 2006). The gene

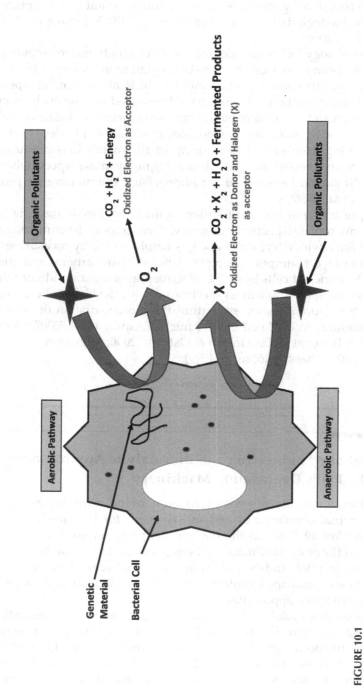

FIGURE 10.1
General mechanism of xenobiotic degradation by bacterial cell using aerobic and anaerobic pathways.

regulation and signal transduction involved in biodegradation can be used to monitor microbial activity during the bioremedial process. The enrichment of the functional genes in response to the pollutant can be optimized to assess the biodegradation efficacy (Kao et al., 2010; Megharaj et al., 2011; Salminen et al., 2008).

Molecular biology techniques coupled with conventional microbiological methods have been used to estimate the biodegradation activity of the microbial community qualitatively and quantitatively. Real-time and competitive polymerase chain reactions of genes have been used to effectively monitor gene regulation and subsequent biodegradation efficacy (El Fantroussi et al., 1997; Lanthier et al., 2000; Robles-González et al., 2008). Besides functional gene monitoring, biochemical monitoring of the metabolites produced in the degradation process by gas chromatography–mass spectrophotometry (GC-MS) should be assessed for improvising the monitoring process (Bachmann et al., 1988).

Tracking of bacterial transport system with ferrographic tracking techniques by microplate spectrofluorometry, ferrographic immunomagnetic capture, or flow cytometry; metabolic gas respirometry by measuring CO_2 production or O_2 consumption using ^{14}C-labeled hydrocarbons; and quantifying the ^{13}C-enriched cells by HPLC-electrospray ionization-MS or whole-cell combustion-isotope ratio mass spectrometry (C/IRMS) have been proven to be valuable approaches for evaluating the bioremediation of persistent organic pollutants in confined environments (Aspray et al., 2007; DeFlaun et al., 2001; Fuller et al., 2000; Holben & Ostrom, 2000; Jørgensen et al., 2000; Lytle et al., 2000; Zhang & Johnson, 1999).

10.5 Metabolic Engineering and Biocatalytic Applications of the POPs Degradation Machinery

Today metabolic engineering has given humankind novel technologies to exploit microbial communities and utilize them for bioremediation. The study of the fate of POPs in the environment has revealed a plethora of microbes and the enzymatic reactions being exploited for a number of potential enzymes in pilot studies and on an industrial scale. Unique catalysts could be accessed utilizing bioinformatics tools, meta-genomic libraries, and DNA-sequence based approaches.

Mutagenesis has broadened the scope of application in bioremediation by matching the increased need and ability in modifying and adapting the catalysts to specific reactions and process requirements. In exceptional cases, these catalysts must be exploited in either fermentations or whole-cell bioconversions in large-scale studies to understand strain physiology and metabolism. On the other hand, the use of genetically modified microorganisms (GMOs) for bioremediation carried out on the industrial scale is not yet very significant. Currently, very few examples exist for applications of genetically engineered microbes (GEMs) in environmental ecosystems (Nagabhushanam et al., 2005). Unfortunately, the only manner in which to fully address the competence of GEMs in bioremediation efforts is through long-term field release studies (Sayler et al., 2000).

10.6 Impact of Metals on the Biodegradation of POPs

The global distribution of POPs in the environment has always been of paramount importance because of their possible hazards to human health as well as whole ecosystems via their transfer from the environment. Numerous microorganisms have been reported in various studies that are found capable of POP degradation; thus bioremediation using microorganisms has been considered as an effective and promising tool for removal of POPs from different contaminated environmental matrices. Metals may inhibit pollutant biodegradation through interaction with enzymes directly involved in biodegradation and interfering with microbial activity. The impact of metals on litter decomposition, methanogenesis, acidogenesis, nitrogen transformation, biomass generation, and enzymatic activity is also known. Metals appear to inhibit bioremediation by affecting both the physiology and ecology of the microorganism responsible for degrading organic pollutants. It is well known that the presence of metals can inhibit and alter a wide range of microbial processes, including growth, nitrogen and methane metabolism, sulfur conversions, reductive processes, and dehalogenation (Baath, 1989). The impact of the toxicity of metals on the biodegradation of organic pollutants with special focus on POPs has not been adequately studied, either qualitatively and quantitatively. This is due to the different characteristics of metals, for example, different physical-chemical forms, transition states, complexed species, and/or ionic solutes.

When a microbial cell faces high concentration of any heavy metal, transportation and accumulation of specific heavy metal ions occurs within the system irrespective of its high concentration. The mechanism behind this process can be attributed to constitutive expression of some unspecific transporters forcing the ions gates to remain always opened, leading to metals toxicity. (Nies, 1999).

Metals inhibit the degradation of organic pollutants by (1) interacting with the enzymes involved directly in biodegradation (an example is pollutant-specific oxygenase enzyme) or (2) interacting with the enzymes involved for the general metabolism within the cell. In both scenarios, metals in the form of ions are found to inhibit the processes (Angle & Chaney, 1989).

Numerous methods have been reported for the biodegradation of POPs, for example, pretreatment before biodegradation, including advanced oxidation processes using various metals as catalysts for reducing their concentration first then treating with microorganism. The efficiency of this method has been reported to be around 70% and above; zero valent iron used for treating PBDE has been reported to have a 74% efficiency for biodegradation in anaerobic conditions (Liu et al., 2001).

Metal exerts a negative effect on microorganisms. Most heavy metals are toxic at their higher concentrations as a result. When added in higher concentrations, they decrease the efficiency of biodegradation by reducing the substrate availability as the excess metal gets adsorbed into the cell surface and consequently reduces the treatment efficiency of organic pollutants (Nies, 1999). A study by Amor et al. (2001) showed that metals at even low concentration had an inhibitory effect on the biodegradation of alkylbenzene by *Bacillus* strain, while some studies reported negligible impact of metals on microbial activity. Hassen et al. (1998) reported that Zn at lower concentrations, between 0.05 to 0.2 mM, was not toxic to the growth of *Bacillus thuringenesis*. Gonzalez-Gil et al. (1999) observed an enhanced effect on metabolic activity and growth rate of aerobic and anaerobic bacteria exposed to low concentrations of Ni. These variable results indicate the complexity of the combined effects of metals on biodegradation of organic pollutants (Ke et al., 2010).

The different inhibitory effects on rates of biodegradation caused by various metals depends on the metal and type of microbial sample. Biodegradation rates were very sensitive to increases in the concentrations of some metals. Even very small concentrations of metal (ppm, ppb) can alter microbial activity and cause resistance to biodegradation (Table 10.4).

The huge differences in the values reported for the inhibition of organic pollutants are also due to different experimental setups and protocols as well as different metal-solution phases and metal concentrations. The pH of the solution and the selected media also have a great impact on the biodegradation of the pollutants. Maximum degradation of the pollutants has been shown to occur at the pH range of 6.5 to 8. A slight change in pH can alter the solubility of metals, reducing the solution-phase metal concentration.

However, research is needed to discover the cost-effective and proper remediation measures needed POP remediation using microorganisms. Attempts have been made to quantify the impact of metals on biodegradation of organic pollutants, but these have shown some inhibitory effects. Many studies have already been conducted, for example, addition of metal-resistant microbes, adjusting pH using chemicals additives, etc.; more field trials are needed to validate the prevailing approaches. New and advanced methods that can define the bioavailability of metals are being developed around the world. The mechanisms that inhibit the biodegradation seem to vary with both the complexity and composition of the ecological system. It also remains clear that metals can act as potent inhibitors on the biodegradation in both aerobic and anaerobic microbial systems.

TABLE 10.4

Inhibition in Biodegradation of Organic Toxicants and Reported Metal Concentration

Metal	Toxicants	Lowest Metal concentration (mg/l)	Micro-organism Studied	Medium	pH	Reference
Cd+2	2,4-D, XYL, BP	25.3–50.6	*Alcaligens, Pesudomonas sp.*	Tris buffer	7.0	Springeal et al. 1993
Cd+2	2,4-D	0.060	*Alcaligenes eutrophus* JMP134	Soil microcosm	8.2	Roane et al. 2001
Cd+2	NAPH	1	*Burkholderia sp.*	Mineral salt with phosphate (1.4 mM)	6.5	Sandrin et al. 2000
Cd+2	TOL	37	*Bacillus sp.*	Mineral salt with phosphate (36 mM)	5.9	Springeal et al. 1993
Cr+3	2,4-DME	0.177	Indigenous community	Aufwuchs (microcosm)	6.1	Said and Lewis, 1991
Cr+6	IPB, NAPH, BP	131	*Alcaligenes sp., Pseudomonas sp.*	Tris buffered plates	7.0	Springeal et al. 1993
Cu+2	Crude oil	6.30	*Pseudomonas sp.*	Mineral salt with phosphate	7.2	Coker and Ekundayo, 1998
Ni+2	TOL	20	*Bacillus sp.*	Mineral salt with phosphate (36 mM)	5.9	Amor et al. 2001
Pb+2	Crude oil	1.41	*Micrococcus sp.*	Mineral salt with phosphate (31 mM)	7.2	Coker and Ekundayo, 1998
Zn+2	PH	10	*Acinetobacter calcoaceticus, AH* strain	Bioreactor with 0.15 mM phosphate	7.8	Nakamura and Swada, 2000
Zn+2	TOL	2.8	*Bacillus sp.*	Mineral salt with phosphate (36 mM)	5.9	Amor et al. 2001

Abbreviations: BP-Biphenyl, TOL-Toluene, XYL-Xylene, NAPH-Naphthalene.

10.7 Microorganisms and GEMs Involved in Biodegradation

Many authors have reported the degradation of pollutants by different bacteria and other microorganisms. Numerous of bacterial species feed exclusively on hydrocarbons for their survival and degrade them into by-products like CO_2, H_2O, etc. Bacteria that can biodegrade hydrocarbons and related compounds are known as hydrocarbon-degrading bacteria. The degradation of hydrocarbons can take place aerobically and anaerobically. But contemporary to this, anaerobic biodegradation is more important. Species of bacteria isolated by Kafilzadeh et al. (2011) belonged to 10 genera: *Staphylococcus, Streptococcus, Shigella, Bacillus, Alcaligenes, Corynebacterium, Enterobacter, Klebsiella, Acinetobacter,* and *Escherichia. Bacillus* was found to be the best for degradation.

Many more species of microorganism are capable of degrading organic pollutants by metabolizing them. Compared to a single bacterium, mixed communities of microorganisms have the most powerful potential for bio-degradation of pollutants because the genetic base of the microorganisms is helpful in degrading the complex mixture of organic compounds present at the contaminated site. Efficient removal of pesticides like atrazine and DDT by adding mixed cultures has also been reported earlier (Struthers et al., 1998). Recent finding on the pesticide-degrading bacteria *Provedencia staurtii* for pesticide chlorpyrifos was isolated from agricultural land, while bacterium species like *Stenotrophomonas, Bacillus,* and *Staphylococcus* were isolated from uncultivated land and found to be capable of degrading DDT (Kanade et al., 2012). Studies have also revealed the degradation of PCBs and their biotransformation both aerobically and anaerobically (Seeger et al., 2001). PCBs with a high degree of chlorination are subjected to dehalogenation with the help of anaerobic microorganisms, while lower chlorinated compounds like biphenyls are broken down by aerobic microorganism. Strains like *Achromobacter, Comamonas, Pseudomonas, Burkholderia,* and *Ralstonia* belonging to gram negative strain have shown surprisingly results for degradation of lower PCBs. All the species showed a similar catabolic pathway catalyzed by enzyme (biphenyl dioxigenase, hydrolase, etc.) activity.

Azo dyes are also found to be degraded by anaerobic and aerobic bacterial strains (Santos et al., 2007). Studies have concluded that decolorization of dyes was more efficient and effective under anaerobic conditions. In contrast to bacteria of mixed cultures, single strains like *Shewanella decolorations* are equally effective for removal of azo dyes, with several advantages (Hong et al., 2007). Heavy metals find another way of degradation as they cannot be degraded biologically (no change in nuclear structure), but they can be transformed into one or more oxidation states or from a complex structure to a simpler one (Garbisu & Alkorta, 2001). Microorganisms have evolved to protect themselves from the toxic effects of heavy metals by different mechanisms, for example, methylation, oxidation, adsorption, uptake, reduction, etc. (Fernández et al., 2012). Microorganisms employ metals as sources of electron donors and use the terminal electron for anaerobic respiration, for example, conversion of Cr(III) by reduction of Cr(VI) in the presence of oxygen (Sayel et al., 2012). Apart from reduction, methylation carried out by microbes also plays a key role for survival against metal toxicity. Mercury, Hg(II) is biomethylated by different species of bacteria like *Bacillus sp.*, *Bacillus pumilus*, and *Alcaligens faecalis* into the gaseous form of methyl mercury (Jaysankar et al., 2006). Apart from methylation and redox chemical reactions, acidophilic bacteria *Acidothiobacillus ferroxidans* and some species of sulfur-oxidizing bacteria have the potential of leaching high concentration of Cd, Cu, Co, Zn, etc., from highly contaminated sites. The sulphate-reducing bacteria are collectively anaerobic bacteria that utilize organic substrates with SO_4^{2-} as electron donor. Heavy metal ions, after transformation, are entrapped into the microorganism cells and subsequently biosorbed in the metal binding site in the cells in a process known as biosorption or passive uptake.

Many authors have suggested the use of pure cultures of microorganism for bioremediation. However, mixed cultures of microorganisms are more advantageous. A study conducted by Adarsh et al. (2007) used a consortium of microorganisms for the removal of heavy metals like Cd, Cr, Ni, Cu, and Pb from wastewater and found it equally effective.

10.7.1 Degradation Using Plant Growth Promoting Bacteria and Plant Growth Stimulating Rhizobecteria

Endophytic bacteria that are nonpathogenic and live in plants and rhizospheric bacteria that live on or near the root of plants contribute significantly to the biodegradation of toxic compounds (Divya & Kumar, 2011). Plant growth stimulating rhizobacteria (PGPR) are naturally occurring bacteria; they live in colonies and help in promoting the growth of plants. *Pseudomonas sp.* have hydrocarbon-degrading potential and are widely known for their PGPR activities. The bacteria aggressively colonize around the root region and promote plant growth. Another species *lysini bacillus* is known to have the potential for PAH degradation. PCB-degrading bacteria are also found in the rhizospheric region in mixed forms and have good potential of PCB degradation. Species like *Rhodococcus*, *Luteibactor*, and *Williamsia* are known for PCB degradation in situ (Liegh et al., 2006). Another species *Azospirullum lipoferum* that is generally found in the rhizosperic region is reported to degrade Malathion, which is the most common organophosphate insecticide used around the world (Kanade et al., 2012). Plant growth promoting bacteria (PGPB) are known as adjuncts for metal phytoremediation and assist plants to grow and survive in areas heavily contaminated with metals. Known as rhizoremediation, biodegradation using a combination of plants and microorganisms is generally useful for increasing the efficiency of toxicant extraction by the plants.

10.7.2 Degradation by Microfungi and Mycorrhiza

Similar to bacteria, fungi are important because they metabolize organic matter and are responsible for carbon decomposition in the biota. Microfungi are a group of microorganisms that are eukaryotic and aerobic. They range from unicellular (yeast) to multicellular (molds).

Like bacteria, fungi can survive and grow in low pH, low moisture, and less light intensity (Spellmen, 2008). Fungi having many enzymes as their extracellular matrix are most efficient for the breakdown of polymeric compounds into simpler ones. Mycorrhiza are symbiotic associations between fungi and the roots of higher plants. In this association, fungi colonize the host plant in two ways: either intracellularly, called arbuscular mycorrizal fungi (AMF), or extacellularly, as ecto mycorrhizal fungi. The bioremediation using Mycorrhiza is collectively called mycorrhizoremediation (Khan, 2006). Fungi are known to have special characteristics and degrading capabilities for the recycling of polymers and phenols (Fritsche & Hofrichter, 2005) and for the removal of hazardous wastes from the environment.

Species like *Trichosporun cutaneum* have an affinity for the uptake of aromatic compounds like phenols from the environment (Mörtberg & Neujahr, 1985). Alkanes having different forms of carbon ranging from C_{10} to C_{20} are the most widely utilized hydrocarbons and are good substrates for microfungi (Bartha, 1986). Species like *Candida lipolytica, C. tropicalsi, Trichosporon,* and *Rhodoturularubra* are widely accepted and known as alkane-utilizing yeasts. Yeasts like *C. methanosorbosa* BP-6 is reported as having the potential to degrade azo dye (Mucha et al., 2010). Many authors have shown that yeasts are better degraders of hydrocarbons than bacteria (Ijah, 1998). Many studies conducted by numerous researchers have shown the degradation of PCB by yeasts. Yeasts known to metabolize PCBs, *C. boidinni* and *C. lipolytica,* along with *Saccharomyces cerevisiae* are reported to degrade PCBs, biphenyls, dibenzofurans, and other toxicants by aerobic fermentation (Cabres et al., 1998).

Several studies have reported that yeasts are superior metal accumulators than bacteria, and they are also capable of accumulating various heavy metals like Cu(II), Co(II), Ni(II), and Cd(II) (Wang and Chen, 2006). Strains of yeasts like *Yarrowialin polytica, Hansenula polymorpha, Rhodotorula pilimanae,* and *Cyberlindnera fabianii* assist in biosorption and reduce Cr(VI) to Cr(III) (Ksheminska et al., 2006). Immobilized yeast cells are more efficient for metal removal; one example is *Schizosaccharomyces pombe* for the removal of copper (SaiSubhashini et al., 2011).

10.7.3 Biodegradation by Filamentous Fungi, Algae, and Protozoa

Filamentous fungi are known as potential biodegraders in both forms: dead or alive. The fungal hyphae rapidly absorb the pollutants from the ambient environment and use them as a substrate. They digest it by secreting a battery of extracellular degrading enzymes. Fungi have a high surface-to-cell ratio, a unique characteristic feature in filamentous fungi that maximizes the mechanical and enzymatic surface contact with the surrounding environment. The extracellular degrading enzymes in the filamentous fungi enable the fungi to tolerate higher concentrations of toxicants for degradation. The members of deuteromycetes of genera *Exophiala, Leptodontium,* and *Cladophialophora* as well as some members of ascomycetes like *Pseudeurotium zonatum* are potential degrades of toluene and other hydrocarbons (Francesc et al., 2001). A group of fungi like *Aspergillus, Penicillium,* and *Cephalosporium* are also reported to degrade PCBs. However, this area of PCB degradation by fungi has not been fully explored. (Tigini et al., 2009). The main mechanism of action of filamentous fungi in the biodegradation of hydrocarbons and other toxicants is due to the extracellular, specific oxidoreductive enzymes that play a key role in the breakdown of hydrocarbons to carbon and energy source for the fungi (Donnelly & Fletcher, 1995). Fungi are also known to detoxify heavy metals by different mechanism like extra- intracellular precipitation, valance transformation, and active uptake of metals. Many species adsorb metals like Cd, Co, Pb, and Cu. A system has been developed to use *Rhiloprzs arrhizus* for uranium and thorium remediation (Treen-Sears et al., 1989).

Species of *Aspergillus niger* AB 10 and *Rhiloprzs arrhizus* M1 act in the biosorption of metals through cell surface adsorption, in which the metal acts as ligands for metal sequestration for removal of heavy metals from aqueous media (Pal et al., 2010). Lignolytic fungi are also equally effective for dye degradation. Filamentous fungi are also reported to degrade some pesticides using intracellular (cytochrome P-450) and exocellular (lignin degradation using lactase and peroxidase) mechanisms (Chaplin et al., 2011).

In addition to filamentous fungi, some algae and protozoans are also well known and equally important in the biodegradation of alkanes, alkenes, hydrocarbons, and PCBs. An alga, *Prototheca zopfi* isolated by Walker et al. (1975), was found to be able to degrade alkanes, iso-alkanes, and aromatic hydrocarbons using crude oil and mixed hydrocarbons as a substrate. A study conducted by Cerniglia and Gibson (1997) reported that nine species of cyanobacteria, including red alga, green alga, brown alga, and diatoms, oxidized naphthalene by using it as a substrate. Some freshwater algae like *Scenedesmus platydiscus, S. capricornutum,* and *Chlorella vulgaris* have been reported to uptake and degrade PAHs. Algae have not shown mush significance in the biodegradation of pesticides, but they are known for the transformation of pesticides into less harmful products (Kobayashi & Rittman, 1982). Species of *Anabaena inacqualis, Chlorella Stigeoclonium* are reported to be able to remove heavy metals, but they are limited to practical application. Metal uptake by brown alga, which includes the adsorption of metals in numerous cell wall constituents like fucoidan and alginate, have been reported for many years (Davis et al., 2003).

Protozoa serve as the main grazer for bacteria in the breakdown of organic materials, so the protozoa work along with bacteria and show prominent results for biodegradation of some organic toxicants. Mattinson and Harayama (2005) have reported that the degradation rate of methylbenzene and benzene has increased 8.5 times when bacteria are mixed with a culture of protozoa. Several mechanism and hypotheses have been proposed by several authors about accelerating the rate of bioremediation of organic chemicals (Chen et al., 2007).

10.7.4 Genetically Engineered Microorganisms

Different methods for biodegradation of toxicants like bioaugmentation, biostimulation, etc., are applied for accelerating the recovery of contaminated sites. Prior to 1980 and in the mid-1970s, researchers started encoding catabolic bacterial genes for characterization of recalcitrant compounds. Since then, genetic engineering has been considered a potential tool for addressing bioremediation (Cases et al., 2005).

GEMs (also known as GMMs) are microorganisms whose genetic material has been transformed using genetic engineering. This technology is referred as recombinant DNA (r-DNA) technology. GEMs have shown a promising potential for the bioremediation of water, sludge, soil etc. Plasmids are the backbone of GEMs that quote for a particular gene. A single plasmid cannot quote and degrade numerous toxicants. Plasmids are grouped into major four categories depending on the coding class (see Table 10.5) (Ramos et al., 1994).

Markandey and Rajvaidya (2004) have demonstrated the potential of different GEMs through genetic manipulation for different types of toxicants. *Pseudomonas putida* containing XML and NAH plasmids, along with recombination parts of OCT and CAM, were reported to degrade camphor, octane, naphthalene, and salicylate (Sayler & Ripp, 2000).

Oil-eating bugs, also known as superbugs, are the products of genetic engineering. They include the plasmid of *Pseudomonas putida* for degradation of chemical compounds such as toluene and xylene, pAC25 (for 3-cne chlorbenxoate), pKF439 (for salicyclate toluene), and RA500 (for 3–5, xylene). This bug was the first living organisms subjected for property rights case.

TABLE 10.5

Different Plasmids for Different Toxicants

S. NO	Toxicants degraded	Plasmid
I.	Hexane, Octane, and decane	OCT plasmid
II.	Toluene and Xylene	XYL plasmid
III.	Napthalene	NAH
IV.	Camphor	CAM plasmid

Genetic engineering has also used as a technique in the removal of heavy metals from highly contaminated sites. *Alcaligens eutrophus* AE104 (pEBz141) was reported for the removal of the heavy metal Cr from industrial wastewater (Srivastava et al., 2010), while a recombinant photosynthetic bacteria *Rhodopseudomonas palustris* was genetically modified to express Hg transport system and metallothionein for mercury in wastewater (Xu and Pie, 2011).

Catabolic genes of *R. eutropha* A5, *A. denitrificans* JB1, and *Achromobacter sp.* LBS1C1 were transformed by genetic engineering into the heavy-metal-resistant strain *R.eutropha* for degradation of PCBs. Endophytic bacteria and rhizospheric bacteria for plant-associated degradation of toxicants are considered the most useful tool developed using genetic engineering. Many bacteria and microorganisms in the rhizospheric region have very limited ability to degrade toxicants; with the help of r-DNA technology, the genetically engineered rhizobacteria that contain toxicant-degrading genes like trichloro-ethylene (TCE) and polychlorinated biphenyls (PCB) are produced and show promising results (Sriprang et al., 2003). A study conducted by Sriprang et al. (2003) introduced gene phytochelatin synthase (PCS;PCSAt) of the species *Arabidopsis thaliana* into the *Mesorhizobium huakuii* strain B3 for studying the accumulation of the heavy metal cadmium and established the symbiotic relationship between *Mesorhizobium huakuii* and *Arabidopsis thaliana* under the control of a bacteriod-specific promoter gene called nifH gene. the nifH gene (Sussman et al., 1988).

However, much controversy exists regarding the development and release of GEMs into the environment. Thus, issues like safety and the risk of ecological damage must remain in the foreground (Wackett, 2004).

10.8 Conclusion

Mineralization of POPs by physiochemical treatment can never be truly comprehending by its norms. Therefore, the literature describes the various aspects of POP biodegradation. Fundamental and applied biomolecular techniques for in-situ and ex-situ biodegradation using native and genetically engineered microorganisms have been explored. Laboratory and field investigations have provided evidence for biodegradation of POPs. The strategies adopted by microbial communities through catabolic evolution and successful metabolic engineering techniques for biodegradation were discussed in this chapter to enable budding researchers to design and strategize better bioremediation processes. It is imperative to seek effective application strategies for minimizing the impact of existing POPs on the environment and on human health.

References

Abbasnezhad, H., Gray, M., & Foght, J. M. (2011). Influence of adhesion on aerobic bio-degradation and bioremediation of liquid hydrocarbons. *Applied Microbiology and Biotechnology.* doi:10.1007/s00253-011-3589-4.

Adams, G. O., Fufeyin, P. T., Okoro, S. E., & Ehinomen, I. (2015). Bioremediation, bio-stimulation and bioaugmention: A review. *International Journal of Environmental Bioremediation & Biodegradation, 3*(1), 28–39. doi:10.12691/ijebb-3-1-5.

Adarsh, V. K., Mishra, M., Chowdhury, S., Sudarshan, M., Thakur, A. R., & Chaudhuri, S. R. (2007). Studies on metal microbe interaction of three bacterial isolates from East Calcutta Wetland. *OnLine Journal of Biological Sciences, 7*(2), 80–88.

Agamuthu, P., Tan, Y. S., & Fauziah, S. H. (2013). Bioremediation of hydrocarbon con-taminated soil using selected organic wastes. *Procedia Environmental Sciences, 18*, 694–702. doi:10.1016/j.proenv.2013.04.094.

Alfreider, A., Pernthaler, J., Amann, R., Sattler, B., Glockner, F. O., Wille, A., & Psenner, R. (1996). Community analysis of the bacterial assemblages in the winter cover and pelagic layers of a high mountain lake by in situ hybridiza-tion. *Applied and Environmental Microbiology, 62*(6), 2138–2144. Retrieved from isi:A1996UP12700042%5Cnhttp://www.pubmedcentral.nih.gov/articlerender.fcgi?artid=1388879&tool=pmcentrez&rendertype=abstract.

Amann, R., Fuchs, B. M., & Behrens, S. (2001). The identification of microorganisms by fluorescence in situ hybridisation. *Current Opinion in Biotechnology, 12*(3), 231–236. doi:10.1016/S0958-1669(00)00204-4.

Amann, R. I., Krumholz, L., & Stahl, D. A. (1990). Fluorescent-oligonucleotide probing of whole cells for determinative, phylogenetic, and environmental studies in microbiology. *Journal of Bacteriology, 172*(2), 762–770. doi:10.1128/JB.172.2.762-770.1990.

Amann, R. I., Ludwig, W., & Schleifer, K. H. (1995). Phylogenetic identification and in situ detection of individual microbial cells without cultivation. *Microbiological Reviews, 59*(1), 143–169. doi:10.1016/j.jip.2007.09.009.

Amor, L., Kennes, C., & Veiga, M. C. (2001). Kinetics of inhibition in the biodegrada-tion of monoaromatic hydrocarbons in presence of heavy metals. *Bioresource Technology 78*, 181–185.

Andreoni, V., & Gianfreda, L. (2007). Bioremediation and monitoring of aromatic-polluted habitats. *Applied Microbiology and Biotechnology, 76*(2), 287–308. doi:10.1007/s00253-007-1018-5.

Angle, J. S., & Chane, R. L. (1989). Cadmium resistance screening in nitrilotriacetate-buffer media. *Applied and Environmental Microbiology, 55*, 2101–2104.

Aprill, W., & Sims, R. C. (1990). Evaluation of the use of prairie grasses for stimu-lating polycyclic aromatic hydrocarbon treatment in soil. *Chemosphere, 20*(1–2), 253–265. doi:10.1016/0045-6535(90)90100-8.

Atekwana, E. A., & Slater, L. D. (2009). Biogeophysics: A new frontier in Earth science research. *Reviews of Geophysics, 47*(4), RG4004. doi:10.1029/2009RG000285.

Atekwana, E. A., Atekwana, E., Legall, F. D., & Krishnamurthy, R. V. (2005). Biodegradation and mineral weathering controls on bulk electrical conduc-tivity in a shallow hydrocarbon contaminated aquifer. *Journal of Contaminant Hydrology, 80*(3–4), 149–167. doi:10.1016/j.jconhyd.2005.06.009.

Atlas, R. M. (1984). Diversity of microbial communities. *Advances in Microbial Ecology, 7,* 1–47. doi:10.1007/978-1-4684-8989-7.

Baath, E. (1989). Effects of heavy metal in soil on icrobial processess and population. *Water, Air and Soil Pollution, 47,* 335–379.

Baboshin, M. A., & Golovleva, L. A. (2012). Aerobic bacterial degradation of polycyclic aromatic hydrocarbons (PAHs) and its kinetic aspects. *Microbiology, 81*(6), 639–650. doi:10.1134/S0026261712060021.

Bachmann, A., de Bruin, W., Jumelet, J. C., Rijnaarts, H. H., & Zehnder, A. J. (1988). Aerobic biomineralization of alpha-hexachlorocyclohexane in contaminated soil. *Applied and Environmental Microbiology, 54*(2), 548–554.

Baldwin, B. R., Peacock, A. D., Park, M., Ogles, D. M., Istok, J. D., McKinley, J. P., & White, D. C. (2008). Multilevel samplers as microcosms to assess microbial response to biostimulation. *Ground Water, 46*(2), 295–304. doi:10.1111/j.1745-6584.2007.00411.x.

Bartha, R. (1986). Biotechnology of petroleum pollutant degradation. *Microbial Ecology, 12,* 155–172.

Bedard, D. L., Ritalahti, K. M., & Löffler, F. E. (2007). The Dehalococcoides population in sediment-free mixed cultures metabolically dechlorinates the commercial polychlorinated biphenyl mixture Aroclor 1260. *Applied and Environmental Microbiology, 73*(8), 2513–2521. doi:10.1128/AEM.02909-06.

Benka Coker, M. O., & Ekundayo, J. A. (1998). Effects of heavy metals on growth of species of Micrococcus and Pseudomonas in a crude oil/mineral salts medium. *Bioresource Technology, 66,* 241–245.

Beškoski, V. P., Gojgić-Cvijović, G., Milić, J., Ilić, M., Miletić, S., Šolević, T., & Vrvić, M. M. (2011). Exsitu bioremediation of a soil contaminated by mazut (heavy residual fuel oil)—A field experiment. *Chemosphere, 83*(1), 34–40. doi:10.1016/j.chemosphere.2011.01.020.

Bhattacharyya, J., Read, D., Amos, S., Dooley, S., Killham, K., & Paton, G. I. (2005). Biosensor-based diagnostics of contaminated groundwater: Assessment and remediation strategy. *Environmental Pollution, 134*(3), 485–492. doi:10.1016/j.envpol.2004.09.002.

Bumpus, J., Tien, M., Wright, D., & Aust, S. (1985). Oxidation of persistent environmental pollutants by a white rot fungus. *Science, 228*(4706), 1434–1436. doi:10.1126/science.3925550.

Bundy, J. G., Campbell, C. D., & Paton, G. I. (2001). Comparison of response of six different luminescent bacterial bioassays to bioremediation of five contrasting oils. *Journal of Environmental Monitoring: JEM, 3*(4), 404–410. doi:10.1039/B103104J.

Bundy, J. G., Paton, G. I., & Campbell, C. D. (2004). Combined microbial community level and single species biosensor responses to monitor recovery of oil polluted soil. *Soil Biology and Biochemistry, 36*(7), 1149–1159. doi:10.1016/j.soilbio.2004.02.025.

Cabras, P., Meloni, M., Pirisi, F. M., Farris, G. A., & Fatichenti, F. (1988). Yeast and pesticide interaction during aerobic fermentation. *Applied Microbiology and Biotechnology, 29*(2–3), 298–301. doi:10.1007/BF01982920.

Caffaro-Filho, R. A., Fantinatti-Garboggini, F., & Durrant, L. R. (2007). Quantitative analysis of Terminal Restriction Fragment Length Polymorphism (T-RFLP) microbial community profiles: Peak height data showed to be more reproducible than peak area. *Brazilian Journal of Microbiology, 38*(4), 736–738. doi:10.1590/S1517-83822007000400027.

Callaghan, A. V., Morris, B. E. L., Pereira, I. A. C., et al. (2012). The genome sequence of Desulfatibacillum alkenivorans AK-01: A blueprint for anaerobic alkane oxidation. *Environmental Microbiology*, 14(1), 101–113. doi:10.1111/j.1462-2920.2011.02516.x.

Cases, I., & de Lorenzo, V. (2005). Genetically modified organisms for the environment: Stories of success and failure and what we have learned from them. *International Microbiology*, 8, 213–222.

Cassidy, N. J. (2007). Evaluating LNAPL contamination using GPR signal attenuation analysis and dielectric property measurements: Practical implications for hydrological studies. *Journal of Contaminant Hydrology*, 94(1–2), 49–75. doi:10.1016/j.jconhyd.2007.05.002.

Cerniglia, C. E., & Gibson, D. T. (1977). Metabolism of naphthalene by Cunninghamella elegans. *Applied and Environmental Microbiology*, 34, 363–370.

Chain, P. S., Denef, V. J., Konstantinidis, K. T., et al. (2006). Burkholderia xenovorans LB400 harbors a multi-replicon, 9.73-Mbp genome shaped for versatility. *Proceedings of the National Academy of Sciences USA*, 103(42), 15280–15287. doi10.1073/pnas.0606924103.

Chakraborty, J., & Das, S. (2016). Molecular perspectives and recent advances in microbial remediation of persistent organic pollutants. *Environmental Science and Pollution Research*, 23, 16883. doi:10.1007/s11356-016-6887-7.

Chaplain, V., Mamy, L., Vieublé-Gonod, L., Mougin, C., Benoit, P., Barriuso, E., & Nélieu, S. (2011). Fate of pesticides in soils: Toward an integrated approach of influential factors. *Biodegradation: Involved Microorganisms and Genetically Engineered Microorganisms*. doi:10.5772/56194315. Pesticides in the Modern World—Risks and Benefits, Margarita Stoytcheva (Ed.) 2011 ISBN: 978-953-307-458-0.

Che-Alota, V., Atekwana, E. A., Atekwana, E. A., Sauck, W. A., & Werkema, D. D. J. (2009). Temporal geophysical signatures from contaminant-mass remediation. *Geophysics*, 74(4), B113–B123. doi:10.1190/1.3139769.

Chen, X., Liu, M., Hu, F., Mao, X., & Li, H. (2007). Contributions of soil micro-fauna (protozoa and nematodes) to rhizosphere ecological functions. *Acta Ecologica Sinica*, 27(8), 3132–3143.

Chikere, C. B. (2013). Application of molecular microbiology techniques in bioremediation of hydrocarbons and other pollutants. *British Biotechnology Journal*, 3(1), 90–115. doi:10.9734/BBJ/2013/2389.

Collin, P. (2004). *Dictionary of Environment and Ecology* (5th ed.) . London, UK: Bloomsbury, 254 pp.

Copley, S. D. (2000). Evolution of a metabolic pathway for degradation of a toxic xenobiotic: The patchwork approach. *Trends in Biochemical Sciences*, 25(6), 261–265. doi:10.1016/S0968-0004(00)01562-0.

Corapcioglu, M. Y., & Baehr, A. L. (1987). A compositional multiphase model for groundwater contamination by petroleum products: 1. Theoretical considerations. *Water Resources Research*, 23(1), 191–200. doi:10.1029/WR023i001p00191.

Cornelissen, G., van Noort, P. C. M., & Govers, H. A. (1997). Desorption kinetics of chlorobenzenes, polycyclic aromatic hydrocarbons, and polychlorinated biphenyls: Sediment extraction with Tenax(R) and effects of contact time and solute hydrophobicity. *Environmental Toxicology and Chemistry*, 16(7), 1351–1357. doi:10.1897/1551-5028(1997)016<1351:dkocpa>2.3.co;2.

Dagley, S. (1986). Biochemistry of aromatic hydrocarbon degradation in pseudomonads. *The Bacteria: A Treatise on Structure and Function*, 10, 527–556.

Das, P., Mukherjee, S., & Sen, R. (2008). Genetic regulations of the biosynthesis of microbial surfactants: An overview. *Biotechnology and Genetic Engineering Reviews*, 25(1), 165–186. doi:10.5661/bger-25-165.

Davis, C. A., Atekwana, E., Atekwana, E., Slater, L. D., Rossbach, S., & Mormile, M. R. (2006). Microbial growth and biofilm formation in geologic media is detected with complex conductivity measurements. *Geophysical Research Letters*, 33(18). doi:10.1029/2006GL027312.

Davis, T. A., Volesky, B., & Mucci, A. (2003). A review of the biochemistry of heavy metal biosorption brown algae. *Water Research*, 37, 4311–4330.

Dawson, J. J. C., Iroegbu, C. O., Maciel, H., & Paton, G. I. (2008). Application of luminescent biosensors for monitoring the degradation and toxicity of BTEX compounds in soils. *Journal of Applied Microbiology*, 104(1), 141–151. doi:10.1111/j.1365-2672.2007.03552.x.

De, J., Ramaiah, N., & Vardanyan, L. (2008). Detoxification of toxic heavy metals by marine bacteria highly resistant to mercury. *Marine Biotechnology*, 10(4), 471–477.

De Beer, D., Schramm, A., Santegoeds, C. M., & Khl, M. (1997). A nitrite microsensor for profiling environmental biofilms. *Applied and Environmental Microbiology*, 63(3), 973–977.

De Beer, D., Van den Heuvel, J. C., & Ottengraf, S. P. P. (1993). Microelectrode measurements of the activity distribution in nitrifying bacterial aggregates. *Applied and Environmental Microbiology*, 59(2), 573–579.

Declercq, I., Cappuyns, V., & Duclos, Y. (2012). Monitored natural attenuation (MNA) of contaminated soils: State of the art in Europe-A critical evaluation. *Science of the Total Environment*, 426, 393–405. doi:10.1016/j.scitotenv.2012.03.040.

DeFlaun, M. F., Fuller, M. E., Zhang, P., Johnson, W. P., Mailloux, B. J., Holben, W. E., & Onstott, T. C. (2001). Comparison of methods for monitoring bacterial transport in the subsurface. *Journal of Microbiological Methods*, 47(2), 219–231. doi:10.1016/S0167-7012(01)00307-4.

Denef, V. J., Park, J., Rodrigues, J. L. M., Tsoi, T. V., Hashsham, S. A., & Tiedje, J. M. (2003). Validation of a more sensitive method for using spotted oligonucleotide DNA microarrays for functional genomics studies on bacterial communities. *Environmental Microbiology*, 5(10), 933–943. doi:10.1046/j.1462-2920.2003.00490.x.

de Voogt, P., & Jansson, B. (1993). Vertical and long-range transport of persistent organics in the atmosphere. In *BT—Reviews of Environmental Contamination and Toxicology: Continuation of Residue Reviews*. G. W. Ware (Ed.), pp. 1–27. New York: Springer. doi:10.1007/978-1-4684-7065-9_1.

Diez, M. C. (2010). Biological aspects involved in the degradation of organic pollutants. *Journal of Soil Science and Plant Nutrition*, 10(3), 244–267. doi:10.4067/S0718-95162010000100004.

Divya, B., & Kumar, M. D. (2011). Plant–microbe interaction with enhanced bioremediation. *Research Journal of Biotechnology*, 6(4), 72–79.

Doick, K. J., & Semple, K. T. (2003). The effect of soil: Water ratios on the mineralisation of phenanthrene: LNAPL mixtures in soil. *FEMS Microbiology Letters*, 220(1), 29–33. doi:10.1016/S0378-1097(03)00056-9.

Donnelly, P. K., & Fletcher, J. S. (1995). PCB metabolism by ectomycorrhizal fungi. *Bulletin of Environmental Contamination and Toxicology*, 54, 507–513.

Dos Santos, A. B., Cervantes, J. F., & Van Lier, B. J. (2007). Review paper on current technologies for decolourisation of textile wastewaters: Perspectives for anaerobic biotechnology. *Bioresource Technology, 98*, 2369–2385.

Dunford, E. A., & Neufeld, J. D. (2010). DNA stable-isotope probing (DNA-SIP). *Journal of Visualized Experiments: JoVE*, (42). doi:10.3791/2027.

Eichner, C. A., Erb, R. W., Timmis, K. N., & Wagner-Dobler, I. (1999). Thermal gradient gel electrophoresis analysis of bioprotection from pollutant shocks in the activated sludge microbial community. *Applied and Environmental Microbiology, 65*(1), 102–109.

El Fantroussi, S., Mahillon, J., Naveau, H., & Agathos, S. N. (1997). Introduction of anaerobic dechlorinating bacteria into soil slurry microcosms and nested-PCR monitoring. *Applied and Environmental Microbiology, 63*(2), 806–811. doi:10.1007/978-94-017-1711-3_37.

Fernández, P. M., Martorell, M. M., Fariña, J. I., & Figueroa, L. I. C. (2012). Removal Efficiency of Cr6+ by Indigenous Pichia sp. Isolated from Textile Factory Effluent. *The Scientific World Journal*, Article ID 708213, 6 pages doi:10.1100/2012/708213.

Fisher, M. M., & Triplett, E. W. (1999). Automated approach for ribosomal intergenic spacer analysis of microbial diversity and its application to freshwater bacterial communities. *Appl Environ Microbiol, 65*(10), 4630–4636.

Francis, A. J., Spanggord, R. J., & Ouchi, G. I. (1975). Degradation of lindane by Escherichia coli. *Applied Microbiology, 29*(4), 567–568. doi:10.1002/cplu.201500098.

Fritsche, W., & Hofrichter, M. (2005) Aerobic degradation of recalcitrant organic compounds by microorganisms. In *Environmental Biotechnology: Concepts and Applications*. H.-J. Jördening and J. Winter (Eds.). Weinheim, Germany: Wiley-VCH Verlag GmbH & Co. KGaA. doi:10.1002/3527604 286.ch7.

Fuentes, S., Méndez, V., Aguila, P., & Seeger, M. (2014). Bioremediation of petroleum hydrocarbons: Catabolic genes, microbial communities, and applications. *Applied Microbiology and Biotechnology, 98*(11), 4781–4794. doi:10.1007/s00253-014-5684-9.

Fuller, M. E., Dong, H., Mailloux, B. J., Onstott, T. C., & DeFlaun, M. F. (2000). Examining bacterial transport in intact cores from Oyster, Virginia: Effect of sedimentary facies type on bacterial breakthrough and retention. *Water Resources Research, 36*(9), 2417–2431. doi:10.1029/2000WR900075.

Garbisu, C., & Alkorta, I. (2001). Phytoextraction: A cost-effective plant based technology for the removal of metals from the environment. *Bioresource Technology, 77*(3), 229–236.

Garon, D., Sage, L., Wouessidjewe, D., & Seigle-Murandi, F. (2004). Enhanced degradation of fluorene in soil slurry by Absidia cylindrospora and maltosyl-cyclodextrin. *Chemosphere, 56*(2), 159–166. doi:10.1016/j.chemosphere.2004.02.019.

Gavrilescu, M. (2005). Fate of pesticides in the environment and its bioremediation. *Journal of the Indian Chemical Society, 88*(10), 1525–1532. doi:10.1002/elsc.200520098.

Ghosal, D., Chatterjee, D. K., & Chakrabarty, A. M. (1985). Microbial degradation of halogenated compounds. *Science.* doi:10.1126/science.228.4696.135.

Glöckner, F. O., Amann, R., Alfreider, A., Pernthaler, J., Psenner, R., Trebesius, K., & Schleifer, K. H. (1996). An in situ hybridization protocol for detection and identification of planktonic bacteria. *Systematic and Applied Microbiology, 19*(3), 403–406. doi:10.1016/S0723-2020(96)80069-5.

Glöckner, F. O., Amann, R., Alfreider, A., Pernthaler, J., Psenner, R., Trebesius, K., & Schleifer, K. H. (1996). An in situ hybridization protocol for detection and identification of planktonic bacteria. *Systematic and Applied Microbiology, 19*(3), 403–406. doi:10.1016/S0723-2020(96)80069-5.

Gonzalez-Gil, G., Kleerebezem, R., & Lettinga, G. (1999). Effects of nickel and cobalt on kinetics of methanol conversion by methanogenic sludge as assessed by on-line CH4 monitoring. *Applied and Environmental Microbiology, 65,* 1789–1793.

Gottesman, S. (1984). Bacterial regulation: Global regulatory networks. *Annual Review of Genetics, 18*(1), 415–441. doi:10.1146/annurev.ge.18.120184.002215.

Grace Liu, P. W., Chang, T. C., Whang, L. M., Kao, C. H., Pan, P. T., & Cheng, S. S. (2011). Bioremediation of petroleum hydrocarbon contaminated soil: Effects of strategies and microbial community shift. *International Biodeterioration and Biodegradation, 65*(8), 1119–1127. doi:10.1016/j.ibiod.2011.09.002.

Gutell, R. R., Larsen, I. N., Woese, C. R., & Larsen, N. (1994). Lessons from an evolving rRNA: 16S and 23S rRNA structures from a comparative perspective. *Microbiological Reviews, 58*(1), 10–26. doi:10.1038/468755a.

Haderlein, S. B., Hofstetter, T. B., & Schwarzenbach, R. P. (2000). Subsurface chemistry of nitroaromatic compounds. In *Biodegradation of Nitroaromatic Compounds and Explosives.* J. C. Spain, J. B. Hughes, & H.-J. Knackmuss (Eds.), pp. 311–356. CRC Press.

Hahn, D., Amann, R. I., Ludwig, W., Akkermans, A. D. L., & Schleifer, K.-H. (1992). Detection of micro-organisms in soil after in situ hybridization with rRNA-targeted, fluorescently labelled oligonucleotides. *Journal of General Microbiology, 138*(5), 879–887. doi:10.1099/00221287-138-5-879.

Harayama, S. (1992). Aerobic biodegradation of aromatic hydrocarbons. *Metal Ions in Biological Systems: Volume 28: Degradation of Environmental Pollutants by Microorganisms and Their Metalloenzymes, 28,* 99.

Heinaru, E., Merimaa, M., Viggor, S., Lehiste, M., Leito, I., Truu, J., & Heinaru, A. (2005). Biodegradation efficiency of functionally important populations selected for bioaugmentation in phenol- and oil-polluted area. *FEMS Microbiology Ecology, 51*(3), 363–373. doi:10.1016/j.femsec.2004.09.009.

Holben, W. E., & Ostrom, P. H. (2000). Monitoring bacterial transport by stable isotope enrichment of cells. *Applied and Environmental Microbiology, 66*(11), 4935–4939. doi:10.1128/AEM.66.11.4935-4939.2000.

Holliger, C., Wohlfarth, G., & Diekert, G. (1999). Reductive dechlorination in the energy metabolism of anaerobic bacteria. *FEMS Microbiology Reviews, 22*(5), 383–398. doi:10.1016/S0168-6445(98)00030-8.

Hong, Y., Xu, M., Guo, J., Xu, Z., Chen, X., & Sun, G. (2007). Respiration and growth of Shewanella decolorations S12 with an azo compound as the sole electron acceptor. *Applied and Environmental Microbiology, 73,* 64–72.

Ijah, U. J. J. (1998). Studies on relative capabilities of bacterial and yeast isolates from tropical soil in degradating crude oil. *Waste Manage, 18,* 293.

Istok, J. D., Humphrey, M. D., Schroth, M. H., Hyman, M. R., & O'Reilly, K. T. (1997). Single-well, "push-pull" test for in situ determination of microbial activities. *Ground Water.* doi:10.1111/j.1745-6584.1997.tb00127.x.

Iwai, S., Johnson, T. A., Chai, B., Hashsham, S. A., & Tiedje, J. M. (2011). Comparison of the specificities and efficacies of primers for aromatic dioxygenase gene analysis of environmental samples. *Applied and Environmental Microbiology.* doi:10.1128/AEM.00331-11.

Iwamoto, T., & Nasu, M. (2001). Current bioremediation practice and perspective. *Journal of Bioscience and Bioengineering*, 92(1), 1–8. doi:10.1016/S1389-1723(01)80190-0.

Iwamoto, T., Tani, K., Nakamura, K., Suzuki, Y., Kitagawa, M., Eguchi, M., & Nasu, M. (2000). Monitoring impact of in situ biostimulation treatment on groundwater bacterial community by DGGE. *FEMS Microbiology Ecology*, 32(2), 129–141. doi:10.1016/S0168-6496(00)00022-2.

Jansson, J. K., Björklöf, K., Elvang, A. M., & Jørgensen, K. S. (2000). Biomarkers for monitoring efficacy of bioremediation by microbial inoculants. *Environmental Pollution*, 107(2), 217–223. doi:10.1016/S0269-7491(99)00140-2.

Jeong, S. W. (2005). Evaluation of the use of capillary numbers for quantifying the removal of DNAPL trapped in a porous medium by surfactant and surfactant foam floods. *Journal of Colloid and Interface Science*, 282(1), 182–187. doi:10.1016/j.jcis.2004.08.108.

Johnsen, A. R., Wick, L. Y., & Harms, H. (2005). Principles of microbial PAH-degradation in soil. *Environmental Pollution*. doi:10.1016/j.envpol.2004.04.015.

Johnson, D. N., Pedit, J. A., & Miller, C. T. (2004). Efficient, near-complete removal of DNAPL from three-dimensional, heterogeneous porous media using a novel combination of treatment technologies. *Environmental Science & Technology*, 38(19), 5149–56. doi:10.1021/es0344765.

Johnson, G. R., Jain, R. K., & Spain, J. C. (2002). Origins of the 2,4-dinitrotoluene pathway. *Journal of Bacteriology*, 184(15), 4219–4232. doi:10.1128/JB.184.15.4219-4232.2002.

Johnson, T. C., Versteeg, R. J., Day-Lewis, F. D., Major, W., & Lane, J. W. (2015). Time-lapse electrical geophysical monitoring of amendment-based biostimulation. *Groundwater*, 53(6), 920–932. doi:10.1111/gwat.12291.

Jones, K. C., & de Voogt, P. (1999). Persistent organic pollutants (POPs): State of the science. *Environmental Pollution*, 100(1–3), 209–221. doi:10.1016/S0269-7491(99)00098-6.

Jørgensen, K. S., Puustinen, J., & Suortti, A. M. (2000). Bioremediation of petroleum hydrocarbon-contaminated soil by composting in biopiles. *Environmental Pollution*, 107(2), 245–254. doi:10.1016/S0269-7491(99)00144-X.

Jørgensen, K. S., Salminen, J. M., & Björklöf, K. (2010). Monitored natural attenuation. In *Bioremediation: Methods and Protocols*. S. P. Cummings (Ed.), pp. 217–233. Totowa, NJ: Humana Press. doi:10.1007/978-1-60761-439-5_14.

Kafilzadeh, F., Sahragard, P., Jamali, H., & Tahery, Y. (2011). Isolation and identification of hydrocarbons degrading bacteria in soil around Shiraz Refinery. *African Journal of Microbiology Research*, 4(19), 3084–3089.

Kanade, S. N., Ade, A. B., & Khilare, V. C. (2012). Malathion degradation by Azospirillum lipoferum Beijerinck. *Science Research Reporter*, 2(1), 94–103.

Kanissery, R. G., & Sims, G. K. (2011). Biostimulation for the enhanced degradation of herbicides in soil. *Applied and Environmental Soil Science*, 2011, 1–10. doi:10.1155/2011/843450.

Kao, C. M., Chen, C. S., Tsa, F. Y., Yang, K. H., Chien, C. C., Liang, S. H., Yang, C. A., & Chen, S. C. (2010). Application of real-time PCR, DGGE fingerprinting, and culture-based method to evaluate the effectiveness of intrinsic bioremediation on the control of petroleum-hydrocarbon plume. *Journal of Hazardous Materials*, 178(1–3), 409–416. doi:10.1016/j.jhazmat.2010.01.096.

Ke, L., Luo, L., Wang, P., Luan, T., & Tam, N. F. (2010). Effects of metals on biosorption and biodegradation of mixed polycyclic aromatic hydrocarbons by a freshwater green alga *Selenastrum capricornutum*. *Bioresource Technology, 101*(2010), 6950–6961.

Kennedy, D. W., Aust, S. D., & Bumpus, J. A. (1990). Comparative biodegradation of alkyl halide insecticides by the white rot fungus, Phanerochaete chrysosporium (BKM-F-1767). *Applied and Environmental Microbiology, 56*(8), 2347–2353.

Kenzaka, T., Yamaguchi, N., & Tani, K. (1998). rRNA-targeted fluorescent in situ hybridization analysis of bacterial community structure in river water. *Microbiology, 144*, 2085–2093. Retrieved from http://mic.sgmjournals.org/content/144/8/2085.short.

Khan, A. G. (2006). Mycorrhizoremediation an enhanced form of phytoremediation. *Journal of Zhejiang University Science, 7*(7), 503–514.

Kivisaar, M. (2011). Evolution of catabolic pathways and their regulatory systems in synthetic nitroaromatic compounds degrading bacteria. *Molecular Microbiology, 82*(2), 265–268. doi:10.1111/j.1365-2958.2011.07824.x.

Kobayashi, H., & Rittman, B. E. (1982). Microbial removal of hazardous organic compounds. *Environmental Science and Technology, 16*, 170A–183A.

Ksheminska, H. P., Taras, M., Galyna, G., & Gonchar, M. (2006). Extracellular chromate-reducing activity of the yeast cultures. *Central European Science Journals, 1*(1), 137–149.

Kumari, K., Sharma, J.K., Kanade, G.S. Kashyap, S.M., Juwarker, A.A., & Wate, S.R. (2014). Investigation of polybrominated diphenyl ethers in old consumer products in India. *Environmental Monitoring and Assessment, 186*, 3001. doi:10.1007/s10661-013-3596-2.

Lane, J. W., Day-Lewis, F. D., & Casey, C. C. (2006). Geophysical monitoring of a field-scale biostimulation pilot project. *Ground Water, 44*(3), 430–443. doi:10.1111/j.1745-6584.2005.00134.x.

Lanthier, M., Villemur, R., Lépine, F., Bisaillon, J. G., & Beaudet, R. (2000). Monitoring of Desulfitobacterium frappieri PCP-1 in pentachlorophenol-degrading anaerobic soil slurry reactors. *Environmental Microbiology, 2*(6), 703–708. doi:10.1046/j.1462-2920.2000.00144.x.

Lebeau, T. (2011). *Bioaugmentation, Biostimulation and Biocontrol* (Vol. 108). doi:10.1007/978-3-642-19769-7.

Lee, K., Park, J. W., & Ahn, I. S. (2003). Effect of additional carbon source on naphthalene biodegradation by Pseudomonas putida G7. *Journal of Hazardous Materials, 105*(1–3), 157–167. doi:10.1016/j.jhazmat.2003.08.005.

Lehman, M. R., Colwell, F. S., & Bala, G. A. (2001). Attached and unattached microbial communities in a simulated basalt aquifer under fracture- and porous-flow conditions. *Applied and Environmental Microbiology, 67*(6), 2799–2809. doi:10.1128/AEM.67.6.2799-2809.2001.

Leigh, B. M., Prouzova, P., Mackova, M., Macek, T., Nagle, D. P., & Fletcher, J. S. (2006). Polychlorinated biphenyl (PCB)-degrading bacteria associated with trees in a PCB-contaminated site. *Applied and Environmental Microbiology, 4*(72), 2331–2342.

Li, E. S. Y., & Liu, W.-T. (2003). DNA microarray technology in microbial ecology studies-principle, applications and current limitations. *Microbes and Environments, 18*(4), 175–187. doi:10.1264/jsme2.18.175.

Liesack, W., & Dunfield, P. F. (2004). T-RFLP Analysis BT - Environmental Microbiology: Methods and Protocols. In J. M. Walker, J. F. T. Spencer, & A. L. Ragout de Spencer (Eds.), (pp. 23–37). Totowa, NJ: Humana Press. doi:10.1385/1-59259-765-3:023.

Liu, C., Jay, J. A., & Ford, T. E. (2001). Evaluation of environmental effect on metal transport from capped contaminated sediment under condition of submarine groundwater discharge. *Environmental Science Technology, 35,* 4549–4555.

Lovley, D. R. (2003). Cleaning up with genomics: Applying molecular biology to bioremediation. *Nature Reviews Microbiology, 1*(1), 35–44. doi:10.1038/nrmicro731.

Luthy, R. G., Aiken, G. R., Brusseau, M. L., Cunningham, S. D., Gschwend, P. M., Pignatello, J. J., Reinhard, M., Traina, S. J., Weber, W. J., & Westall, J. C. (1997). Sequestration of hydrophobic organic contaminants by geosorbents. *Environmental Science & Technology, 31*(12), 3341–3347. doi:10.1021/es970512m.

Lytle, C. A., Gan, Y. D., & White, D. C. (2000). Electrospray ionization/mass spectrometry compatible reversed-phase separation of phospholipids: Piperidine as a post column modifier for negative ion detection. *Journal of Microbiological Methods, 41*(3), 227–234. doi:10.1016/S0167-7012(00)00156-1.

Manefield, M., Whiteley, A. S., Griffiths, R. I., & Bailey, M. J. (2002). RNA stable isotope probing, a novel means of linking microbial community function to phylogeny. *Applied and Environmental Microbiology, 68*(11), 5367–5373. doi:10.1128/AEM.68.11.5367.

Manz, W., Szewzyk, U., Ericsson, P., Amann, R., Schleifer, K. H., & Stenstrom, T. A. (1993). In situ identification of bacteria in drinking water and adjoining biofilms by hybridization with 16S and 23S rRNA-directed fluorescent oligonucleotide probes. *Applied and Environmental Microbiology, 59*(7), 2293–2298.

Markandey, D. K., & Rajvaidya, N. (2004). *Environmental Biotechnology,* 1st ed. APH Publishing Corporation, p. 79.

Marsh, T. L., Saxman, P., Cole, J., & Tiedje, J. (2000). Terminal restriction fragment length polymorphism analysis program, a web-based research tool for microbial community analysis. *Applied and Environmental Microbiology, 66*(8), 3616–3620. doi:10.1128/AEM.66.8.3616-3620.2000.

Maruyama, A., & Sunamura, M. (2000). Simultaneous direct counting of total and specific microbial cells in seawater, using a deep-sea microbe as target. *Applied and Environmental Microbiology, 66*(5), 2211–2215. doi:10.1128/AEM.66.5.2211-2215.2000.

Mattison, R. G., Taki, H., & Harayama, S. (2005). The soil flagellate Heteromita globosa accelerates bacterial degradation of alkylbenzenesthrough grazing and acetate excretion in batch culture. *Microbial Ecology, 49,* 142–150.

McGlashan, M. A., Tsoflias, G. P., Schillig, P. C., Devlin, J. F., & Roberts, J. A. (2012). Field GPR monitoring of biostimulation in saturated porous media. *Journal of Applied Geophysics, 78,* 102–112. doi:10.1016/j.jappgeo.2011.08.006.

McGowan, C., Fulthorpe, R., Wright, A., & Tiedje, J. M. (1998). Evidence for interspecies gene transfer in the evolution of 2,4-dichlorophenoxyacetic acid degraders. *Applied and Environmental Microbiology, 64*(10), 4089–4092.

Megharaj, M., Ramakrishnan, B., Venkateswarlu, K., Sethunathan, N., & Naidu, R. (2011). Bioremediation approaches for organic pollutants: A critical perspective. *Environment International, 37*(8), 1362–1375. doi:10.1016/j.envint.2011.06.003.

Mohan, S. V., Kisa, T., Ohkuma, T., Kanaly, R. A., & Shimizu, Y. (2006). Bioremediation technologies for treatment of PAH-contaminated soil and strategies to enhance process efficiency. *Reviews in Environmental Science and Biotechnology, 5*(4), 347–374. doi:10.1007/s11157-006-0004-1.

Mörtberg, M., & Neujahr, H. Y. (1985). Uptake of phenol in Trichosporon cutaneum. *Journal of Bacteriology, 161,* 615–619.

Mougin, C., Pericaud, C., Dubroca, J., & Asther, M. (1997). Enhanced mineralization of lindane in soils supplemented with the white rot basidiomycete Phanerochaete Chrysosporium. *Soil Biology and Biochemistry, 29*(9–10), 1321–1324. doi:10.1016/S0038-0717(97)00060-6.

Mrozik, A., & Piotrowska-Seget, Z. (2010). Bioaugmentation as a strategy for cleaning up of soils contaminated with aromatic compounds. *Microbiological Research, 165*(5), 363–375. doi:10.1016/j.micres.2009.08.001.

Mucha, K., Kwapisz, E., Kucharska, U., & Okruszeki, A. (2010). Mechanism of aniline degradation by yeast strain Candida methanosorbosa BP-6. *Polish Journal of Microbiology, 59*(4), 311–315.

Mulligan, C. N., & Yong, R. N. (2004). Natural attenuation of contaminated soils. *Environment International, 30*(4), 587–601. doi:10.1016/j.envint.2003.11.001.

Muyzer, G., De Waal, E. C., & Uitterlinden, A. G. (1993). Profiling of complex microbial populations by denaturing gradient gel electrophoresis analysis of polymerase chain reaction-amplified genes coding for 16S rRNA. *Applied and Environmental Microbiology, 59*(3), 695–700.

Nagasawa, S., Kikuchi, R., Nagata, Y., Takagi, M., & Matsuo, M. (1993). Aerobic mineralization of γ-HCH by Pseudomonas paucimobilis UT26. *Chemosphere, 26*(9), 1719–1728. doi:10.1016/0045-6535(93)90115-L.

Nagata, Y., Miyauchi, K., & Takagi, M. (1999). Complete analysis of genes and enzymes for γ-hexachlorocyclohexane degradation in Sphingomonas paucimobilis UT26. *Journal of Industrial Microbiology and Biotechnology, 23*(4–5), 380–390. doi:10.1038/sj.jim.2900736.

Nakajima, F., Baun, A., Ledin, A., & Mikkelsen, P. S. (2005). A novel method for evaluating bioavailability of polycyclic aromatic hydrocarbons in sediments of an urban stream. *Water Science and Technology, 51*(3–4), 275–281.

Nakamura, Y., & Sawada, T. (2000). Biodegradation of phenol in the presence of heavy metals. *Journal of Chemical Technology & Biotechnology, 75,* 137–142.

National Research Council (NRC). (1993). *In Situ Bioremediation: When Does it Work?* Washington, DC: National Academic Press.

Neufeld, J. D., Vohra, J., Dumont, M. G., Lueders, T., Manefield, M., Friedrich, M. W., & Murrell, C. J. (2007). DNA stable-isotope probing. *Nature Protocols, 2*(4), 860–866. doi:10.1038/nprot.2007.109.

Nies, D. H. (1999). Microbial heavy metal resistance. *Applied Microbiology and Biotechnology, 51,* 730–750.

Nzila, A. (2013). Update on the cometabolism of organic pollutants by bacteria. *Environmental Pollution, 178,* 474–482.

Pal, T. K., Bhattacharyya, S., & Basumajumdar, A. (2010). Cellular distribution of bioaccumulated toxic heavy metals in Aspergillus niger and Rhizopus arrhizus. *International Journal of Pharma and Bio Sciences, 1*(2), 1–6.

Pandey, J., Sood, S. S., & Jain, R. K. (2007). Terminal restriction fragment length polymorphism (T-RFLP) analysis: Characterizing the unseen. *Indian Journal of Microbiology, 47*(1), 90–91. doi:10.1007/s12088-007-0017-7.

Perelo, L. W. (2010). Review: In situ and bioremediation of organic pollutants in aquatic sediments. *Journal of Hazardous Materials, 177*(1–3), 81–89. doi:10.1016/j.jhazmat.2009.12.090.

Phillips, T. M., Seech, A. G., Lee, H., & Trevors, J. T. (2005). Biodegradation of hexachlorocyclohexane (HCH) by microorganisms. *Biodegradation, 16*(4), 363–392. doi:10.1007/s10532-004-2413-6.

Pieper, D. H., Martins Dos Santos, V. A. P., & Golyshin, P. N. (2004). Genomic and mechanistic insights into the biodegradation of organic pollutants. *Current Opinion in Biotechnology, 15*(3), 215–224. doi:10.1016/j.copbio.2004.03.008.

Pieper, D. H., & Seeger, M. (2008). Bacterial metabolism of polychlorinated biphenyls. *Journal of Molecular Microbiology and Biotechnology.* doi:10.1159/000121325.

Pignatello, J. J., Ferrandino, F. J., & Huang, L. Q. (1993). Elution of aged and freshly added herbicides from a soil. *Environmental Science & Technology, 27*(8), 1563–1571. doi:10.1021/es00045a010.

Prasanna, D., Venkata Mohan, S., Purushotham Reddy, B., & Sarma, P. N. (2008). Bioremediation of anthracene contaminated soil in bio-slurry phase reactor operated in periodic discontinuous batch mode. *Journal of Hazardous Materials, 153*(2), 244–251. doi:10.1016/j.ibiod.2008.01.006.

Prenafeta-Boldu, F. X., Kuhn, A., Luykx, D., Anke, H., van Groenestijn, J. W., & de Bont, J. A. M. (2001). Isolation and characterisation of fungi growing on volatile aromatic hydrocarbons as their sole carbon and energy source. *Mycological Research, 105*(4), 477–484.

Purohit, H. J. (2003). Biosensors as molecular tools for use in bioremediation. *Journal of Cleaner Production, 11*(3), 293–301. doi:10.1016/S0959-6526(02)00072-0.

Ramos, J. L., Díaz, E., Dowling, D., de Lorenzo, V., Molin, S., O'Gara, F., Ramos, C., & Timmis, K. N. (1994). The behavior of bacteria designed for biodegradation. *Biotechnology, 12*(13), 1349–1356.

Ranjard, L., Poly, F., Lata, J., Mougel, C., Thioulouse, J., & Nazaret, S. (2001). Characterization of bacterial and fungal soil communities by automated ribosomal intergenic spacer analysis fingerprints: Biological and methodological variability. *Applied and Environmental Microbiology, 67*(10), 4479–4487. doi:10.1128/AEM.67.10.4479.

Reid, B. J., Jones, K. C., & Semple, K. T. (2000). Bioavailability of persistent organic pollutants in soils and sediments—A perspective on mechanisms, consequences and assessment. *Environmental Pollution, 108*(1), 103–112. doi:10.1016/S0269-7491(99)00206-7.

Rhee, S. K., Liu, X., Wu, L., Chong, S. C., Wan, X., & Zhou, J. (2004). Detection of genes involved in biodegradation and biotransformation in microbial communities by using 50-mer oligonucleotide microarrays. *Applied and Environmental Microbiology, 70*(7), 4303–4317. doi:10.1128/AEM.70.7.4303-4317.2004.

Rieger, P. G., Meier, H. M., Gerle, M., Vogt, U., Groth, T., & Knackmuss, H. J. (2002). Xenobiotics in the environment: Present and future strategies to obviate the problem of biological persistence. *Journal of Biotechnology, 94*(1), 101–123. doi:10.1016/S0168-1656(01)00422-9.

Roane, T. M., Josephson, K. L, & Pepper, I. L. (2001). Microbial cadmium detoxification allows remediation of co-contaminated soil. *Applied and Environmental Microbiology, 67*, 3208–3215.

Robles-González, I. V., Fava, F., & Poggi-Varaldo, H. M. (2008). A review on slurry bioreactors for bioremediation of soils and sediments. *Microbial Cell Factories, 7*(1), 5. doi:10.1186/1475-2859-7-5.

Rochkind-Dubinsky, M. L., Sayler, G. S., & Blackburn, J. W. (1987). Metabolism of non-chlorinated compounds. In *Microbiological Decomposition of Chlorinated Aromatic Compounds*, pp. 48–72. New York: Marcel Dekker.

Rouse, J. D., Sabatini, D. A., Brown, R. E., & Harwell, J. H. (1996). Evaluation of ethoxylated alkylsulfate surfactants for use in subsurface remediation. *Water Environment Research, 68*(2), 162–168. doi:10.2175/106143096×127343.

Said, W. A., & Lewis, D. L. (1991). Quantitative assessment of the effects of metals on microbial degradation of organic chemicals. *Applied and Environmental Microbiology, 57,* 1498–1503.

SaiSubhashini, S., Kaliappan, S., & Velan, M. (2011). Removal of heavy metal from aqueous solution using Schizosaccharomyces pombe in free and alginate immobilized cells. *Second International Conference on Environmental Science and Technology,* 6107–6111.

Salminen, J. M., Tuomi, P. M., & Jørgensen, K. S. (2008). Functional gene abundances (nahAc, alkB, xylE) in the assessment of the efficacy of bioremediation. *Applied Biochemistry and Biotechnology, 151*(2–3), 638–652. doi:10.1007/s12010-008-8275-3.

Sanchez, M. A., Campbell, L. M., Brinker, F. A., & Owens, D. (2000). Attenuation the natural way. A former wood-preserving site offers a case study for evaluating the potential of monitored natural attenuation. *Industrial Wastewater, 5,* 37–42. Retrieved from https://www.researchgate.net/publication/285915650_Attenuation_the_natural_way_A_former_wood-preserving_site_offers_a_case_study_for_evaluating_the_potential_of_monitored_natural_attenuation.

Sandrin, T. R., Chech, A. M, & Maier, R. M. (2000). A rhamnolipid biosurfactant reduces cadmium Toxicity during biodegradation of napthalene. *Applied Environmental Microbiology, 66,* 4585–4588.

Sarkar, D., Ferguson, M., Datta, R., & Birnbaum, S. (2005). Bioremediation of petroleum hydrocarbons in contaminated soils: Comparison of biosolids addition, carbon supplementation, and monitored natural attenuation. *Environmental Pollution, 136*(1), 187–195. doi:10.1016/j.envpol.2004.09.025.

Satoh, H., Okabe, S., Yamaguchi, Y., & Watanabe, Y. (2003). Evaluation of the impact of bioaugmentation and biostimulation by in situ hybridization and microelectrode. *Water Research, 37*(9), 2206–2216. doi:10.1016/S0043-1354(02)00617-6.

Sayel, H., Bahafid, W., Tahri-Joutey, N., Derraz, K., Fikri-Benbrahim, K., Ibnsouda-Koraichi, S., & El Ghachtouli, N. (2012). Cr(VI) reduction by Enterococcus gallinarum isolated from tannery waste-contaminated soil. *Annals of Microbiology.* doi:10.1007/s13213-011-0372-9.

Sayler, G. S., & Ripp, S. (2000). Field applications of genetically engineered microorganisms for bioremediation processes. *Current Opinion in Biotechnology, 11,* 286–289.

Sayler, G. S., Sherrill, T. W., & Perkins, R. E. (1982). Impact of coal-coking effluent on sediment microbial communities: A multivariate approach. *Applied and Environmental Microbiology, 44*(5), 1118–1129.

Schlömann, M. (1994). Evolution of chlorocatechol catabolic pathways—Conclusions to be drawn from comparisons of lactone hydrolases. *Biodegradation, 5*(3–4), 301–321. doi:10.1007/BF00696467.

Schwartz, R., Gerth, J., Neumann-Hensel, H., Bley, S., & Forstner, U. (2006). Assessment of highly polluted fluvisol in the Spittelwasser floodplain—Based on national guideline values and MNA-Criteria. *Journal of Soils and Sediments, 6*(3), 145–155. doi:10.1065/jss2006.06.166.

Scow, K. M. & Hicks, K. A. (2005). Natural attenuation and enhanced bioremediation of organic contaminants in groundwater. *Current Opinion in Biotechnology.* doi:10.1016/j.copbio.2005.03.009.

Scragg, A. H. (2005). *Environmental Biotechnology* (2nd ed.). New York: Oxford University Press.

Seeger, M., Cámara, B., & Hofer, B. (2001). Dehalogenation, denitration, dehydroxylation, and angular attack on substituted biphenyls and related compounds by a biphenyl dioxygenase. *Journal of Bacteriology, 183,* 3548–3555.

Seeger, M., Timmis, K. N., & Hofer, B. (1997). Bacterial pathways for the degradation of polychlorinated biphenyls. *Marine Chemistry, 58*(3–4), 327–333. doi:10.1016/S0304-4203(97)00059-5.

Sheehan, D. (1997). *Bioremediation Protocols.* Totowa, NJ: Humana Press.

Singh, B. K., & Kuhad, R. C. (1999). Biodegradation of lindane (γ-hexachlorocyclohexane) by the white-rot fungus Trametes hirsutus. *Letters in Applied Microbiology, 28*(3), 238–241. doi:10.1046/j.1365-2672.1999.00508.x.

Smejkal, C. W., Vallaeys, T., Seymour, F. A., Burton, S. K., & Lappin-Scott, H. M. (2001). Characterization of (R/S)-mecoprop [2-(2-methyl-4-chlorophenoxy) propionic acid]-degrading Alcaligenes sp. CS1 and Ralstonia sp. CS2 isolated from agricultural soils. *Environmental Microbiology, 3*(4), 288–293.

Snieder, R., Hubbard, S., Haney, M., Bawden, G., Hatchell, P., Revil, A., & Geophysical Monitoring Working Group. (2007). Advanced noninvasive geophysical monitoring techniques. *Annual Review of Earth and Planetary Sciences, 35*(1), 653–683. doi:10.1146/annurev.earth.35.092006.145050.

So, C. M., Phelps, C. D., & Young, L. Y. (2003). Anaerobic transformation of alkanes to fatty acids by a sulfate-reducing bacterium, strain Hxd3. *Applied and Environmental Microbiology, 69*(7), 3892–3900. doi:10.1128/AEM.69.7.3892-3900.2003.

Spellman, F.R. (2008). *Ecology for Non-Ecologists,* 1st ed., Government Institutes, p. 364.

Springael, D., & Top, E. M. (2004). Horizontal gene transfer and microbial adaptation to xenobiotics: new types of mobile genetic elements and lessons from ecological studies. *Trends in Microbiology, 12*(2), 51–53. doi:10.1016/j.tim.2003.12.001.

Springeal, D., Diels, L., Hooybergh, L., Krepsk, S., & Mergeay, M. (1993). Construction and characterization of heavy metal resistant haloaromatic dergrading Alcaligens eutrophus strains. *Applied and Environmental Microbiology, 59,* 334–339.

Sriprang, R., Hayashi, M., Ono, H., Takagi, M., Hirata, K., & Murooka, Y. (2003). Enhanced accumulation of Cd2+ by a Mesorhizobium sp. transformed with a gene from Arabidopsis thaliana coding for phytochelatin synthase. *Applied and Environmental Microbiology, 69,* 1791–1796.

Srivastava, N. K., Jha, M. K., Mall, I. D., & Singh, D. (2010). Application of genetic engineering for chromium removal from industrial wastewater. *International Journal of Chemical and Biological Engineering, 3,* 3.

Stiner, L., & Halverson, L. J. (2002). Development and characterization of a green fluorescent protein-based bacterial biosensor for bioavailable toluene and related compounds. *Applied and Environmental Microbiology, 68*(4), 1962–1971. doi:10.1128/AEM.68.4.1962-1971.2002.

Strathdee, F., & Free, A. (2013). Denaturing gradient gel electrophoresis. *Current Opinion in Biotechnology, 4*(1), 37–40. doi:10.1016/0958-1669(93)90029-V.

Struthers, J. K., Jayachandran, K., & Moorman, T. B. (1998). Biodegradation of atrazine by Agrobacterium radiobacter J14a and use of this strain in bioremediation of contaminated soil. *Applied of Environmental Microbiology, 64,* 3368–3375.

Stockholm Convention on Persistent Organic Pollutants. (2017). Link- chm.pops.int/tabid/208/Default.aspx.

Suarez, M. P., & Rifai, H. S. (2002). Evaluation of BTEX remediation by natural attenuation at a coastal facility. *Groundwater Monitoring & Remediation*. Retrieved from http://info.ngwa.org/gwol/pdf/020273050.pdf.

Sussman, M. C., Collins, H., Skinner, F. A., & Stewart-Tull, D. E. (ed.). (1988). *Release of Genetically Engineered Micro-Organisms*. London, UK: Academic Press.

Suzuki, M. T., & Giovannoni, S. J. (1996). Bias caused by template annealing in the amplification of mixtures of 16S rRNA genes by PCR. *Applied and Environmental Microbiology, 62*(2), 625–630.

Thompson, I. P., Van Der Gast, C. J., Ciric, L., & Singer, A. C. (2005). Bioaugmentation for bioremediation: The challenge of strain selection. *Environmental Microbiology*. doi:10.1111/j.1462-2920.2005.00804.x.

Tigini, V., Prigione, V., Di Toro, S., Fava, F., & Varese, G.C. (2009). Isolation and characterisation of polychlorinated biphenyl (PCB) degrading fungi from a historically contaminated soil. Microbial Cell Factories, 8, 5. doi:10.1186/1475-2859-8-5.

Timmis, K. N., & Pieper, D. H. (1999). Bacteria designed for bioremediation. *Trends in Biotechnology, 17*(5), 201–204. doi:10.1016/S0167-7799(98)01295-5.

Tiquia, S. M. (2010). Using terminal restriction fragment length polymorphism (T-RFLP) analysis to assess microbial community structure in compost systems. *Methods in Molecular Biology (Clifton, NJ), 599*, 89–102. doi:10.1007/978-1-60761-439-5_6.

Top, E. M., & Springael, D. (2003). The role of mobile genetic elements in bacterial adaptation to xenobiotic organic compounds. *Current Opinion in Biotechnology, 14*(3), 262–269. doi:10.1016/S0958-1669(03)00066-1.

Top, E. M., Springael, D., & Boon, N. (2002). Catabolic mobile genetic elements and their potential use in bioaugmentation of polluted soils and waters. *FEMS Microbiology Ecology, 42*(2), 199–208. doi:10.1016/S0168-6496(02)00370-7.

Treen-Sears, M. E., Martin, S. M., & Volesky, B. (1989). Propagation of Rhiloprzs juvanicus biosorbent. *Applied and Environmental Microbiology, 448*, 137–141.

Trejo-Hernandez, M. R., Lopez-Munguia, A., & Quintero Ramirez, R. (2001). Residual compost of Agaricus bisporus as a source of crude laccase for enzymic oxidation of phenolic compounds. *Process Biochemistry, 36*(7), 635–639. doi:10.1016/S0032-9592(00)00257-0.

Tu, C. M. (1976). Utilization and degradation of lindane by soil microorganisms. *Archives of Microbiology, 108*(3), 259–263. doi:10.1007/BF00454850.

Uhlik, O., Leewis, M. C., Strejcek, M., Musilova, L., Mackova, M., Leigh, M. B., & Macek, T. (2013). Stable isotope probing in the metagenomics era: A bridge towards improved bioremediation. *Biotechnology Advances, 31*(2), 154–165. doi:10.1016/j.biotechadv.2012.09.003.

United Nations Environment Programme Chemicals 2011. UNEP Persistent Organic Pollutants. Accessed July 5, 2010, http://www.chem.unep.ch/pops/.

USEPA. Use of monitored natural attenuation at superfund, RCRA corrective action and underground storage tank sites. OSWER Directive Number 9200.4-17P. Washington, DC: Office of Solid Waste and Emergency Response; 1999.

Vallaeys, T., Courde, L., Mc Gowan, C., Wright, A. D., & Fulthorpe, R. R. (1999). Phylogenetic analyses indicate independent recruitment of diverse gene cassettes during assemblage of the 2,4-D catabolic pathway. *FEMS Microbiology Ecology, 28*(4), 373–382. doi:10.1016/S0168-6496(98)00123-8.

Van de Peer, Y., Chapelle, S., & De Wachter, R. (1996). A quantitative map of nucleotide substitution rates in bacterial rRNA. *Nucleic Acids Research, 24*(17), 3381–3391. doi:10.1093/nar/24.17.3381.

Van der meer, J., & Sentchilo, V. (2003). Genomic islands and the evolution of catabolic pathways in bacteria. *Current Opinion in Biotechnology, 14*(3), 248–254. doi:10.1016/S0958-1669(03)00058-2.

Van der Meer, J. R., Werlen, C., Nishino, S. F., & Spain, J. C. (1998). Evolution of a pathway for chlorobenzene metabolism leads to natural attenuation in contaminated groundwater. *Applied and Environmental Microbiology, 64*(11), 4185–4193.

Van der Meer, J. R., Zehnder, A. J., & de Vos, W. M. (1992). Identification of a novel composite transposable element, Tn5280, carrying chlorobenzene dioxygenase genes of Pseudomonas sp. strain P51. *Journal of Bacteriology, 173,* 7077–7083.

Vasile Pavel, L., & Gavrilescu, M. (2008). Overview of Ex situ decontamination techniques for soil cleanup. *Environmental Engineering and Management Journal, 7*(6), 815–834.

Venkata Mohan, S., Purushotham Reddy, B., & Sarma, P. N. (2009). Ex situ slurry phase bioremediation of chrysene contaminated soil with the function of metabolic function: Process evaluation by data enveloping analysis (DEA) and Taguchi design of experimental methodology (DOE). *Bioresource Technology, 100*(1), 164–172. doi:10.1016/j.biortech.2008.06.020.

Vidali, M. (2001). Bioremediation—An overview. *Pure and Applied Chemistry, 73*(7), 1163–1172. doi:10.1351/pac200173071163.

Voparil, I. M., & Mayer, L. M. (2004). Commercially available chemicals that mimic a deposit feeder's (Arenicola marina) digestive solubilization of lipids. *Environmental Science and Technology, 38*(16), 4334–4339. doi:10.1021/es049506y.

Wackett, L. P. (2004). Stable isotope probing in biodegradation research. *Trends Biotechnology, 22,* 153–154.

Walker, J. D., Colwell, R. R., Vaituzis, Z., & Meyer, S. A. (1975). Petroleum-degrading achlorophyllous alga Prototphecazopfi. *Nature (London), 254,* 423–424.

Wang, J., & Chen, C. (2006). Biosorption of heavy metals by Saccharomyces cerevisiae: A Review. *Biotechnology Advances, 24,* 427–451.

Wang, L., Wang, W., Lai, Q., & Shao, Z. (2010). Gene diversity of CYP153A and AlkB alkane hydroxylases in oil-degrading bacteria isolated from the Atlantic Ocean. *Environmental Microbiology, 12*(5), 1230–1242. doi:10.1111/j.1462-2920.2010.02165.x.

Wentzel, A., Ellingsen, T. E., Kotlar, H. K., Zotchev, S. B., & Throne-Holst, M. (2007). Bacterial metabolism of long-chain n-alkanes. *Applied Microbiology and Biotechnology.* doi:10.1007/s00253-007-1119-1.

West, C. C., & Harwell, J. H. (1992). Surfactants and subsurface remediation. *Environmental Science and Technology, 26*(12), 2324–2330. doi:10.1021/es00036a002.

Whiteley, A. S., Manefield, M., & Lueders, T. (2006). Unlocking the "microbial black box" using RNA-based stable isotope probing technologies. *Current Opinion in Biotechnology.* doi:10.1016/j.copbio.2005.11.002.

Whyte, L. G., Smits, T. H. M., Labbé, D., Witholt, B., Greer, C. W., & Van Beilen, J. B. (2002). Gene cloning and characterization of multiple alkane hydroxylase systems in Rhodococcus strains Q15 and NRRL B-16531. *Applied and Environmental Microbiology, 68*(12), 5933–5942. doi:10.1128/AEM.68.12.5933-5942.

Xu, D., & Pei, J. (2011). Construction and characterization of a photosynthetic bacterium genetically engineered for Hg2+ uptake. *Bioresource Technology, 102,* 3083–3088.

Yamaguchi, N., Inaoka, S., Tani, K., Kenzaka, T., & Nasu, M. (1996). Detection of specific bacterial cells with 2-hydroxy-3-naphthoic acid-2'-phenylanilide phosphate and fast red TR in situ hybridization. *Applied and Environmental Microbiology, 62*(1), 275–278.

Yule, W. N., Chiba, M., & Morley, H. V. (1967). Fate of insecticide residues. Decomposition of lindane in soil. *Journal of Agricultural and Food Chemistry, 15*(6), 1000–1004. doi:10.1021/jf60154a037.

Zhang, P., & Johnson, W. P. (1999). Rapid selective ferrographic enumeration of bacteria. *Journal of Magnetism and Magnetic Materials, 194*(1), 267–274. doi:10.1016/S0304-8853(98)00582-4.

Zhang, S., Xia, X., Xia, N., Wu, S., Gao, F., & Zhou, W. (2013). Identification and biodegradation efficiency of a newly isolated 2,2',4,4'-tetrabromodiphenyl ether (BDE-47) aerobic degrading bacterial strain. *International Biodeterioration & Biodegradation, 76,* 24–31.

Zhang, T. C., & Bishop, P. L. (1996). Evaluation of substrate and pH effects in a nitrifying biofilm. *Water Environment Research, 68*(7), 1107–1115. doi:10.2175/106143096×128504.

Zhou, J. (2003). Microarrays for bacterial detection and microbial community analysis. *Current Opinion in Microbiology.* doi:10.1016/S1369-5274(03)00052-3.

Zoeteman, B. C. J., Harmsen, K., Linders, J. B. H. J., Morra, C. F. H., & Slooff, W. (1980). Persistent organic pollutants in river water and ground water of the Netherlands. *Chemosphere, 9*(4), 231–249. doi:10.1016/0045-6535(80)90080-6.

11

Heavy Metal Immobilization during Bio-Processing of Solid Waste

Ronghua Li

CONTENTS

11.1 Global Perspective of Heavy Metal Contamination and Conventional Approaches for Removal

11.1.1 The Potential for the Use of Organic Solid Waste in Agriculture

Organic wastes, especially sewage sludge, farmland plant residue, and live-stock and poultry droppings, are of growing environmental concern around the world due to extended organic waste control, advanced wastewater treatments, and extensive food demand resulting from urbanization and economic growth, as well as the growing environmental concerns of governments and citizens. It was estimated that the annual output of sewage sludge in the

United States and Europe is around 8 and 2 billion tons, respectively (Pathak et al. 2009). The collected sewage capacity in China increased dramatically in the past 30 years, with the sewage treatment capacity about $1.7 \times 10^8 \, m^3/d$ by 2013 (Jin et al. 2014; Ministry of Environmental Protection of the PRC, 2015; Yang et al. 2017). Consequently, the production of sewage sludge in China had an average annual growth rate of 13%, and the dry sewage sludge production reached 6.25×10^6 tons at the end of 2013; it was projected that the production will increase even more in the near future (Yang et al. 2015a; Zhang et al. 2018b). Meanwhile, with the increasing food demand, the total straw yield increased sharply in 1978–2009 with the improvement of agricultural production capacity. Among the 0.7 billion tons of agricultural straw production in 2009, the straw collected from rice, wheat, and maize was the main source of crop straw resource (Zhu et al. 2012). Except for food crop planting, the animal husbandry industry in China also developed very fast from 1978 to 2011. According to the data collected from the *Statistical Yearbook of China* and *China Animal Husbandry Yearbook*, the production amount of animal manure was mainly from cattle, pig, sheep, and poultry. The animal manure production and derived COD chemical oxygen demand (COD), is an important measure of organic pollution in water had increased by 1.35 and 0.91 times from 1978 to 2011, reaching 2.545 billion tons and 0.233 billion tons in 2011, respectively. Amounts of nitrogen and phosphorous from animal manure were 14.20 million and 2.48 million tons in 2011, increasing by 1.39 and 1.66 times from 1978 to 2011, respectively. The number of animals from the poultry breeding industry in most regions or provinces had surpassed 50% of the holding capacity of the environment (Zhu et al. 2014). Therefore, resource utilization of organic solid waste (mainly sewage sludge, livestock manure, and plant residue) and reduction in the related non-point-source pollution in China is an issue of concern and one in need of an urgent solution.

The development of agriculture is very important to China because of its large population. Organic wastes, including sewage sludge, livestock manure, and plant residue, have great potential in agricultural application since they are the common traditional fertilizer resource. The application of sewage sludge, livestock manure, and plant residue has been a worldwide agricultural practice for many years. However, because of environmental protection concerns about the disposal of this huge amount of organic wastes, utilization of these organic resources as fertilizer in China shrank dramatically. The insufficient utilization of organic fertilizer and excessive application of chemical fertilizers had led to a decrease in the water-retention capability and soil quality of farmland (Environment Protection Administration, China, 2001). Therefore, increasing the production of organic fertilizer made from organic wastes improves the utilization of organic fertilizer in agriculture, and optimizing the nutrients in soil is essential for sustainable development in China.

11.1.2 Global Perspective of Heavy Metal Contamination from Solid Waste

Until now, the main approaches for disposal of organic solid waste can be classified into five categories: landfill, incineration, industrial application, energy recovery, and agricultural use (Fijalkowski et al. 2017). Among these conventional methods of sewage sludge management, landfilling is one of the most widely employed methods in many countries. However, managing degradable sewage sludge by landfilling often requires open land or space occupation, which limits its application in areas that are short of land resources (Huang et al. 2018). Rising environmental and health concerns related to landfilling pollution, such as the release of highly contaminated leachate, odor gases, and greenhouse gases from the landfilling site, also make it less practicable (Syed-Hassan et al. 2017). Compared to landfilling (20%–30%), most of the sewage sludge in the United States and Europe were deposited in landfills (50%–60%) or incinerated (approximately 20%) (Syed-Hassan et al. 2017). Meanwhile, industrial application and incineration or thermochemical conversion of solid waste are energy-intensive processes that still cannot be used widely in developing countries. It is well known that most of the organic solid wastes (mainly sewage sludge and livestock manure) contain N, P, K, and several microelements such as Fe, Mg, and Ca that are essential to plant growth. Hence, these solid wastes have great potential in agricultural nutrient recycling practice (Fijalkowski et al. 2017). Although the amount of sewage sludge in land application (2.4%) is very limited in China at present, it expected that these practices will change in the very near future due to increasing concerns about sustainable development and recourse recycling (Syed-Hassan et al. 2017).

However, despite the fact that the application of organic solid waste may benefit cultivated soil and improve crop production, concerns about possible environmental contamination, such as unpleasant odor and greenhouse gas emissions, microbial infections, heavy metal distribution, etc., have to be addressed (Antoniadis et al. 2008, 2010; Awasthi et al. 2016; Li et al. 2012, 2015; Mendoza et al. 2010; Oleszczuk and Hollert 2011; Wang et al. 2016a, 2016b; Wang et al. 2017a; Yang et al. 2011; Zhu et al. 2014). Today, the major routes of exogenous heavy metal input to farmland soil are believed to be sewage sludge and animal manure utilization, inorganic fertilizer application, wastewater irrigation, and atmospheric deposition (Nicholson et al. 1998).

Many evaluation pointed out that the source of heavy metals in sewage sludge was highly dependent on the environmental condition of the area where municipal wastewater treatment plants were located. Duan et al. (2015) found that metallic pollutants such as Cd, Hg, As, Pb, Cr, Cu, Zn, etc., exist in sewage sludge, with Cu and Zn being the major contaminants. They also found that about 35.7% of the contaminations was related to transportation, coking factories and smelting facilities; 29.0% came from water supply pollution; and 16.2% originated from leather tanning, textile processing, chemical manufacturing industries, etc. Human and animal excretions, drinking water, and atmospheric deposition are also important sources of heavy metal in municipal sewage and contribute more than 50% of the Cu, Zn, and Pb content of municipal sewage sludge for soil application (Koch and Rotard 2001).

Similar to studies about sewage sludge, various researchers declared that the source of heavy metals in animal excretion were highly dependent on the feed supplement because metal additives are commonly overused and incorporated in animal feed worldwide (Cang et al. 2004; Hossain et al. 1998; Li et al. 2005, 2010; Li and Chen, 2005; Nicholson et al. 1999; Odstrci et al. 2010; Solaiman et al. 2007; Wang et al. 2016a, 2017a; Xiong et al. 2010; Zhu et al. 2013). Recently, an investigation conducted by Zhu et al. (2013) showed that Cu, Zn, Cr, Ni, As, Pb, and Cd occurred widely in the samples of pig manure and feed. The presence of heavy metals in pig feed was the main source of heavy metals in pig manure. These results were in line with other researchers' findings, for example, Cang et al. (2004) and Xiong et al. (2010). It has been reported that the Cu content in pig manure was around 399 and 699 mg/kg in Jiangsu and Beijing, China, respectively (Cang et al. 2004; Xiong et al. 2010). Except for the contamination of Cu, the detected maximum contents of Cd were 129.76 mg/kg in the pig manure and 27.60 mg/kg in pig feed samples from Beijing and Fuxin, China, respectively (Li et al. 2010). In addition, Nicholson et al. (1999) also revealed that in 12 pig slurry samples from England and Wales, the mean content of Cu, Zn, Ni, Pb, Cd, As, and Cr in pig manures was 351, 575, 10.4, 2.48, 0.30, 1.68, and 2.82 mg/kg, respectively. All these studies demonstrate that the possible sink for heavy metals are organic solid waste like sewage sludge and animal manure (Table 11.1).

The heavy metal content varies from less than 1 mg/kg to more than 10000 mg/kg in sewage sludge and animal manure. However, the continuous application of these organic wastes to the soil might pose a threat to the environment due to the possible transfer of heavy metal from sludge to soil, from which it may enter food chain (Soon et al. 1980; Zhu et al. 2013). The hazards of heavy metal soil pollution may be aggravated if the toxic heavy metal pollutant in the soil is mobile and can thus be taken up easily by plants or transported in the ecosystem (Kelly et al. 1999; Ma et al. 2011). In order to promote the land application of these organic wastes, heavy metal immobilization, potential toxicity reduction, and metal chemical forms adjustment must be made in the sewage sludge and animal manure (Duan et al. 2015).

TABLE 11.1

Levels of Heavy Metals (mg/kg) in Sewage Sludge and Animal Manure Reported in the Literature

Sample	Regions	Cu	Pb	Zn	Cd	Ni	Cr	As	References
Sewage sludge	China	62.75–796.63	86.25–136.75	290.38–831.0	9.63–15.13	98.63–2180.13	50.00–212.5	NR	Yang et al. (2017)
	Ireland	520	252	0.08	12	18	25	NR	Healy et al. (2016)
	Italy	90–206	80–126	0.02–0.09	0.3–0.9	11–15	18–65	NR	Gianico et al. (2013)
	Russia	200–300	34.7	0.07–0.08	NR	75–77	305–310	NR	Nikovski and Kalinichenko (2014)
	Canada	180–2300	26–465	354–640	2.3–10	37–179	66–2021	NR	Pathak et al. (2009)
	Hong Kong	112–255	52.5–57	1009–2823	2.3–10	NR	663	NR	Pathak et al. (2009)
	USA	616	170	1285	25	71	178	NR	Pathak et al. (2009)

(Continued)

TABLE 11.1 *(Continued)*

Levels of Heavy Metals (mg/kg) in Sewage Sludge and Animal Manure Reported in the Literature

Sample	Regions	Cu	Pb	Zn	Cd	Ni	Cr	As	References
Pig manure	Northeast China	77.62–1521.43	0–5.08	63.37–1622.8	0–203.4	NR	0–43.45	0.61–33.48	Zhang (2004)
	Shandong, China	46.1–1310.6	1.9–5.5	151.1–14679.8	0.6–1.5	NR	0.6–258.8	0.5–373.8	Pan et al. (2013)
	Zhejiang, China	96.58–1788.04	0.37–7.78	112.17–10056.68	0.02–4.87	2.14–23.18	0.43–86.58	2.45–76.43	Zhang et al. (2012)
	Shannxi, China	78.99–1543.28	0.05–35.81	68.72–3011.72	0.08–50.19	0.66–28.36	1.99–115.53	0.04–117.01	Zhu et al. (2013)
	Guangxi, China	123.3–1361.7	NR	370.4–2078.0	0.7–1.7	NR	10.8–40.6	NR	Huang et al. (2007)
	Jiangsu, China	35.7–1726.3	4.22–82.91	113.6–1505.6	1.13–4.35	3.62–22.10	23.21–64.67	4.00–78.00	Cang et al. (2004)
	England, Wales	17.5–780	1.01–9.79	68–716	0.10–0.84	1.00–49.8	0.67–15.7	0.10–6.7	Nicholson et al. (1999)
Dairy cattle slurry	England, Wales	1–352	0.10–16.9	5–727	0.10–1.74	0.1–11.40	0.20–21.40	0.10–4.83	Nicholson et al. (1999)
Beef cattle slurry	England, Wales	10.5–48.7	1.00–18.0	41–274	0.10–0.53	0.2–20.4	0.79–15.70	0.39–10.8	Nicholson et al. (1999)

NR means not reported.

11.1.3 Conventional Approaches for Heavy Metal Removal or Toxicity Reduction from Organic Solid Waste

Currently, various methods have been applied for heavy metal removal and reduction from organic solid waste. The main conventional clean-up approaches reported are chemical extraction, bioleaching, heat treating (combustion and thermochemical conversion), and bioprocess disposal (composting) techniques.

Chemical extraction is a chemical leaching process used for removal of heavy metals from solid waste. Various chemicals, including inorganic acids (e.g., H_2SO_4, HNO_3, HCl, etc.) (Naoum et al. 2001; Deng et al. 2009), organic acids (e.g., oxalic acid, citric acid, etc.) (Marchioretto 2003; Wang et al. 2015), soluble salts (e.g., $Fe_2(SO_4)_3$, $FeCl_3$, $MgCl_2$, etc.) (Strasser et al. 1995; Ito et al. 2000), and chelating agents (e.g., ethylenediamine, ethylene diamine tetraacetic acid, nitrilotriacetic acid, etc.) (Lo and Chen, 1990; Suanon et al. 2016; Veeken and Hamelers 1999) had been used for heavy metal extraction from sludge. Heavy metals can be removed efficiently from the solid waste with these extraction reagents. Many disadvantages associated with the chemical extraction process have been addressed, such as the high operating cost, the excessive requirement of chemicals, the operational difficulties in solid-liquid separation, the loss of valuable nutrient elements, and the unavoidable secondary pollution problems (Pathak et al. 2009). These drawbacks limit the utilization of chemical leaching in practice.

Consequently, research interests were gradually shifted toward bioleaching, which can achieve the reduction of heavy metals from solid waste efficiently and economically (Gu et al. 2017; Wong et al. 2004; Wong and Henry 1983). Bioleaching is the dissolution of metals from solid substrates by the metabolic activity of microorganisms (Wong et al. 2004). In practice, some sulfur-containing substances like FeS_2 and S^0 are often added to the organic solid waste leaching system at aerobic condition (Pathak et al. 2009; Wong et al. 2004). Heavy metals are removed by the catalytic effect of the metabolic activities of microorganisms because the acidic environment (sulfuric acid or organic acids) are often generated and result in an acceleration of metal removal during metabolic oxidation of insoluble iron and sulfur compounds by the iron- and sulfur-oxidizing bacteria (Pathak et al. 2009; Wong et al. 2004). It is believed that bioleaching is an environment-friendly technique due to the biogenic acidic environment in nature. Bioleaching is also one of the low-cost technologies in terms of chemical consumption because it needs relatively fewer chemicals in practice compared to traditional chemical leaching applied for metal leaching and recovery from sludge (Tyagi et al. 1988; Xiang et al. 2000). Furthermore, the chemical and bioleaching processes can also be facilitated by the assistance of electrokinetic improvement and ultrasonic or microwave disposal in practice (Peng et al. 2011; Oh et al. 2016). Although the chemical and bioleaching approaches have proved to be feasible technologies for cleaning up solid waste, the leaching operational

practice often includes several steps such as sludge conditioning and dehy-
dration, solid-liquid separation, decontaminated solid sludge neutralization,
disposal of leachate containing metals, etc. (Pathak et al. 2009). These operational difficulties make it impossible for quick reduction of solid waste volume in a very short time by using bioleaching technology.

Heat treatment appears to be a promising method for the management
of sewage sludge, crop residue, and animal manure (Li et al. 2018a). Heat
treatment involves the decomposition of most of the organic parts of the
solid waste by the application of controlled heating and/or oxidation, which
reduces the toxicity of metals and allows the conversion of solid waste into
energy, fuel, or other valuable materials in a relatively short time, making it
more advantageous than the bioleaching and chemical extraction processes
(Huang et al. 2018). The main conventional heat treatments are combustion
and thermochemical conversion (pyrolysis and gasification) (Li et al. 2018a,
2018b). Burning solid organic waste into ash in an incinerator with oxygen is
called incineration or combustion. It can reduce the volume of solid organic
waste efficiently and has been widely used in Germany (more than 54% of
the sewage sludge was burned) (Fijalkowski et al. 2017). The organic substance in organic solid waste was burned during the incineration, with the
heavy metal mainly accumulated in the ash (Huang et al. 2018). The bioavailability and mobility of heavy metal in ash was remarkably inhibited in the
combustion treatment process (Li et al. 2017b). Thermochemical conversion
is an emerging technology in organic waste disposal. It often pyrolyzes or
gasifies the organic solid waste by thermal destruction of organic material
at a high temperature in oxygen-limited conditions (Dias et al. 2010). The
thermochemical conversion of organic solid waste results in the production of liquid bio-oil, solid biochar, and syngas mixture (H_2, CO, CO_2, H_2O,
some light C1–C4 hydrocarbons, etc.) (Huang et al. 2018; Syed-Hassan et al.
2017). Although the composition of these three products depend mainly on
the feedstock of organic solid waste, peak temperature, heating rate, and
residence time, most of the heavy metals are retained and immobilized in
the solid biochar, with the bioavailability and mobility strongly inhibited
after thermal disposal (Huang et al. 2018; Xiao et al. 2015). It should be noted
that there are some similarities between the thermochemical conversion and
incineration in the aspect of reactions taking place during both processes (Li
et al. 2018a). However, the incineration was aimed at converting stored chemical energy in organic solid waste directly into heat as a source of thermal
energy (e.g., for house heating or electricity generation), while the thermochemical conversion was mainly focused on producing the carbon-rich solid
biochar or energy-rich gaseous mixture but not a hot flue gas (Syed-Hassan
et al. 2017). The need for efficient volume reduction and harmless disposal
and the concern of sustainable utilization of the solid waste sludge gave rise
to the methods of heating treatment and biochemical processing disposal (Li
et al. 2017a, 2017b, 2018b).

Composting is a kind of aerobic biochemical processing method for the degradable organic solid waste disposal (Bernal et al. 2009; Li et al. 2012). The composting process involves the concentration of nitrogen (N), phosphorus(P) and potassium (K) nutrients, the decomposition of organic waste into humus, the destruction of pathogens or unwanted seeds, and the transformation of heavy metals from a labile form to a stable form (Guerra-Rodriguez et al., 2006; Zhang et al. 2018a). The need for efficient volume reduction and harmless disposal and the concern of sustainable utilization of the solid waste sludge gave rise to the methods of heating treatment and bioprocessing disposal. The bioprocessing disposal of sewage sludge appears to be one of the most promising methods compared to heat treatment in terms of operation and economy (Xiao et al. 2017). Composting of solid waste such as sewage sludge and animal feces is not only an economical approach to reduce the vast volume but also a way of producing fertilizer for agriculture (Li et al. 2012, 2015). The organic solid waste composting disposal process can be improved with the help of additives in engineering practice (Awasthi et al. 2017ab).

11.2 Immobilization of Heavy Metals Using Different Additives as Amendments

As summarized in Figure 11.1, composting is one of the potential routes for the organic solid waste disposal. The process is often affected by many factors, including the properties of the composting mixture and the external environmental conditions (Xiao et al. 2017). In composting, all the parameters affecting the composting could be controlled or adjusted by engineering approaches and by the application of additives (Yu and Huang. 2009; Wong and Fang 2000). It is believed that the additives would facilitate the solid waste disposal (Awasthi et al. 2017ab; Shaheen et al. 2014). The bioavailability of heavy metals in compost could be significantly reduced with the utilization of additives (Li et al. 2015). Utilization of these additives could create feasible conditions for use of the manure and sewage sludge as fertilizers (Li et al. 2015; Shaheen et al. 2014). Due to the operational convenience in practice and the improvement of composting processes with the help of additives, many additives had been used for heavy metal immobilization during composting (Awasthi et al. 2017ab; Chen et al. 2010; Hua et al. 2009; Li et al. 2015; Wang et al. 2017ab; Wong et al. 2000; Xiao et al. 2017; Yang et al. 2015b; Zorpas et al. 2000ab). These reported additives applied in composting could be divided into two categories: organic additives and inorganic additives. The primary physiochemical properties of some organic and inorganic additives employed extensively in composting are listed in Table 11.2.

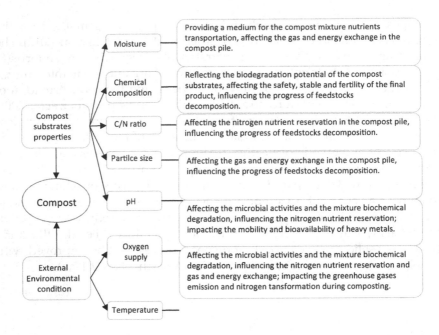

FIGURE 11.1
Parameters affecting composting processes.

TABLE 11.2

The Primary Physiochemical Property of Some Reported Inorganic Additives Employed in Composting

Additives	Types of Source	Physiochemical Property
Sawdust, rice straw, wheat straw, water hyacinth, sawdust, etc.	Wood processing waste, crop waste, wild plant residue, etc.	Easy degradable organic material rich in carbon element.
Biochar or charcoal	A by-product of organic material (crop residue, woody waste, sewage sludge, animal waste, etc.) thermal conservation treatment	The relatively stable property, high carbon content, enriched inorganic compounds, existing surface functional groups, developed porous structures, acceptable adsorptive capacity, and favorable biocompatibility.
Commercial humic acid, peat	Natural or artificial humic substance	Relatively stable organic aromatization substance with plenty of functional groups (–OH, –COOH, –COO, –NH, and –NH$_2$, etc.) in the structure.

(Continued)

TABLE 11.2 (*Continued*)

The Primary Physiochemical Property of Some Reported Inorganic Additives Employed in Composting

Additives	Types of Source	Physiochemical Property
Sodium sulfide	Chemical	Strong basicity, soluble, easy react with transit cation metal ion.
$FeCl_3$, $AlCl_3$	Chemical	Strong acidity, soluble, easy hydrolyzation, forms coprecipitation with transit cation metal ion.
Medical stone, Bentonite, Zeolite	Natural mineral	Porous mineral with ion exchangeability in solution.
Lime	Natural mineral	Strong basicity, easily reacts with transit cation metal ion.
Flyash	Coal combustion by-product	Strong basicity, easily reacts with transit cation metal ion.

11.2.1 Use of Organic Additives as Amendments

Table 11.3 summarizes the types and main impacts of reported organic additives used in composting heavy metal stabilization. The widely applied organic additives in composting are the organic bulking agents, such as sawdust, crop straw, grass, and so on. These organic additives are often added to the compost feedstock with the aim of adjusting the C:N ratio of compost substrate and providing a porous medium for the gas and energy exchange (Li et al. 2015). The organic waste decomposition process is enhanced with the help of these additives. The heavy metals gradually transform from a labile form into a stable form and consequently result in the reduction of heavy metal bioavailability (Awasthi et al. 2016; Li et al. 2012; Shaheen et al. 2014). A study conducted by Hua et al. (2009) showed that, by amending the sewage sludge compost mixture with rapeseed marc (10% w/w), the DTPA (diethylenetriaminepentaacetic acid) extractable Cu and Zn contents in the mixture decreased 29.2% and 12.0%, respectively after 42 days of aerobic composting.

Apart from the organic bulking agents, biochar is nothing but another widely employed organic additive in composting (Xiao et al. 2017). As mentioned earlier, biochar is the kind of solid product of organic waste thermochemical conservation like pyrolysis and gasification (Li et al. 2015, 2018a, 2018b), and nearly all organic materials can be employed as feedstock for biochar production (Dias et al., 2017; Xiao et al. 2017; Huang et al. 2017; Li et al. 2018a, 2018b). Many researchers have demonstrated the multifunctional roles of biochar in solid organic waste composting, contaminant immobilization, nutrient conservation, wastewater purification, and soil amelioration because of its relatively stable properties, highly porous structure, high carbon content, enriched

TABLE 11.3

Types and Main Impacts of Reported Organic Additives Used in Heavy Metal Immobilization during Solid Waste Disposal

Organic Additives	Adding Dosage	Substrates Mixture	Main Impacts	References
Corn straw	8.3% (dry w/w)	Pig manure	After composting, bioavailabilities of Pb and Cd reduced 42.79% and 73.07% in compost, while those of Zn and Cu in compost were activated slightly (0.18%–3.75% %). The content of exchangeable Pb speciation significantly reduced dramatically.	Hou et al. (2014)
Rapeseed marc	10% (fresh w/w)	Sewage sludge	Compared with the initial value before composting, extractable Cu and Zn contents in the composting material after composting decreased 29.2% and 12.0%, respectively.	Hua et al. (2009)
Peat	2.5% (dry w/w)	Pig manure + corn straw	After composting, bioavailabilities of Cu, Pb, and Zn reduced 25.35%, 4.08% and 50.28% in compost, respectively; while about 6.90% Cd was activated in compost. Contents of exchangeable Cu, Zn, and Pb speciations significantly reduced.	Hou et al. (2014)
Biological humic acid (Fujian)	2.5% (dry w/w)	Pig manure + corn straw	After composting, bioavailabilities of Cu and Cd decreased 14.70% and 47.39% in compost, while about 17.13% Zn and 1.25% Pb were activated,	Hou et al. (2014)
Biological humic acid (Jiabowen)	2.5% (dry w/w)	Pig manure + corn straw	Bioavailabilities of Cu, Pb, Zn, and Cd were reduced 47.78%, 47.54%, 64.94%, and 87.36% in compost.	Hou et al. (2014)
Rice straw	25% (dry w/w)	Sewage sludge	Stabilizing the sewage sludge increased the Sorghum *vulgare* L. and *Eurica sativa* growth and decreased the Cu and Zn bioavailabilities as compared to the non-stabilized sewage sludge.	Shaheen et al. (2014)
Water hyacinth	25% (dry w/w)	Sewage sludge	Stabilizing the sewage sludge increased the Sorghum *vulgare* L. and *Eurica sativa* growth and decreased the Cu and Zn bioavailabilities as compared to the non-stabilized sewage sludge.	Shaheen et al. (2014)
Bamboo biochar (600°C)	0%, 1%, 3%, 5%, 7%, and 9% (fresh w/w)	Sewage sludge + rapeseed marc	The effects of Cu and Zn immobilization in compost depend on the level of biochar addition. In comparison with the control treatment, amendment with 9% biochar is recommended in composting practice.	Hua et al. 2009)

(Continued)

TABLE 11.3 (Continued)

Types and Main Impacts of Reported Organic Additives Used in Heavy Metal Immobilization during Solid Waste Disposal

Organic Additives	Adding Dosage	Substrates Mixture	Main Impacts	References
Wheat straw biochar (350°C–550°C)	Biochar (0%, 1%, 3%, 5% and 7%) + 0.4% microbial agents	Sewage sludge + wheat straw	Addition of biochar benefits the passivation of Pb, Ni, Cu, Zn, As, Cr and Cd. The available content of Pb, Cu and As decreased 0.46%–51.5%, 7.25%–59.54% and 2.73%–56.31%, respectively.	Liu et al. (2017a)
Bamboo biochar	3%, 6% and 9% (fresh w/w)	Pig manure + sawdust	The increase of bamboo biochar addition significantly decreased Cu and Zn mobility in composting.	Chen et al. (2010)
Wood biochar	2.5% (dry w/w)	Pig manure + corn straw	After composting, bioavailabilities of Pb, Zn and Cd were reduced about 13.09%, 55.42% and 94.67% in compost respectively.	Hou et al. (2014)
Corn stalks biochar	2.5% (dry w/w)	Pig manure + corn straw	After composting, bioavailabilities of Cd, Zn, Pb, and Cu were decreased more than 49.93%, 18.08%, 57.20%, and 24.31% in compost, respectively.	Hou et al. (2014)
Peanut shell biochar	2.5% (dry w/w)	Pig manure + corn straw	After composting, bioavailabilities of Cd, Zn, Pb, and Cu were decreased 25.09%, 13.66%, 46.20%, 65.71% in compost, respectively.	Hou et al. (2014)
Corn stalk biochar (250°C–300°C, 450°C–500°C, 600°C–700°C, 750°C–900°C)	2.5% (dry w/w)	Pigmanure + corn stalk	In the control, 250°C–300°C, 450°C–500°C, 600°C–700°C, and 750°C–900°C biochar amended treatments, the DTPA extractable Zn reductions were 2.0%, 5.8%, 9.9%, 8.6% and 8.7%, respectively; the DTPA-extractable Cu reductions were 11.5%, 22.3%, 24.8%, 23.1%, and 22.1%, respectively.	Li et al. (2015)
Lime, wheat straw biochar	1% lime, 12% biochar +1% lime (w/w dry basis)	Sewage sludge + wheat straw	Mixing biochar and lime as additive reduced the bioavailabilities of Cu (34.81%), Zn (56.74%), Pb (87.96%), and Ni (86.65%) and improved the maturity and the quality of final compost product.	Awasthi et al. (2016)

inorganic mineral content, acceptable adsorptive capacity, and favorable bio-compatibility (Awasthi et al. 2016; Chen et al. 2010; Dias et al. 2010; Hua et al. 2009; Li et al. 2015, 2016a, 2016b, 2017c, 2018a, 2018b; Steiner et al. 2010). In the market, biochar produced from feedstock pyrolysis is very popular and the most cost-effective due to the relatively low temperatures (300°C–800°C) and the lack of need for oxygen supply during the preparation (Jeong et al. 2012; Li et al. 2018a, 2018b, 2018c). However, the costs of feedstock collecting, trans-portation, and processing, as well as the related energy requirement, always need to be considered during biochar production (Li et al. 2018a, 2018b, 2018c; Jeong et al. 2012; Xiao et al. 2017).

Previous studies have pointed out that most of the reported additives are effective for heavy metal immobilization during organic solid waste disposal and reduction in the environmental risk of solid waste farmland utilization (Table 11.3). For instance, Shaheen et al. (2014) successfully proved that using rice straw and water hyacinth as amendments could decrease the content of DTPA-extractable Cu and Zn in comparison with the original sewage sludge. In addition, using the amended sewage sludge as fertilizer improved *Sorghum vulgare L.* and *Eurica sativa* growth and increased the yield of plant dry biomass significantly.

The impact of the additives on heavy metal immobilization is strongly influenced by the type of additive. Even for biochar and humic acid material, the effects on heavy metal immobilization are strongly dependent on the feedstock and preparation conditions. For example, a series of comparison composting tests were conducted by Hou et al. (2014) to investigate the effect of different additives on the status of heavy metals during pig manure com-posting. Pig manure with different additives (peanut shell biochar, wood bio-char, corn stalks charcoal, Fujian brand humic acid, Jiabowen brand humic acid, peat) and a blank manure (no additive condition) were mixed and aero-bically composted in reactors over 30 days. The investigation demonstrated that peanut shell biochar, wood biochar, corn stalks charcoal, and Jiabowen brand humic acid had the best effects on compost quality improvement and reduced the content of exchangeable Cu, Pb, Zn, and Cd in pig manure com-post. The peanut shell biochar did not result in a mature product during pig manure composting (Hou et al. 2014). Based on the evidence of these additives on compost maturity, facilitating heavy metal immobilization, and improving compost quality, peanut shell biochar was not recommended for use as an amendment for pig manure composting. It was reported by Hou et al. (2014) that not only were the wood biochar and corn stalks charcoal more effective than the peanut shell biochar but also the Jiabowen brand biological humic acid was more effective than the Fujian brand biological humic acid and peat. These results demonstrate that the feedstock and bio-char types have an important impact on heavy metal immobilization.

In addition to the feedstock, the additive processing procedure is consid-ered to pose some non-negligible effect on heavy metal stabilization. Li et al. (2015) revealed the Cu and Zn immobilization capacities (DTPA-extractable

metal content reduction) of corn straw biochar in pig manure composting were strongly dominated by the biochar pyrolysis temperature. They
noticed that the percentages of Zn and Cu immobilization were in line with
the increase of corn straw biochar carbonization temperatures (250°C–300°C,
450°C–500°C, 600°C–700°C, and 750°C–900°C). The Cu immobilization capacity of corn biochar was higher than that of Zn, and similar observations were
also noticed in other biochar-amended aerobically composted pig manure
and sewage sludge (Awasthi et al. 2016; Chen et al. 2010; Hua et al. 2009).
These studies indicate that additives like biochar prepared from different
conditions also had a non-negligible effect on heavy metal stabilization.

The amounts of the additives in the compost mixture are also crucial to the
heavy metal immobilization (Camps and Tomlinson 2015; Liu et al. 2017a).
For example, Chen et al. (2010) and Hua et al. (2009) studied the amount of
bamboo biochar amendment on immobilization of Cu and Zn in sewage
sludge compost. They discovered significant effects of bamboo biochar on
Cu and Zn immobilization in sewage sludge compost and also noticed that
the heavy metal immobilization strongly depended on the biochar amount.
In comparison with the control treatment, extractable Cu and Zn content
in compost with 9% (w/w) bamboo biochar addition reduced by 27.5% and
8.2%, respectively. In addition, Liu et al. (2017a) also reported that the biochar amount had a great influence on heavy metal immobilization during
sewage sludge composting, but considering compost pile temperature and
nitrogen transformation, the treatment with 5% biochar was suggested for
sludge composting application. Some researchers reported that the biochar
amounts varied from 1% to 20% (w/w) and even up to 50% (w/w) (Awasthi
et al. 2017a; Chen et al. 2010; Dias et al. 2010; Hua et al. 2009; Li et al. 2015;
Sonoki et al. 2013; Steiner et al. 2010; Xiao et al. 2017; Zhang et al. 2014). In
terms of benefit and economic cost, biochar amounts lower than 20% are
acceptable in composting (Xiao et al. 2017). Additives with a high dosage
may not only increase the organic solid waste disposal cost but could also
possibly reduce the nutrient value of the final products (Li et al. 2012; Wong
et al. 2000; Xiao et al. 2017).

11.2.2 Effect of Inorganic Additive as an Amendment

Stabilization or safe disposal of organic solid wastes like sewage sludge
and animal manure prior to their land application not only helps to control the mobility of solid waste–borne contaminants but also transforms
the waste into valuable fertilizer. Among the additives used in the practice
of heavy metal stabilization, inorganic additives have been extensively
investigated due to their diverse physiochemical properties (Table 11.2).
The types and main impacts of some reported typical inorganic additives used in heavy metal immobilization during composting are listed
in Table 11.4. These inorganic additives are mainly salts, minerals, and
industrial by-products.

TABLE 11.4

Overview of the Types and Main Impacts of Reported Inorganic Additives Used in Heavy Metal Immobilization

Inorganic Additives	Adding Dosage	Substrates Mixture	Main Impacts	References
Sodium based bentonite	0%, 2.5%, 5%, 7.5%, and 10% in dry weight	Pig manure	Increasing the amount of bentonite in pig manure compost feasibly reduced the contents of extractable Cu and Zn.	Li et al. (2012)
Calcium-based bentonite	0%, 2.5%, 5%, 7.5%, and 10% in dry weight	Pig manure	Increasing the amount of Calcium based bentonite improved the reduction of extractable Cu and Zn content in pig manure compost.	Wang et al. (2016a)
Medical stone	0%, 2.5%, 5%, 7.5%, and 10% in dry weight	Pig manure	Medical stone addition promoted the organic waste degradation and prolonged the thermophilic phase as well as enhanced the immobilization of heavy metals Cu and Zn.	Wang et al. (2016b)
Bentonite	25% of dry weight	Sewage sludge	Stabilize the sewage sludge increased the Sorghum *vulgare L.* and *Eurica sativa* growth and decreased the metal bioavailability (DTPA-extractable Cu and Zn) as compared to the non-stabilized sewage sludge.	Shaheen et al. (2014)
Brick factory fly ash	25% of dry weight	Sewage sludge	Stabilization the sewage sludge increased the plant growth and decreased the metal bioavailability (DTPA-extractable Cu and Zn) of sewage sludge.	Shaheen et al. (2014)
Sugar beet factory lime	25% of dry weight	Sewage sludge	Immobilization of the sewage sludge increased the plant growth and decreased the contents of DTPA-extractable Cu and Zn in sewage sludge.	Shaheen et al. (2014)
Fishpond sediment and rock phosphate	Fishpond sediment (at 0, 25, and 35%) and/or rock phosphate (at 0, 10, and 15%) in dry weight	Green waste	The combined addition of fish pond sediment and rock phosphate accelerated green waste degradation. The amendment improved compost parameters variation like mixture temperature, windrow volume, pH, electrical conductivity, cation exchange capacity, NH_3 emission, microbial biomass, enzyme activities, nutrient content, and seed germination etc. Addition of 25% fishpond sediment and 15% rock phosphate yielded the highest quality compost product and shortened the compost mature time to 22 days.	Zhang and Sun (2017)

(Continued)

TABLE 11.4 (*Continued*)

Overview of the Types and Main Impacts of Reported Inorganic Additives Used in Heavy Metal Immobilization

Inorganic Additives	Adding Dosage	Substrates Mixture	Main Impacts	References
Sodium sulfide + Lime	Sodium sulfide and lime mixture (Na$_2$S/ CaO=1:1) at 3% in dry weight	Sewage sludge and sawdust mixture (4:1 /w fresh weight)	The mobile fractions of Cu, Zn, and Ni in exchangeable and carbonate bound forms in sewage sludge were transformed into low availability fractions (organic matter and sulfide, Fe-Mn oxides bound and residual forms). The addition of sodium sulfide and lime enhanced the transformation of Cu, Zn, and Ni during composting of sewage sludge.	Wang et al. (2008)
Lime	0%, 0.63%, 1.0%, and 1.63% in dry weight	Sewage sludge + sawdust (2:1, w/w fresh weight)	Lime addition decreased the transformation of Ni from reducible to residual form and the transformation of Zn from residual into oxidizable form but increased the content of the residual form of Pb. Addition of lime to sewage sludge produced compost with lower DTPA extractable Cu, Mn, Ni, Pb, and Zn contents.	Wong and Selvam (2006)
Lime	0%, 0.63%, 1.0%, and 1.63% in dry weight	Sewage sludge + sawdust (2:1, w/w fresh weight)	Lime adding reduced water-soluble and DTPA-extractable Cu, Mn, Zn, and Ni contents.	Fang and Wong (1999)
Zeolite (clinoptilolite)	0.99%, .91%, 4.76%, 6.54%, and 9.09% in dry weight	Sewage sludge	Metals in the sewage sludge are bound mainly (over 50%) in the residual fraction. The metals in mobile form was in the order of Ni > Cd ≥ Cr > Cu ≥ Pb. Addition of the clinoptilolite to the sludge leads to the stabilization of metals. Concentrations of mobile forms of Cd, Cr, Cu and Ni decrease by 87%, 64%, 35%, and 24%, respectively, as a result of the addition of 9.09% of the clinoptilolite. The reduction of Cu, Ni, Cr, Cd and Pb in sludge were 11%, 15%, 25%, 41%, and 51% in 9.09% clinoptilolite added treatment, respectively.	Sprynskyy et al. (2007)

(Continued)

TABLE 11.4 (Continued)

Overview of the Types and Main Impacts of Reported Inorganic Additives Used in Heavy Metal Immobilization

Inorganic Additives	Adding Dosage	Substrates Mixture	Main Impacts	References
Zeolite (clinoptilolite)	20% of dry weight	Sewage sludge mixed with sawdust in 10%, 30% and 40% (w/w)	The leachabilities of Cu, Cr, Fe, Ni, Mn, Pb, and Zn decreased after 75-day composting. Most of the number of metals was bound in residual fraction.	Zorpas et al. (2008)
Zeolite (clinoptilolite)	0%, 5%, 10%, 15%, 20%, 25%, and 30% in dry weight	Sewage sludge	During the 150 days composting the more zeolite added the more heavy metals immobilized. In the 25% and 30% zeolite treatment, about 100% Cd, 28%–45% Cu, 10%–15% Cr, 41%–47% Fe, 9%–24% Mn, 50%–55% Ni and Pb, and 40%–46% Zn were removed.	Zorpas et al. (2000b)
Zeolite (clinoptilolite)	25% zeolite in dry weight with particle size of 0.161, 0.161–1.0, 1.1–2.5, 2.6–3.2, and 3.3–4.0 mm	Sewage sludge	Amount of heavy metal uptake by zeolite was increased with particles size increase. Up to 12% of Co, 27% of Cu, 14% of Cr, 30% of Fe, 40% of Zn, 55% of Pb, and 60% of Ni was uptaken by 3.3–4.0 mm clinoptilolite.	Zorpas et al. (2002)
Zeolite	Three kinds of zeolite (all in size of 5–6 mm and pore size less than 2 nm) were added in a ratio of 10%, 25%, and 40% in dry weight	Sewage sludge	At the end of composting, zeolites retained 100% of the Ni, Cr, and Pb that was present in the sludge. Among the three zeolites (mordenite, clinoptilolite (Klinolith) and clinoptilolite (clinoptilolite E568)), clinoptilolite E568 was the most effective zeolite for the removal of Cu, Zn, and Hg.	Villaseñor et al. (2011)
Zeolite (clinoptilolite)	Ratio of 0%, 0.5%, 1.0%, 2.0%, 5.%, and 10% in dry weight; Particle size of 0.5–0.3, 1.0–0.7, and 2.0–1.4 mm.	Sewage sludge	The optimized additive for immobilization of Pb and Cr was zeolite in particle size of 0.7–1.0 mm, mixed for 5 h with zeolite/sewage sludge ratio of 2/98.	Kosobucki et al. (2008)
Coal fly ash	0%, 10%, 25%, and 35% in dry weight	Sewage sludge + sawdust (2:1, w/w fresh weight)	Adding of coal fly ash raised the pH (above 9.0), decreased the soluble organic carbon and total carbon content of the sludge compost, and inhibited the heavy metal mobility in sludge.	Wong et al. (1997)

It is easy to understand that the heavy metal immobilization capacity of inorganic additives depends considerably on the physiochemical properties of the inorganic materials. Shaheen et al. (2014) compared the differences of some additives on Cu and Zn immobilization of sewage sludge by selecting easily discomposed organic material (rice straw and water hyacinth), a mineral with a favorable ion exchange capacity (bentonite), and industrial by-products with alkaline features (sugar beet factory lime and brick factory fly ash) as the typical inorganic additives. The application of these additives in 25% (w/w) ratio to stabilize the sewage sludge increased *Sorghum vulgare L.* and *Eurica sativa* growth and decreased the bioavailability of Cu and Zn. Of these additives, sugar beet factory lime and bentonite are the recommended additives for sewage sludge immobilization (Shaheen et al. 2014).

The efficiency of lime as an additive for heavy metal immobilization in sewage sludge had been extensively proved. However, optimizing the adding ratio of lime was one of the important factors that affecting the biosolid disposal cost and composting efficiency. In an in-vessel 100-day aerobiotic composting test, Wong and Selvam (2006) treated the sewage sludge and sawdust mixture with lime at dry weight ratios of 0%, 0.63%, 1%, and 1.63%, respectively. They found the composting transformed the residual form of Cu into an oxidizable form, the residual form of Mn into a reducible form, the reducible form of Ni into a residual form, and the predominant residual form of Zn into an oxidizable form during the composting process, respectively. Except for the increase of the residual form of Pb content with an increase in the lime additive, no other significant transformation of different forms of Pb was observed. Similar observations were also reported in the study conducted by Fang and Wong (1999), who composted lime-amended sewage sludge in an aerobic reactor that produced safer organic fertilizer with lower water-extractable and DTPA-extractable metal contents. They found, for the control treatments of compost treatment, that a maximum reduction of Cu, Mn, Zn, and Ni were 40%, 40%, 10%, and 20%, respectively, percentages that were lower than those of the lime-treated sludge. After 63 days of composting, the DTPA-extractable Cu, Mn, Zn, and Ni were reduced by 60%, 80%, 55%, and 20%, respectively. Judging from the levels of maturity and nutrient levels of the final compost product, an additive with lime lower than 1% was suggested in practice.

Zeolite is another type of inorganic additive being widely studied in composting. Generally, zeolite is a type of aluminosilicate mineral that is commonly used as a commercial adsorbent and catalyst in industry. Zeolite occurs naturally and can also be produced industrially on a large scale. The porous structure of zeolite allows it to accumulate a wide variety of readily exchangeable positive ions (Na^+, K^+, Ca^{2+}, Mg^{2+}, and others) in its structure. The positive ion exchangeability makes zeolite a promising additive for heavy metal immobilization during composting (Zorpas et al. 2000a). One of the most common types of the natural zeolite, clinoptilolite has been used as an additive by Zorpas et al. (2000b) during the amendment of dewatered, anaerobically stabilized primary sewage sludge, with clinoptilolite in dry weight adding ratio of 0%, 5%, 10%, 15%, 20%, 25%, and 30%, respectively. The researchers noticed that the more zeolite added to the initial compost mixture, the more metals (Cd, Cu, Cr, Fe, Mn, Ni, Pb, Zn, etc.) immobilized in the final product after 150 days of composting. In the 25% and 30% zeolite treatments, the maximum metal reductions were 100% for Cd, 28%–45% for Cu, 10%–15% for Cr, 41%–47% for Fe, 9%–24% for Mn, 50%–55% for Ni, 50%–55% for Pb, and 40%–46% for Zn, respectively. The leachability of Cu, Cr, Fe, Ni, Mn, Pb, and Zn in sewage sludge decreased after composting. The later Tessier sequential extraction analysis proved nearly all metals in zeolite-treated sludge were predominantly bound in residual form rather than other forms (e.g. exchangeable, carbonate, reducible, organic bound, residual fraction, etc.). Aiming to declare the mechanism of clinoptilolite addition on heavy metals immobilization during the sewage sludge composting, Zorpas et al. (2008) first separated the clinoptilolite from the mixture of compost product and then analyzed the clinoptilolite sample. They observed that the metal bound in the exchangeable and carbonate forms was most readily taken up by clinoptilolite. The effect of zeolite in composting can be increased when using a mixture of clinoptilolite and sawdust or waste paper as an amendment for sewage sludge composting (Zorpas et al. 2003; Zorpas and Loizidou 2008).

To promote the utilization of zeolite amendment on sewage sludge composting disposal, Zorpas et al. (2002) investigated the influence of the particle size of zeolite (<0.161, 0.161–1.0, 1.1–2.5, 2.6–3.2, and 3.3–4.0 mm) on metal accumulation. Theoretically, the smaller particle size of zeolite meant a greater surface area and a higher ion-exchange capacity. In their study, they noticed the amount of heavy metal uptake by zeolite *increased* with a particle size *increase*. With the 25% (w/w, dry basis) clinoptilolite addition, in respect to total metal content, up to 12%, 27%, 14%, 30%, 40%, 55%, and 60% of Co, Cu, Cr, Fe, Zn, Pb, and Ni was taken up by clinoptilolite, respectively. Despite the particle size and concentration of zeolite, the type of zeolite might also influence the performance of heavy metal immobilization during composting. Villaseñor et al. (2011) mixed three kinds of zeolite (mordenite, klinolith, and clinoptilolite E568, all in particle size of 5–6 mm with a pore size smaller than 2 nm) with sewage sludge in ratios of 10%, 25%, and 40% (w/w wet basis). At the end of 82 days of composting, the metals Ni, Cr, and Pb presented in sludge were all retained by zeolites. Among these zeolites, clinoptilolite E568 was most effective for the adsorbing of Cu, Zn, and Hg. On the whole, the leachability of heavy metals of sewage sludge was effectively decreased with the co-composting of clinoptilolite with sewage sludge (Zorpas, 2009).

11.3 Mechanism of Heavy Metal Immobilization with Different Additives

Extensively produced organic wastes like sewage sludge and livestock and poultry droppings drew growing environmental concern around the world due to the release of odor gases and the presence of toxic heavy metals. In addition to carbon, some of these organic wastes contained other nutrient elements such as nitrogen, phosphorus, and potassium. As a potential approach for converting organic waste into valuable products, composting is a biochemical process that biologically decomposes and converts degradable solid wastes into organic fertilizer. In composting, the entire biochemical process occurs in an aqueous medium with a solid interface within the compost substrate matrix. As summarized in Tables 11.2 through 11.4, most of the additives used in composting are solid materials that have been used as low-cost adsorbents for the adsorptive removal of heavy metal ions from aqueous solution or wastewater as well (Agarwal and Rani 2017; Balaib et al. 2014; Fan and Zhang 2018; Li et al. 2016a, 2016b, 2018a, 2018b; Lin et al. 2018; Salazar-Rabago and Leyva-Ramos 2016; Tian et al. 2011; Zhou et al. 2017). Therefore, the mechanisms of heavy metal ions immobilization in compost is believed to be related to their detoxification behavior in the liquid-solid interface of adsorbents (additives or amendments) (Zorpas et al. 2008). The physiochemical property of the additive controlled the heavy metal ions' environmental behavior within the surface of the adsorbent (Zorpas 2009).

For example, biochar produced by thermal decomposition of biomass under oxygen-limited conditions has achieved increasing attention as a low-cost adsorbent material to treat wastewater and as a low-cost additive to improve composting quality (Li et al. 2015, 2016a, 2016b, 2017c, 2018a, 2018b). Generally, biochar's properties varied considerably with the feedstock material and pyrolysis temperature. A high pyrolysis temperature often produces biochar with high porosity, large surface area, high pH value and mineral content, and less surface functional group (Huang et al. 2018; Li et al. 2018a). Many studies have demonstrated that cation metal ions such as Cd, Pb, Hg etc., could be immobilized by biochar through the processes of complexation, cation exchange, precipitation, and so on (Li et al. 2017b). For instance, Harvey et al. (2011) found that, for the plant biochar with high cation exchange capacities, cation exchange was the major mechanism for Cd adsorption. Zhang et al. (2015) proved that released cations (K^+, Ca^{2+}, Na^+, and Mg^{2+}) from water hyacinth biochar were almost equal to the amount of sorbed Cd, indicating that cation exchange was the dominant mechanism. The FTIR spectra investigation of biochar before and after Cd adsorption revealed Cd complexation with carboxyl groups in the biochar was another minor mechanism for Cd sorption (Chen et al. 2015). The relatively high concentrations of soluble carbonate and phosphate in biochar implied that chemical precipitation also contributed to the cations' (Cd^{2+}, Pb^{2+}, etc.) removal (Li et al. 2018b).

Mineral-like zeolites are basically naturally occurring crystalline aluminosilicates. During the 1970s, natural zeolites drew attention in science and technology due to their ion-exchange capability to preferentially remove toxic heavy metal ions (Vaca et al. 2001). This unique characteristic makes zeolites favorable for detoxification of heavy metal ions in wastewater treatment. Some researchers selected the most abundant zeolite, clinoptilolite, as an additive for the removal of heavy metals (Cd, Cr, Cu, Fe, Mn, Ni, Pb, Zn) from dewatered anaerobic sewage sludge (Zorpas et al. 2000a). They noticed during the 150 days of composting, with increasing amounts of clinoptilolite in the compost (from 0% to 30% [dry w/w]), the concentration of all heavy metals in the compost decreased while the concentration of sodium and potassium increased. Later, they separated and analyzed the clinoptilolite from the compost mixture and found the metal uptake in clinoptilolite was in line with the sodium and potassium release from clinoptilolite. The clinoptilolite had the ability to stabilize heavy metals associated with the mobile fractions such as the exchangeable and the carbonate forms from organic waste (Zorpas et al. 2008). The high cation exchange and adsorption capacity of clinoptilolite contributed to the metal uptake from sewage sludge subjected to the composting process (Stylianou et al. 2008).

Unlike zeolite, lime and fly ash are more alkaline substances. The interactions of these alkaline substances with heavy metal cations were probably governed by the chemical precipitation process (Marchioretto et al. 2005). In addition, the existence of some anion such as phosphate, sulfate, and carbonate could also help in the formation of heavy metal precipitation during composting (Sugiyama et al. 2003). During the composting process, the increasing amount of humic substance also plays a crucial role in the heavy metal ions binding through coordination reaction (He et al. 2014).

11.4 Physiochemical Changes of Solid Waste during Heavy Metal Immobilization Bio-Processing

Composting is a biochemical process for biological transformation of solid organic wastes into safe and stable materials (compost) that can be employed as fertilizer and soil conditioner in agricultural applications (Li et al. 2012). Most of the combined additives can accelerate biosolid waste degradation, shorten the compost maturity time, and improve compost quality (Zhang and Sun 2017). Therefore, additive-assisted composting technology has made composting easier and opened new methods for waste management sector

(Gabhane et al. 2012). The applied additives are various nutrient-supplying or pH-balancing compounds, minerals, readily available forms of carbon, and inoculums of enzymes/microorganisms that could enhance microbial activity and optimize the composting medium condition when the additives are in contact with the waste materials (Himanen and Hänninen 2009). As summarized in Tables 11.3 and 11.4, various additives like sodium acetate, coal fly ash, lime, wood ash, bauxite, bentonite, zeolites, biochar, kaoline, medical stone, and phosphogypsum have been used in the composting of municipal solid waste, animal breeding waste, sewage sludge, and food waste by many scientists (Belyaeva and Haynes 2009; Himanen and Hänninen 2009; Koivula et al. 2004; Kurola et al. 2011; Li et al. 2012, 2015; Liu et al. 2017a; Villaseñor et al. 2011; Wang et al. 2017a, 2017b; Wong et al. 1997; Yang et al. 2015ab). All these additives, including coal fly ash, lime, biochar, and commercial additives (mixture of Fe, Mg, Mn, and Ca-sulfate and oxide compounds), have acceptable effects on heavy metal immobilization during composting (Himanen and Hänninen 2009; Liu et al. 2017b; Wong et al. 1997; Wong and Fang 2000). Because of the various physiochemical and biochemical properties of these additives, however, their effects on the physiochemical changes in solid waste during bioprocessing had vast differences. In this section, we only discuss the effects of several typical additives on the physiochemical changes of solid waste bioprocessing disposal.

As one of the widely examined organic additives, biochar often has an alkaline nature. The addition of biochar in a compost mixture inevitably affects the initial pH of the mixture at the beginning stages but not the final compost product. Li et al. (2015) examined the effect of biochar from different pyrolysis temperature (from 400°C to 900°C) on pig manure composting, and they found that the initial pH of composting mixture increased with increasing pyrolysis temperature of the corn biochar. The initial pH value could reach 9.4 with the application of biochar produced at 900°C. Dias et al. (2010) and Vandecasteele et al. (2016) also observed the application of alkaline biochar as an additive in poultry manure compost. They obtained the highest pH of all composting processes. As a carbon-rich porous material, biochar added to the compost mixture also affects the moisture content of the compost. Owing to the developed porous structure of biochar, the introduction of biochar with small size can increase the moisture content and improve the moisture-holding capacity of compost (Lehmann and Joseph 2015), while the addition of large-size biochar might decrease the compost mixture moisture content due to the overdeveloped pile porosity, which favors gas exchange and vapor transport (Liu et al. 2017a). As a carbon-rich and stable solid material, biochar introduced into compost will change the C:N ratio of the starting compost substrates (Dias et al. 2010; Jindo et al. 2012); the high recalcitrance of carbon in the biochar would exist for a much longer time (Lehmann and Joseph 2015). The addition of biochar decreases the compost pile density and particle size, facilitates oxygen supply, avoids the formation of anaerobic sites, and reduces the odorous gas release. For example, adding 10% (w/w) biochar into the starting compost substrates could reduce the compost pile density from 0.35 to 0.26 kg/L (Liu et al. 2017a). Addition of wood biochar at 12% to 18% (w/w) could dramatically change the sludge microstructure because of the formation of channels or voids on the sludge surface (Zhang et al. 2014). These channels, voids, and porous structures favor O_2 diffusion, aerobic biodegradation, and humification of the compost substrate (Jindo et al. 2016). Sánchez-García et al. (2015) noticed that the addition of biochar could prevent the formation of large clumps and facilitated O_2 diffusion into the composting mixture.

However, basic minerals have been proven again and again to have the greatest potential for biosolid waste treatment. Although alkaline additives such as lime and coal fly ash exhibited excellent heavy metal immobilization capacities, the alkaline environment often causes other adverse problems during biosolid waste disposal. Hence, the additive adding ratio should be addressed carefully. For example, Fang and Wong (1999) noticed that the application of lime as an additive not only raised the pH value of the compost mixture to above 9.2 at the initial stage of composting but also caused a decrease in electrical conductivity through precipitation of soluble ions. Lime amendment significantly reduced water-soluble and DTPA-extractable metal contents. The adverse impacts of the lime amendment on all biological parameters increased with an increase in lime application rates, although these effects were generally restricted to the early stage of the thermophilic phase. Addition of optimized lime amount (0.63% lime in dry weight) to a sludge and sawdust mixture slightly improved the microbial activity during the 100-day aerobic composting (Wong and Fang 2000). An increase in the amount of lime additive inhibited the biological decomposition activity of sewage sludge and resulted in more nitrogen loss (through ammonia volatilization) at 1.63% lime amendment (Fang and Wong 1999). Hence, the lime amendment at a rate of lower than 1.0% (w/w) was recommended for lime-sludge co-compost by Fang and Wong (1999) and Wong and Fang (2000). Like lime, coal fly ash is another kind of alkaline solid additive used in biosolid waste treatment. Fang et al. (1999) declared that adding coal fly ash at a rate of 35% (w/w) raised the pH (above 9.0) of the sludge compost and inhibited the sludge decomposition activity. The addition of coal fly ash decreased the soluble organic carbon and total carbon content of sludge compost as the prolonging of composting time, whereas the total nitrogen content was increased with an increase in the composting time. The coal fly ash amendment could cause ammonia nitrogen loss during the thermophilic phase and inhibited the nitrification process and phosphorus transformation of the treated sludge compost. Therefore, the utilization of these alkaline additives in biosolid waste management should be carefully monitored (Fang et al. 1999; Wong et al. 1995; Wong et al. 1997; Wong and Fang 2000).

Unlike lime and fly ash, zeolite has no strong alkalinity. Many reports had proved the benefits of zeolite as an additive for heavy metal immobilization in the biosolid waste disposal process. In addition to this ability of zeolite for heavy metal immobilization, zeolite addition has many benefits for compost-ing progress and final product quality improvement. Stylianou et al. (2008) tested the changes of physiochemical characteristics of 20% clinoptilolite (in dry weight) in sewage sludge during 120 days of composting in pilot bioreac-tors. They noticed that the addition of zeolite neutralized the sludge, with the pH of the raw sludge changed from 8.46 to 7.10 for control treatment and to 7.60 for the zeolite-added treatment. The addition of zeolite allowed the com-posting sample to retain more moisture, and it decreased the conductivity of the zeolite-mixed compost from 3.83 to 1.53 mS/cm and to 2.63 mS/cm for the blank sample. The introduction of zeolite also enhanced the decomposition of organic matter (from 25% to 15%) due to its ability to increase the poros-ity of the compost substrate. After 120 days of composting for the zeolite-amended sewage sludge compost, the total nitrogen content decreased from 8.2% to lower than 0.6%, while the C:N ratio of the zeolite-amended sewage sludge compost (C:N ratio = 18) was lower than that of blank compost sam-ple (C:N ratio = 24), indicating that the zeolite benefited the decomposition of carbon and nitrogen-containing organic matter during composting. The addition of zeolite in composting could not only improve compost maturity but also significantly reduce the water extractable metal contents (Cu, Cr, Zn, Ni, and Mn) in sludge compost samples. The latest research conducted also demonstrated that the application of zeolite or the zeolite and other minerals or additives to the mixture could reduce the NH_3, CH_4 and N_2O emissions during the composting (Awasthi et al. 2016, 2017a; Turan and Ergun 2007, 2008). The use of zeolite in organic waste stabilization reduces the number of contaminants and improves the quality of the compost product (Turan 2008).

11.5 Conclusion

Composting is one of the most effective and economical bioprocessing approaches for organic solid waste management. Incorporating an addi-tive or a mixture of additives, either in organic or inorganic form, into the sewage sludge and animal manure not only benefits the composting prog-ress but also facilitates the reduction of the bioavailable heavy metals in organic solid wastes. Consequently, some effort has recently been devoted to researching additives to composting, especially their effect on compost heavy metal immobilization and potential toxicity reduction. Additive composting technology is developing. To date, several different addi-tives, including organic plant residue and inorganic minerals, have been used. Among them, biochar is one of the most promising additives; it has great potential for organic waste management, compost nutrient conserva-tion, heavy metal immobilization, and organic fertilizer quality improve-ment. Based on the source and character of biochar, the potential of this material may be greater than so far shown in composting. However, the performance of biochar depends heavily on its feedstock and preparation. Further research is needed to better understand the effect of the addition of biochar on composting, including biochar from different feedstocks and production methods. The integrated utilization of minerals and biochar is another emerging technology in composting. Thus, amendment of compost with minerals and biochar mixture must also be addressed and tested in the field.

References

Agarwal, S., Rani, A. 2017. Adsorption of resorcinol from aqueous solution onto CTAB/NaOH/flyash composites: Equilibrium, kinetics and thermodynamics. *Journal of Environmental Chemical Engineering* 5: 526–538.

Antoniadis, V., Robinson, J.S., Alloway, B.J. 2008. Effects of short-term pH fluctuations on cadmium, nickel, lead, and zinc availability to ryegrass in a sewage sludge-amended field. *Chemosphere* 71: 759–764.

Antoniadis, V., Tsadilas, C.D., Samaras, V. 2010. Trace element availability in a sewage sludge-amended cotton grown Mediterranean soil. *Chemosphere* 80: 1308–1313.

Awasthi, M.K., Wang, M., Chen, H., Wang, Q., Zhao, J., Ren, X., Li, D., et al. 2017a. Heterogeneity of biochar amendment to improve the carbon and nitrogen sequestration through reduce the greenhouse gasses emissions during sewage sludge composting. *Bioresource Technology* 224: 428–438.

Awasthi, M.K., Wang, Q., Huang, H., Li, R., Shen, F., Lahori, A.H., Wang, P., Guo, D., Guo, Z., Jiang, S. 2016. Effect of biochar amendment on greenhouse gas emission and bio-availability of heavy metals during sewage sludge co-composting. *Journal of Cleaner Production* 135: 829–835.

Awasthi, M.K., Wang, M., Pandey, A., Chen, H., Awasthi, S.K., Wang, Q., Ren, X., et al. 2017b. Heterogeneity of zeolite combined with biochar properties as a function of sewage sludge composting and production of nutrient-rich compost. *Waste Management* 68: 760–773.

Balaib, F., Azzedine, M., Boubeker, B., Abdeslam-Hassen, M. 2014. Experimental study of oxytetracycline retention by adsorption onto polyaniline coated peanut shells. *International Journal of Hydrogen Energy* 39: 1511–1515.

Belyaeva, O.N., Haynes, R.J. 2009. Chemical, microbial and physical properties of manufactured soils produced by co-composting municipal green waste with coal fly ash. *Bioresource Technology* 100: 5203–5209.

Bernal, M.P., Alburquerque, J., Moral, R. 2009. Composting of animal manures and chemical criteria for compost maturity assessment. A review. *Bioresource Technology* 100: 5444–5453.

Camps, M., Tomlinson, T. 2015. The use of biochar in composting. International Biochar Initiative. http://www.biochar-international.org/sites/default/files/Compost_biochar_IBI_final.pdf.

Cang, L., Wang, Y.J., Zhou, D.M., Dong, Y.H. 2004. Heavy metals pollution in poultry and livestock feeds and manures under intensive farming in Jiangsu Province, China. *Journal of Environmental Sciences* 16: 371–374.

Chen, T., Zhou, Z.Y., Han, R., Meng, R.H., Wang, H.T., Lu, W.J. 2015. Adsorption of cadmium by biochar derived from municipal sewage sludge: Impact factors and adsorption mechanism. *Chemosphere* 134: 286–293.

Chen, Y., Huang, X., Han, Z., Huang, X., Hu, B., Shi, D., Wu, W. 2010. Effects of bamboo charcoal and bamboo vinegar on nitrogen conservation and heavy metals immobility during pig manure composting. *Chemosphere* 78: 1177–1181.

Deng, J., Feng, X., Qiu, X. 2009. Extraction of heavy metal from sewage sludge using ultrasound-assisted nitric acid. *Chemical Engineering Journal* 152: 177–182.

Dias, B.O., Silva, C.A., Higashikawa, F.S., Roig, A., Sánchez-Monedero, M.A. 2010. Use of biochar as bulking agent for the composting of poultry manure: Effect on organic matter degradation and humification. *Bioresource Technology* 101: 1239–1246.

Dias, D., Lapa, N., Bernardo, M., Godinho, D., Fonseca, I., Miranda, M., Pinto, F., Lemos, F. 2017. Properties of chars from the gasification and pyrolysis of rice waste streams towards their valorisation as adsorbent materials. *Waste Management* 65: 186–194.

Duan, B., Liu, F., Zhang, W., Zheng, H., Zhang, Q., Li, X., Bu, Y. 2015. Evaluation and source apportionment of heavy metals (HMs) in sewage sludge of municipal wastewater treatment plants (WWTPs) in Shanxi, China. *International Journal of Environmental Research and Public Health* 12: 15807–15818.

Environment Protection Administration, China. 2001. Chinese Bulletin of Environment of 2000 [N]. China Environment News, 2001–09–30 (in Chinese). http://www.envir.gov.cn/info/2001/6/616569.htm.

Fan, C., Zhang, Y. 2018. Adsorption isotherms, kinetics and thermodynamics of nitrate and phosphate in binary systems on a novel adsorbent derived from corn stalks. *Journal of Geochemical Exploration* 188: 95–100.

Fang, M., Wong, J.W.C. 1999. Effects of lime amendment on availability of heavy metals and maturation in sewage sludge composting. *Environmental Pollution* 106: 83–89.

Fang, M., Wong, J.W.C., Ma, K.K., Wong, M.H. 1999. Co-composting of sewage sludge and coal fly ash: Nutrient transformations. *Bioresource Technology* 67: 19–24.

Fijalkowski, K., Rorat, A., Grobelak, A., Kacprzak, M.J. 2017. The presence of contaminations in sewage sludge–The current situation. *Journal of Environmental Management* 203: 1126–1136.

Gabhane, J., Prince William, S.P.M., Bidyadhar, R., Bhilawe, P., Anand, D., Vaidya, A.N., Wate, S.R. 2012. Additives aided composting of green waste: Effects on organic matter degradation, compost maturity, and quality of the finished compost. *Bioresource Technology* 114: 382–388.

Gianico, A., Braguglia, C.M., Mascolo, G., Mininni, G. 2013. Partitioning of nutrients and micropollutants along the sludge treatment line: A case study. *Environmental Science and Pollution Research* 20: 6256–6265.

Gu, X.Y., Wong, J.W.C., Tyagi, R.D. 2017. Bioleaching of heavy metals from sewage sludge for land application. In *Current Developments in Biotechnology and Bioengineering, Solid Waste Management.*, ed J.W.C.Wong, R.D.Tyagi, A.Pandey. Amsterdam, the Netherlands: Elsevier, pp. 241–265.

Guerra-Rodriguez, E., Alonso, J., Melgar, M.J., Vazquez, M. 2006. Evaluation of heavy metal contents in co-composts of poultry manure with barley wastes or chestnut burr/leaf litter. *Chemosphere* 65: 1801–1805.

Harvey, O.R., Herbert, B.E., Rhue, R.D., Kuo, L.J. 2011. Metal interactions at the biochar-water interface: Energetics and structure-sorption relationships elucidated by flow sorption microcalorimetry. *Environmental Science & Technology* 45: 5550–5556.

He, X.-S., Xi, B.-D., Li, D., Guo, X.-J., Cui, D.-Y., Pan, H.-W., Ma, Y. 2014. Influence of the composition and removal characteristics of organic matter on heavy metal distribution in compost leachates. *Environmental Science and Pollution Research* 21(12): 7522–7529.

Healy, M.G., Fenton, O., Forrestal, P.J., Danaher, M., Brennan, R.B., Morrison, L. 2016. Metal concentrations in lime stabilised, thermally dried and anaerobically digested sewage sludge. *Waste Management* 48: 404–408.

Himanen, M., Hänninen, K. 2009. Effect of commercial mineral-based additives on composting and compost quality. *Waste Management* 29: 2265–2273.

Hossain, S.M., Barreto, S.L., Silva, C.G. 1998. Growth performance and carcass composition of broilers fed supplemental chromium from chromium yeast. *Animal Feed Science and Technology* 71: 217–228.

Hou, Y., Zhao, L., Meng, H. 2014. Passivating effect of biochar and humic acid materials on heavy metals during composting of pig manure. *Transactions of the Chinese Society of Agricultural Engineering* (Transactions of the CSAE) 30(11): 205–215. (in Chinese with English abstract)

Hua, L., Wu, W., Liu, Y., McBride, M.B., Chen, Y. 2009. Reduction of nitrogen loss and Cu and Zn mobility during sludge composting with bamboo charcoal amendment. *Environmental Science Pollution Research* 16: 1–9.

Huang, H., Yao, W., Li, R., Ali, A., Du, J., Guo, D., Xiao, R., Guo, Z., Zhang, Z., Awasthi, M.K. 2018. Effect of pyrolysis temperature on chemical form, behavior and environmental risk of Zn, Pb and Cd in biochar produced from phytoremediation residue. *Bioresource Technology* 249: 487–493.

Huang, H.-J., Yang, T., Lai, F.-Y., Wu, G-Q. 2017. Co-pyrolysis of sewage sludge and sawdust/rice straw for the production of biochar. *Journal of Analytical and Applied Pyrolysis* 125: 61–68.

Huang, Y., Liu, B., Chen, G. 2007. Contents of heavy metals in formulated feed for pig and pig manure of scaled piggery. *Guangxi Agriculture Science* 38(5): 544–546. (In Chinese with English abstract.)

Ito, A., Umita, T., Aizawa, J., Takachi, T., Morinaga, K. 2000. Removal of heavy metals from anaerobically digested sewage sludge by a new chemical method using ferric sulfate. *Water Research* 34 (3): 751–758.

Jeong, C.Y., Dodla, S.K., Wang, J.J. 2012. Fundamental and molecular composition characteristics of biochars produced from sugarcane and rice crop residues and by-products. *Chemosphere* 142: 4–13.

Jin, L.Y., Zhang, G.M., Tian, H.F. 2014. Current state of sewage treatment in China. *Water Research* 66: 85–98.

Jindo, K., Sánchez-Monedero, M.A., Hernández, T., García, C., Furukawa, T., Matsumoto, K., Sonoki, T., Bastida, F. 2012. Biochar influences the microbial community structure during manure composting with agricultural wastes. *Science of the Total Environment* 416: 476–481.

Jindo, K., Sonoki, T., Matsumoto, K., Canellas, L., Roig, A., Sanchez-Monedero, M.A. 2016. Influence of biochar addition on the humic substances of composting manures. *Waste Management* 49: 545–552.

Kelly, J.J., Häggblom, M., Tate Iii, R.L.T. 1999. Effects of the land application of sewage sludge on soil heavy metal concentrations and soil microbial communities. *Soil Biology & Biochemistry* 31(31): 1467–1470.

Koch, M., Rotard, W. 2001. On the contribution of background sources to the heavy metal content of municipal sewage sludge. *Water Science and Technology* 43: 67–74.

Koivula, N., Raikkonen, T., Urpilainen, S., Ranta, J., Hanninen, K. 2004. Ash in composting of source-separated catering waste. *Bioresource Technology* 93: 291–299.

Kosobucki, P., Kruk, M., Buszewski, B. 2008. Immobilization of selected heavy metals in sewage sludge by natural zeolites. *Bioresource Technology* 99: 5972–5976.

Kurola, J.M., Arnold, M., Kontro, M.H., Talves, M., Romantschuk, M. 2011. Wood ash for application in municipal biowaste composting. *Bioresource Technology* 102: 5214–5220.

Lehmann, J., Joseph, S. 2015. *Biochar for Environmental Management: Science, Technology and Implementation*. UK Routledge.

Li, H., Dong, X., da Silva, E.B., de Oliveira, L.M., Chen, Y., Ma, L.Q. 2017a. Mechanisms of metal sorption by biochars: Biochar characteristics and modifications. *Chemosphere* 178: 466–478.

Li, J.-S., Xue, Q., Fang, L., Poon, C.S. 2017b. Characteristics and metal leachability of incinerated sewage sludge ash and air pollution control residues from Hong Kong evaluated by different methods. *Waste Management* 64: 161–170.

Li, R., Liang, W., Wang, J.J., Gaston, L.A., Huang, H., Lei, S., Awasthi, M.K., Zhou, B., Xiao, R., Zhang, Z. 2018b. Facilitative capture of As(V), Pb(II) and methylene blue from aqueous solutions with MgO hybrid sponge-like carbonaceous composite derived from sugarcane leafy trash. *Journal of Environmental Management* 212, 77–87.

Li, R., Wang, J.J., Gaston, L.A., Zhou, B., Li, M., Xiao, R., Wang, Q., Zhang, Z., Huang, H., Liang, W., Huang, H., Zhang, X. 2018a. An overview of carbothermal synthesis of metal–biochar composites for the removal of oxyanion contaminants from aqueous solution. *Carbon* 129: 674–687.

Li, R., Wang, J.J., Zhang, Z., Awasthi, M.K., Du, D., Dang, P., Huang, Q., Zhang, Y., Wang, L. 2018c. Recovery of phosphate and dissolved organic matter from aqueous solution using a novel CaO-MgO hybrid carbon composite and its feasibility in phosphorus recycling. *Science of The Total Environment* 642: 526–536.

Li, R., Wang, J.J., Zhang, Z., Shen, F., Zhang, G., Qin, R., Li, X., Xiao, R. 2012. Nutrient transformations during composting of pig manure with bentonite. *Bioresource Technology* 121: 362–368.

Li, R., Wang, J.J., Zhou, B.M., Awasthi, M.K., Ali, A., Zhang, Z., Lahori, A.H., Mahar, A. 2016b. Recovery of phosphate from aqueous solution by magnesium oxide decorated magnetic biochar and its potential as phosphate-based fertilizer substitute. *Bioresource Technology* 215: 209–214.

Li, R., Wang, J.J., Zhou, B., Zhang, Z., Liu, S., F., Lei, S., Xiao, R. 2017c. Simultaneous capture removal of phosphate, ammonium and organic substances by MgO impregnated biochar and its potential use in swine wastewater treatment. *Journal of Cleaner Production* 147: 96–107.

Li, R., Wang, J.J., Zhou, B.M., Awasthi, M.K., Ali, A., Zhang, Z., Gaston, L.A., Lahori, A.H., Mahar, A. 2016a. Enhancing phosphate adsorption by Mg/Al layered double hydroxide functionalized biochar with different Mg/Al ratios. *Science of the Total Environment* 559: 121–129.

Li, R., Wang, Q., Zhang, Z., Zhang, G., Li. Z., Wang, L., Zheng, J. 2015. Nutrient transformation during aerobic composting of pig manure with biochar prepared at different temperatures. *Environmental Technology* 36: 815–826.

Li, Y.X., Chen, T.B. 2005. Concentrations of additive arsenic in Beijing pig feeds and the residues in pig manure. *Resources, Conservation and Recycling*, 45: 356–367.

Li, Y.X., Xiong, X., Lin, C.Y., Zhang, F.S., Li, W., Han, W. 2010. Cadmium in animal production and its potential hazard on Beijing and Fuxin farmlands. *Journal of Hazardous Materials* 177: 475–480.

Lin, J., Jiang, B., Zhan, Y. 2018. Effect of pre-treatment of bentonite with sodium and calcium ions on phosphate adsorption onto zirconium-modified bentonite. *Journal of Environmental Management* 217: 183–195.

Liu, N., Zhou, J., Han, L., Ma, S., Sun, X., Huang, G. 2017a. Role and multi-scale characterization of bamboo biochar during poultry manure aerobic composting. *Bioresource Technology* 241: 190–199.

Liu, W., Huo, R., Xu, J., Liang, S., Li, J., Zhao, T., Wang, S. 2017b. Effects of biochar on nitrogen transformation and heavy metals in sludge composting. *Bioresource Technology* 235: 43–49.

Lo, K.S.L., Chen, Y.H. 1990. Extracting heavy metals from municipal and industrial sludges. *Science of the Total Environment* 90: 99–116.

Ma, L., Sun, H., Chen, L., Zhao, J. 2011. Relation between heavy metal fraction in soils and plants enrichment in pilot scale experiment on land application of sewage sludge. *Journal of Food Agriculture & Environment* 9: 967–973.

Marchioretto, M.M. 2003. Heavy metals removal from anaerobically digested sludge. PhD thesis, Wageningen University, Wageningen, the Netherlands.

Marchioretto, M.M., Bruning, H., Rulkens, W. 2005. Heavy metals precipitation in sewage sludge. *Separation Science & Technology* 40(16): 393–405.

Mendoza, H.R., Gallmann, E., Zheng, K. 2010. Pig husbandry and solid manures in a commercial pig farm in Beijing, China. *International Journal of Biological and Life Sciences* 6(2): 107–116.

Ministry of Environmental Protection of the PRC. 2015. Announcement on national urban sewage treatment facilities in 2014. http://www.mep.gov.cn/gkml/hbb/bgg/201506/t20150609_303209.htm. Accessed July 4, 2017, in Chinese.

Naoum, C., Fatta, D., Haralambous, K.J., Loizidou, M. 2001. Removal of heavy metals from sewage sludge by acid treatment. *Journal of Environmental Science and Health, Part A* 36: 873–881.

Nicholson, F., Chambers, B., Alloway, B., Hird, A., Smith, S., Carlton-Smith, C. 1998. An inventory of heavy metal inputs to agricultural soils in England and Wales. In: *Proceedings of the 16th World Congress of Soil Science*. Montpellier, France.

Nicholson, F.A., Chambers, B.J., Williams, J.R., Unwin, R.J. 1999. Heavy metal contents of livestock feeds and animal manures in England and Wales. *Bioresource Technology* 70(1): 23–31.

Nikovski, G.N., Kalinichenko, K.V. 2014. Biotechnology of utilization of municipal wastewater sediments. *Biotechnologia Acta* 7: 21–32.

Odstrci A del, C.A., Carino, S.N., Ricci, J.C.D., Mandalunis, P.M. 2010. Effect of arsenic in endochondral ossification of experimental animals. *Experimental and Toxicologic Pathology* 62: 243–249.

Oh, J.-Y., Choi, S.-D., Kwon, H.-O., Lee, S.-E. 2016. Leaching of polycyclic aromatic hydrocarbons (PAHs) from industrial wastewater sludge by ultrasonic treatment. *Ultrasonics Sonochemistry* 33: 61–66.

Oleszczuk, P., Hollert, H. 2011. Comparison of sewage sludge toxicity to plants and invertebrates in three different soils. *Chemosphere* 83: 502–509.

Pan, X., Han, Z., Ben, W. 2013. Heavy metal contents in pig manure and pig feeds from intensive pig farms in Shandong Province, China. *Journal of Agro-Environment Science* 32(1): 160–165. (in Chinese)

Pathak, A., Dastidar, M.G., Sreekrishnan, T.R. 2009. Bioleaching of heavy metals from sewage sludge: A review. *Journal of Environmental Management* 90: 2343–2353.

Peng, G., Tian, G., Liu, J., Bao, Q., Zang, L. 2011. Removal of heavy metals from sewage sludge with a combination of bioleaching and electrokinetic remediation technology. *Desalination* 271: 100–104.

Salazar-Rabago, J.J., Leyva-Ramos, R. 2016. Novel biosorbent with high adsorption capacity prepared by chemical modification of white pine (Pinus durangensis) sawdust. Adsorption of Pb(II) from aqueous solutions. *Journal of Environmental Management* 169: 303–312.

Sánchez-García, M., Alburquerque, J., Sánchez-Monedero, M., Roig, A., Cayuela, M. 2015. Biochar accelerates organic matter degradation and enhances N mineralisation during composting of poultry manure without a relevant impact on gas emissions. *Bioresource Technology* 192: 272–279.

Shaheen, S.M., Shams, M.S., Ibrahim, S.M., Elbehiry, F.A., Antoniadis, V. 2014. Stabilization of sewage sludge by using various by-products: Effects on soil properties, biomass production, and bioavailability of copper and zinc. *Water Air & Soil Pollution* 225 (7): 2014.

Solaiman, S.G., Craig Jr., T.J., Reddy, G., Shoemaker, C.E. 2007. Effect of high levels of Cu supplement on growth performance, rumen fermentation, and immune responses in goat kids. *Small Ruminant Research* 69: 115–123.

Sonoki, T., Furukawa, T.,Jindo, K., Suto, K., Aoyama, M., Sánchez-Monedero, M.Á. 2013. Influence of biochar addition on methane metabolism during thermophilic phase of composting. *Journal of Basic Microbiology* 53: 617–621.

Soon, Y.K., Bates, T.E., Moyer, J.R. 1980. Land application of chemically treated sewage sludge: III. Effects on soil and plant heavy metal content. *Journal of Environmental Quality* 3: 497–404.

Sprynskyy, M., Kosobucki, P., Kowalkowski, T., Buszewski, B. 2007. Influence of clinoptilolite rock on chemical speciation of selected heavy metals in sewage sludge. *Journal of Hazardous Materials* 149: 310–316.

Steiner, C., Das, K., Melear, N., Lakly, D. 2010. Reducing nitrogen loss during poultry litter composting using biochar. *Journal of Environmental Quality* 39: 1236–1242.

Strasser, H., Brunner, H., Schinner, F. 1995. Leaching of iron and toxic heavy metals from anaerobically-digested sewage sludge. *Journal of Microbiology and Biotechnology* 14: 281–287.

Stylianou, M.A., Inglezakis, V.J., Moustakas, K.G., Loizidou, M.D. 2008. Improvement of the quality of sewage sludge compost by adding natural clinoptilolite. *Desalination* 224: 240–249.

Suanon, F., Sun, Q., Dimon, B., Mama, D., Yu, C.-P. 2016. Heavy metal removal from sludge with organic chelators: Comparative study of N, N-bis(carboxymethyl) glutamic acid and citric acid. *Journal of Environmental Management* 166: 341–347.

Sugiyama, S., Ichii, T., Fujisawa, M., Kawashiro, K., Tomida, T. 2003. Heavy metal immobilization in aqueous solution using calcium phosphate and calcium hydrogen phosphates. *Journal of Colloid & Interface Science* 259(2): 408–410.

Syed-Hassan, S.S.A., Wang, Y., Hu, S., Su, S., Xiang, J. 2017. Thermochemical processing of sewage sludge to energy and fuel: Fundamentals, challenges and considerations. *Renewable and Sustainable Energy Reviews* 80: 888–913.

Tian, Y., Wu, M., Lin, X., Huang, P., Huang, Y. 2011. Synthesis of magnetic wheat straw for arsenic adsorption. *Journal of Hazardous Materials* 193: 10–16.

Turan, N.G. 2008. The effects of natural zeolite on salinity level of poultry litter compost. *Bioresource Technology* 99: 2097–2101.

Turan, N.G., Ergun, O.N. 2007. Ammonia uptake by natural zeolite in municipal solid waste compost. *Environmental Progress* 26: 149–156.

Turan, N.G., Ergun, O.N. 2008. Improving the quality of municipal solid waste compost by using expanded perlite and natural zeolite. *Clean* 36: 330–334.

Tyagi, R.D., Coullard, D., Fran, F.T. 1988. Heavy metal removal from anaerobically digested sludge by chemical and microbiological methods. *Environmental Pollution* 50: 295–316.

Vaca, M.M., López, C.R., Gehr, R., Jiménez, C.B., Alvarez, P. 2001. Heavy metal removal with Mexican clinoptilolite: Multi-component ionic exchange. *Water Research* 35: 373–378.

Vandecasteele, B., Sinicco, T., D'Hose, T., Nest, T.V., Mondini, C. 2016. Biochar amendment before or after composting affects compost quality and N losses, but not P plant uptake. *Journal of Environmental Management* 168: 200–209.

Veeken, A.H.M., Hamelers, H.V.M. 1999. Removal of heavy metals from sewage sludge by extraction with organic acids. *Water Science & Technology* 40: 129–136.

Villaseñor, J., Rodríguez, L., Fernández, F.J. 2011. Composting domestic sewage sludge with natural zeolites in a rotary drum reactor. *Bioresource Technology* 102: 1447–1454.

Wang, M., Awasthi, M.K., Wang, Q., Chen, H., Ren, X., Zhao, J., Li, R., Zhang, Z. 2017a. Comparison of additives amendment for mitigation of greenhouse gases and ammonia emission during sewage sludge co-composting based on correlation analysis. *Bioresource Technology* 243: 520–527.

Wang, Q., Awasthi, M.K., Zhao, J., Ren, X., Li, R., Wang, Z., Wang, M., Zhang, Z. 2017b. Improvement of pig manure compost lignocellulose degradation, organic matter humification and compost quality with medical stone. *Bioresource Technology* 243: 771–777.

Wang, Q., Li, R., Cai, H., Awasthi, M.K., Zhang, Z., Wang, J.J., Ali, A., Amanullah, M. 2016a. Improving pig manure composting efficiency employing Ca-bentonite. *Ecological Engineering* 87: 157–161.

Wang, Q., Wang, Z., Awasthi, M.K., Jiang, Y., Li, R., Ren, X., Zhao, J., Shen, F., Wang, M., Zhang, Z. 2016b. Evaluation of medical stone amendment for the reduction of nitrogen loss and bioavailability of heavy metals during pig manure composting. *Bioresource Technology* 220: 297–204.

Wang, X., Chen, J., Yan, X., Wang, X., Zhang, J., Huang, J., Zhao, J. 2015. Heavy metal chemical extraction from industrial and municipal mixed sludge by ultrasound-assisted citric acid. *Journal of Industrial and Engineering Chemistry* 27: 368–372.

Wang, X., Chen, L., Xia, S., Zhao, J. 2008. Changes of Cu, Zn, and Ni chemical speciation in sewage sludge co-composted with sodium sulfide and lime. *Journal of Environmental Science* 20: 156–160.

Wong, J.W.C., Fang, M. 2000. Effect of lime addition on sewage sludge composting process. *Water Research* 34(15): 3691–3698.

Wong, J.W.C., Fang, M., Li, G.X., Wong, M.H. 1997. Feasibility of using coal ash residues as co-composting materials for sewage sludge. *Environmental Technology* 18: 563–568.

Wong, J.W.C., Fung, S.O., Selvam, A. 2009. Coal fly ash and lime addition enhances the rate and efficiency of decomposition of food waste during composting. *Bioresources Technology* 100: 3324–3331.

Wong, J.W.C., Li, S.W.Y., Wong, M.H. 1995. Coal fly ash as a composting material for sewage sludge: Effects on microbial activities. *Environmental Technology Letters* 16: 527–537.

Wong, J.W.C., Selvam, A. 2006. Speciation of heavy metals during co-composting of sewage sludge with lime. *Chemosphere* 63: 980–986.

Wong, J.W.C., Xiang, L., Gu, X.Y., Zhou, L.X. 2004. Bioleaching of heavy metals from anaerobically digested sewage sludge using FeS$_2$ as an energy source. *Chemosphere* 55: 101–107.

Wong, L., Henry, J.G. 1983. Bacterial leaching of heavy metals from anaerobically digested sewage sludges. *Water Quality Research Journal of Canada* 18: 151–162.

Xiang, L., Chan, L.C., Wong, J.W.C. 2000. Removal of heavy metals from anaerobically digested sewage sludge by isolated indigenous iron-oxidizing bacteria. *Chemosphere* 41: 283–287.

Xiao, R., Awasthi, M.K., Li, R., Park, J., Pensky, S.M., Wang, Q., Wang, J. J., Zhang, Z. 2017. Recent developments in biochar utilization as an additive in organic solid waste composting: A review. *Bioresource Technology* 246: 203–213.

Xiong, X., Li, Y.X., Li, W., Lin, C.Y., Han, W., Yang, M. 2010. Copper content in animal manures and potential risk of soil copper pollution with animal manure use in agriculture. *Resources, Conservation and Recycling* 54: 985–990.

Yang, C., Meng, X.Z., Chen, L., Xia, S. Q. 2011. Polybrominated diphenyl ethers in sewage sludge from Shanghai, China: Possible ecological risk applied to agricultural land. *Chemosphere* 85: 418–423.

Yang, F., Li, G., Shi, H., Wang, Y. 2015b. Effects of phosphogypsum and superphosphate on compost maturity and gaseous emissions during kitchen waste composting. *Waste Management* 36: 70–76.

Yang, G., Zhang, G., Wang, H. 2015a. Current state of sludge production, management, treatment and disposal in China. *Water Research* 78: 60–73.

Yang, T., Huang, H., Lai, F. 2017. Pollution hazards of heavy metals in sewage sludge from four wastewater treatment plants in Nanchang, China. *The Transactions of Nonferrous Metals Society of China* 27:2249–59.

Yu, H., Huang, G.H. 2009. Effects of sodium acetate as a pH control amendment on the composting of food waste. *Bioresource Technology* 100: 2005–2011.

Zhang, C.S., Xu, Y., Zhao, M.H., Rong, H.W., Zhang, K.F. 2018a. Influence of inoculating white-rot fungi on organic matter transformations and mobility of heavy metals in sewage sludge-based composting. *Journal of Hazardous Materials* 344: 163–168.

Zhang, F., Li, Y., Yang, M., Li, W. 2012. Content of heavy metals in animal feeds and manures from farms of different scales in Northeast China. *International Journal of Environmental Research and Public Health* 9: 2 658–668.

Zhang, F., Wang, X., Yin, D.X., Peng, B., Tan, C.Y., Liu, Y.G., Tan, X.F., Wu, S.X. 2015. Efficiency and mechanisms of Cd removal from aqueous solution by biochar derived from water hyacinth (*Eichornia crassipes*). *Journal of Environmental Management* 153: 68–73.

Zhang, J., Lü, F., Shao, L., He, P. 2014. The use of biochar-amended composting to improve the humification and degradation of sewage sludge. *Bioresource Technology* 168: 252–258.

Zhang, L., Sun, X. 2017. Addition of fish pond sediment and rock phosphate enhances the composting of green waste. *Bioresource Technology* 233: 116–126.

Zhang, S. 2004. Analysis of main harmful components in manures of livestock and poultry from intensive feedlots and the harmless disposal. PhD thesis, Beijing, China: Chinese Academy of Agricultural Sciences. (In Chinese with English abstract.)

Zhang, X., Xiang, N., Wang, W., Liao, W., Yang, X., Shui, W., Wu, J., Deng, S. 2018b. An emergy evaluation of the sewage sludge treatment system with earthworm compositing technology in Chengdu, China. *Ecological Engineering* 110: 8–17.

Zhou, Y., Zhang, R., Chen, K., Zhao, X., Gu, X., Lu, J. 2017. Enhanced adsorption and photo-degradation of bisphenol A by β-cyclodextrin modified pine sawdust in an aquatic environment. *Journal of the Taiwan Institute of Chemical Engineers* 78: 510–516.

Zhu, J., Li, R., Yang, X., Zhang, Z., Fan, Z. 2012. Spatial and temporal distribution of crop straw resources in 30 years in China. *Journal of Northwest A&F University* 40: 139–145. (in Chinese with English abstract)

Zhu, J., Li, R., Zhang, Z., Mao, H., Fan, Z. 2013. Heavy metal contents in pig manure and feeds under intensive farming and potential hazard on farmlands in Shaanxi Province, China. *Transactions of the Chinese Society for Agricultural Machinery* 44(11): 98–104. (In Chinese.)

Zhu, J., Zhang, Z., Fan, Z., Li, R. 2014. Biogas potential, cropland load and total amount control of animal manure in China. *Journal of Agro-Environment Science* 33: 435–445. (In Chinese with English abstract.)

Zorpas, A.A. 2009. Heavy metals leachability before, during and after composting of sewage sludge with natural clinoptilolite. *Desalination & Water Treatment* 8: 256–262.

Zorpas, A.A., Arapoglou, D., Panagiotis, K. 2003. Waste paper and clinoptilolite as a bulking material with dewatered anaerobically stabilized primary sewage sludge (DASPSS) for compost production. *Waste Management* 23: 27–35.

Zorpas, A.A., Constantinides, T., Vlyssides, A.G., Haralambous, I., Loizidou, M. 2000b. Heavy metal uptake by natural zeolite and metals partitioning in sewage sludge compost. *Bioresource Technology* 72: 113–119.

Zorpas, A.A., Inglezakis, V.J., Loizidou, M. 2008. Heavy metals fractionation before, during and after composting of sewage sludge with natural zeolite. *Waste Management* 28: 2054–2060.

Zorpas, A.A., Kapetanios, E., Zorpas, G.A., Karlis, P., Vlyssides, A., Haralambous, I., Loizidou, M. 2000a. Compost produced from organic fraction of municipal solid waste, primary stabilized sewage sludge and natural zeolite. *Journal of Hazardous Materials* 77: 149–159.

Zorpas, A.A., Loizidou, M. 2008. Sawdust and natural zeolite as a bulking agent for improving quality of a composting product from anaerobically stabilized sewage sludge. *Bioresource Technology* 99: 7545–552.

Zorpas, A.A., Vassilis, I., Loizidou, M., Grigoropoulou, H. 2002. Particle size effects on uptake of heavy metals from sewage sludge compost using natural zeolite clinoptilolite. *Journal of Colloid & Interface Science* 250: 1–4.

12

Recent Developments in the Treatment of Petroleum Hydrocarbon and Oily Sludge from the Petroleum Industry

Surendra Sarsaiya, Sanjeev Kumar Awasthi, Archana Jain, Saket Mishra, Qi Jia, Fuxing Shu, Jiao Li, Yumin Duan, Ranjan Singh, Mukesh Kumar Awasthi, Jingshan Shi, and Jishuang Chen

CONTENTS

12.1 Introduction

The power of the petroleum industry in the global market creates a situation in which diverse types of hydrocarbons are distributed in massive quantities as hydrocarbon sludge waste (HSW). Land and water are harmed by hydrocarbon pollution in equal measure because of the accidental or intentional release of unfinished hydrocarbon sludge waste into the environment. Oil spills are not discharged directly into the surrounding areas, which is dangerous as they harm the nearby ecosystems. Petroleum hydrocarbons were present in the environment before humans utilized it as a supply of energy. Natural deep-sea leaks have released hydrocarbons into ecosystems with microorganisms capable of r exploiting the hydrocarbons. However, this natural biological equilibrium in the environments cannot assimilate huge quantities of hydrocarbons from oil spills and other human activities.

Since the mid-1980s, hydrocarbon contamination has become a critical environmental issue because of its adverse environmental and health effects. In recent decades, the worldwide rise in petroleum exploration and invention has resulted in an ever-growing degree of ecological harm from hydrocarbons from activities like exploration, drilling, processing, and refining. This increase in the discharge of petroleum sludge waste has thus led to an increase the number of polluted wastewater sites (Mandal et al. 2011). The substances in hydrocarbon waste are highly carcinogenic and immunotoxic agents. Hence, growing attention is being given to the study and progress of techniques for cleaning up this contamination. To understand the environmental impacts that can occur during an oil spill, the molecular structure of petroleum hydrocarbons must be considered. The petroleum diesel composition is normally extremely complex. Petroleum hydrocarbons are presently used as the chief source of power; therefore, they constitute a significant worldwide environmental pollutant. Despite accidental pollution of bionetwork, the enormous quantity of oily sludge produces in refineries from water oil handling division and build-up of oily waste materials (OWMs) in crude oil storeroom tank bottoms create great troubles because of costly disposal practices (Ferrari et al. 1996; Vasudevan and Rajaram 2001). In spite of decades of study, safe petroleum- and hydrocarbon-contaminated soil biotreatment remains a great challenge to society. Inorganic substances restrict petroleum hydrocarbons biodegradation in aquatic regions. Atlas and Bartha (1972) showed that microbial mineralization and degradation were not amplified by phosphate or nitrate only, but they were increased noticeably when phosphate and nitrate were mixed together.

Petroleum unfinished oil is an extremely complex combination of organic mixture of a number of 1.3 million liters that mix into the ecosystem every year. It is an in nature occurring multipart combination of hydrocarbons with nonhydrocarbon compounds at which suitable quantity possess an assessable toxicity to organisms. The toxicity of crude petroleum oil products varies extensively depending on their concentration, their composition, ecological factors, and the natural state of the organisms that are contaminated. Oily sludge waste (OSW) is a viscous combination of sediments, oil, water, and elevated hydrocarbon concentration encountered throughout crude oil decontamination sites, the clean-up of oil storage vessels, and processing plant wastewater management. The chemical composition of oily sludge (OS) is quite complex because of its source. OS is chiefly composed of aromatics, alkanes, resin, and phaltenes (Ubani et al. 2013). It contains equally inorganic and organic oily pollutants. The inorganic parts include a variety of metals, for example, lead, zinc, copper, chromium, nickel, and mercury. Discharge of heavy metals into water supplies can cause serious toxic harm to human health and to water organisms. These metals are toxic and do not biodegrade.

Management of petro hydrocarbon oily sludge (PHOS) is possible via biological, physical, and chemical processes (Udotong et al. 2011). The natural processes are the main environmentally sustainable courses of action (Figure 12.1). A general biological approach is the exploitation of potential and effective microorganisms for the degradation of OS in the environment (Paulauskiene et al. 2009). Biodegradation by microbes is an effective and cost-efficient method for the management of petroleum hydrocarbons. This is achievable when the microbial inhabitants are present in the environment for degradation and the conditions are adequate for the activity of microbial growth (Phillips et al. 2000). Microbes are shown extended record in the oil along with gas production. Now, molecular investigation process, united with increases information of microbe discovery and chemistry, has led to advances in fighting microbiologically subjective corrosion and reservoir harm. Land farming methods have many restrictions in terms of their bulky space, time limits, and air pollution because of the volatile emission through organic matters. Consequently, notwithstanding decades of investigation, successful scientific and cost-efficient remediation of petroleum hydrocarbon pollution remains complex.

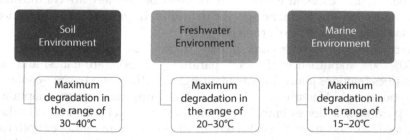

FIGURE 12.1
The optimal conditions for hydrocarbon removal through the soil mixture, fresh water system, and marine environments.

12.1.1 Overview of Petroleum Hydrocarbon and Oily Sludge

Crude oil source has dramatically changed humanity's way of life. It is a quick and accessible source of energy. Petroleum oil is the basis for the developments in numerous technologically advanced countries. The use of unfinished oil as a basis of energy has allowed many countries to build up a higher standard of living. Continued commercial growth in turn leads to increased demand for oil, which in turn leads to newer technologies or more exploration for original sources of oil (Helmy et al. 2010).

Hydro-petroleum, which in its unpolished/or raw state, is recognized as crude/raw oil. It is a fossil fuel that has accrued below the surface of the ground over many centuries. The Chinese first used primitive unpurified oil in the pre-Christian period because of the discovery of oil that had leaked to the earth's surface. Colonel E. A. Drake built up a petroleum business in Pennsylvania in the United States in 1859 (Alloway and Ayres 1993). The initial use of petroleum goods was to substitute expensive oil that has been applied for lighting. Today, raw oil is also applied as an energy source, a lubricant, as well as a raw substantial for a variety of miscellaneous industries, manufacture crude oil dominance in the global market. The United States Environmental Protection Agency (USEPA) has categorized OSW as a hazardous waste organic substance (Liu et al. 2010). The oily hazardous leftover materials are collected of entire petro-hydrocarbons (PHs), sediments, and water. As a result, management of this pollutant is important for countering the harmful effects of OS pollution in the environment.

12.1.1.1 History and Classification

The word *petroleum* comes from the Latin words *petra* ("rock") and *oleum* ("oil") (Testa and Winegardner 1991). Hydro-petroleum is chiefly produced biologically at temperatures below 200°C from materials deposited in deep seas and consequently compacted by the overload of deposited clays (Alloway and Ayres 1997). Each petroleum reserve is a unique combination of biomass breakdown products. Hence, each petroleum reserve has a unique compositional complexity, with additional variations occurring within each individual petroleum reservoir (Overton et al. 1994). Since there are compositional differences in petroleum, no specific definition of petroleum composition is available for every type of crude oil.

Petroleum raw oil contains mostly hydrocarbons, which is composed of 30%–60% naphthenes; 15%–60% paraffins; 3%–30% aromatics; and other asphaltics fractions jointly by way of nitrogen-, sulfur-, and oxygen-containing compounds (Al-Hawash et al. 2018). The aliphatic portion (naphthenes and paraffins) includes branched or linear chain cycloalkanes and alkanes (Figure 12.2). The aromatic parts contain alkyl side chains either or both fused cycloalkanes are possessed mono, di and poly ring/nuclear aromatic hydrocarbons (AHs). The asphaltenes as well as resins contain additional polar compounds consisting of oxygenated heterocyclic hydrocarbons and elevated

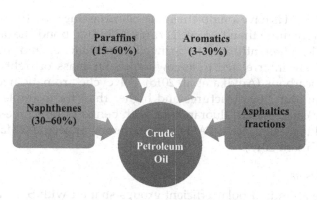

FIGURE 12.2
Crude petroleum oil hydrocarbon combination.

molecular weight aggregates (Harmsen and Rietra 2018; Brown et al. 2017). Asphaltenes and resins have extremely complex indefinite carbon assemblies that include sulfur, nitrogen, as well as oxygen molecules. Every segment has incomparable organic trials that directly affect their biodegradation process. In structural prearrangement of four chief hydrocarbon crude oil mechanisms, saturates framework the outmost coating, while asphalthenes existence larger molar form part constitute the inmost part of oil (Al-Hawash et al. 2018; Varjani 2014). Crude oil is not only used as energy but also as a feedstuff for the petrochemical industries to manufacture commercial energy sources (diesel, petrol, kerosene etc.), synthetic rubbers, plastics, and many other chemicals (Harayama et al. 2004).

12.1.1.1.1 Aliphatics

Aliphatic hydrocarbons may be saturated or unsaturated and branched or linear uncluttered-chain forms, for instance, iso-alkanes, n-alkanes, steranes, terpenes, in addition to cyclo-alkanes (Zabbey et al. 2017); n-alkanes are categorized into four clusters based on their weight of molecules:

  a. Gaseous/fumy alkanes

  b. Lesser molecular weight aliphatic hydrocarbons (C8–C16)

  c. Average molecular weight aliphatic hydrocarbons (C17–C28)

  d. Higher molecular weight aliphatic hydrocarbons (>C28) (Abbasian et al. 2015; Varjani 2017)

12.1.1.1.2 Aromatics

The aromatics are hydrocarbon molecules along with ring structure (Costa et al. 2012). They are chiefly categorized as:

  a. Aromatic hydrocarbons (AHs) in the form of monocyclic type, for example xylenes, toluene, benzene, and ethylbenzene

  b. AHs in the variety of polycyclic category (Anyika et al. 2015)

The polycyclic AHs have a supplementary benzene ring structure (one), while those that have more than two cyclic rings with dual bonds hexagon chains (for example, phenanthrene, two-ringed naphthalene, and triple-ringed anthracene) are interrelated to as low-molecular mass or lightweight aromatic hydrocarbons (Anyika et al. 2015). Polycyclic aromatic hydrocarbons (PAHs) form four-ring structures and higher than for example, chrysenes (four-ringed) and pyrene, fluoranthene with benzo pyrene (five-ringed) are interrelated to as significant PAHs mass of heavy molecules (Macaulay and Rees 2014; Varjani 2017).

12.1.1.1.3 Resins

Resins have abundant polar efficient groups shaped with S, N, and O, and metals like Ni, Fe, and V. Resins are shapeless and are essentially liquified in oil (Al-Hawash et al. 2018). Resins include aromatic compounds by way of extended alkyl chain in addition to mixtures of n-pentane and n-heptane (Zabbey et al. 2017). There forms are exceptionally differing to surface-active substance in rough oils and be used as peptizing mediators (Zabbey et al. 2017; Varjani 2017).

12.1.1.1.4 Asphaltenes

The asphaltenes consist of numerous different polar efficient groups. They are huge, complex, dark brown colloidal particles that are disperse in aromatic and saturate. They dissolve in low aromatic hydrocarbons, for example, toluene as well as benzene. Asphaltenes are high-molecular-weight viscid compounds made up of variably substituted polycyclic clusters by means of alkyl groups that contribute to their opposition of biodegradation. Peptizing agents retain asphaltenes in suspense, thus indorsing the constancy of unfinished oil (Varjani 2017).

12.1.2 Nature, Magnitude, and Distribution of Source

Crude/raw oil is a multipart, heterogeneous fluid combination of many hydrocarbon segments and carbon-based compounds, of which carbon (C: 85%–90%) and hydrogen (H: 10%–14%) are the chief elements. Other components include nearly non-elemental hydrocarbon, for example, 0.2%–3% sulfur, $\leq$ 0.1%–2%nitrogen, 1%–1.5% oxygen, and other trace components (parts/million) of metallic complexes together with phosphorus, arsenic, lead, nickel, and vanadium compounds (Okoh 2006). The petroleum crude/raw oil organic compounds possess linear with branched chain volatile along with nonvolatile aromatic compounds (26%–30%) and aliphatic group (up to 50%) fractions (Lal and Khanna 1996), ranging from light fumes (C1–C4) to heavy remains (C35–C40), with resins forming the residual (Kadali et al. 2012). The configuration of crude/raw oil is not stable and may vary depending on the stage and position of an oil arena, and on the deepness of each separate oil well. In view of the earlier 1850s fact, once Edwin Drake penetrated the foremost oil well, insist for oil has continual to increase. It is predicted that the day-to-day worldwide consumption of petroleum will increase from 85 million barrels in 2006 to 106.6 million barrels in 2030. The consumption of petroleum produces vast quantities of waste: by wastewater rationale for in excess of 80% of fluid waste and as much as 95% in aged oilfields (Igunnu and Chen 2012).

Over 2 million tons of oil are produced in the world. Soil pollution and the presence of groundwater hydrocarbons in the region of oil refineries and the surrounding area of fuel transportation services are substantial pollution issues. Various chemicals are present in the naturally occurring petroleum hydrocarbons that are refined for use as energy sources, for example, vehicle fuel and home heating oil (Vaziri et al. 2013). The magnitude of ordinary unpurified oil outflow was expected to reach 600,000 metric tons (MT) per year, plus or minus 200,000 metric tons (MT) per year (Das and Chandran 2011). Crude petroleum oil is a highly composite combination of carbon-based complexes of which a sum of 1.3 million liters arrives into the neighboring areas per year (Brooijmans 2009). In the European Union (EU), it is estimated that 3.5 million places are polluted, and 0.5 million places are actually polluted and require remediation (Perez-Sanz et al. 2012). Annually, approximately 1.7 to 8.8 million metric tons of hydrocarbon oil are released into the water and food chains, more than 90% of which can be traced directly to accidents caused by humans (Megharaj et al. 2011).

12.2 The Toxicity of Oil and Sludge Hydrocarbons

Hydrocarbon pollutants are harmful because they are highly toxic. They cause destruction in the environment (Al-Hawash et al. 2018). Extensive free discharge of petro-carbon contaminants from sources like discharges and seepage, subversive tanks, steamers, unblocking of oily wells, discarded oil industry places are caused natural properties of surface soil, groundwater, and ocean water sources (Brown et al. 2017). Numerous parts of crude petroleum oils are recalcitrant and extremely toxic because of their role in carcinogenic, blood toxic, and teratogenic mechanisms. These parts include benzene, toluene, xylenes, ethyl benzene, and PAHs (Meckenstock et al. 2016; Al-Hawash et al. 2018).

Various indirect or direct acute/chronic properties of petroleum contaminants have already been documented (Meckenstock et al. 2016). The many direct and indirect effects include anoxia, suffocation, stunted evolution, conflict in metabolic actions, and hormonal inequity in existence formulae (Walker 2006; Brown et al. 2017). Serious necrosis death rate, smothering, hypothermia, sinking and intake of deadly compounds at some stage in preening are a number of short-term effects (Desforges et al. 2016). Extended-term effects consist of developmental abnormalities in aquatic animals, for example, jaw reductions, unfused skulls, and lack of pigmentation (Van Meter et al. 2006; Varjani 2014). These effects cause variations in inhabitants and communities, and in this manner reason variations to a whole flora and fauna (Walker 2006; Brown et al. 2017). Crude oil containing volatile organic compound is predominantly harmful to humans. PAHs in particular may encourage malignant tumors that first and foremost have an effect on skin and supplementary epithelial material as they have a large similarity for nucleophilic interior of function molecules like protein, RNA, and DNA (Perez-Cadahia et al. 2007).

Petroleum carbons are also categorized into individual series of correspondent carbon figure or segment(s) to measure environmental and human health danger (CCME 2008, 2010; Anyika et al. 2015). These parts can be further categorized and described in accordance with their physical, chemical, and toxicological properties. Part 1 explains the range of the same carbon quantity from C6 to C10, instead of volatile segment of maximum hydrocarbon combinations. Part 2 can be clearly explained as series of same carbon quantity from C10 to C16, in lieu of partial-volatile part. Part 3 incorporates variety of correspondent carbon number from C16 to C34, which is measured as nonvolatile parts. Part 4 mentions the complexes with same carbon quantity of C35+; they are measured as part with the lowermost solubility and volatility (Table 12.1) (CCME 2008; Desforges et al. 2016).

TABLE 12.1

Major Effects of Petroleum Hydrocarbon and OS Contaminants on the Environment

Parameter	Source	Effects	Reference
Rural land areas	Accidental oil spills, accumulation of oil at oil storage areas, fugitive equipment leaks	Reduce soil fertility, Affects soil biota, reduce seed germination	Singh and Chandra (2014)
Aquatic life	Accidental oil spills Unscientific methods Pipe line and well leakage	Death of aquatic life including flora and founa, negative effect on the aquatic birds,	
Human	Accidents and equipment failures etc.	Skin cancer, Skin erythema, gastrointestinal cancer, Effect on CNS, Irregular heartbeat	
Ecosystem		Imbalance in marine ecosystem, damage to the natural habitat, affect the food chain	
Plants		Affects the plant photosynthesis	
Animals		Damage animal lungs, kidney, intestine and other tissue	

12.3 Recent Advancements in the Petro-Hydrocarbon Contamination Remediation Process

12.3.1 Fenton Method

The Fenton system is useful for the breakdown of H_2O_2 along with hydroxyl radicals formed in the presence Fe^{3+} ions acting as a catalyst (Yoon et al. 2001). Various investigations have revealed that released hydroxyl radicals can decompose petro organic contaminants, for instance, oil and sludge hydrocarbons (Sung-Ho et al. 1998; Yun and Kyung 2000; Chamarro et al. 2001; Neyens and Baeyens 2003; Yeh et al. 2003; Ahad and Slater 2008; Lin et al. 2012). Lu et al. (2001) demonstrated the bioremediation of natural petroleum pollutants from the soil surface with the help of the Fenton method. They showed that the Fenton method amplified the effectiveness of the natural course of action (Mang et al. 2010). In other investigations, petroleum-polluted soil was preserved by consuming the phosphate which upsurge the efficacy of the Fenton method but the not greater than 40% (Watts and Dilly 1996). The elimination effectiveness of PAHs was in the range of 70%–98% (and depended on the chemical character of the PAHs) with the collective biological interruption by the modified way of Fenton method (Nam et al. 2001).

With the Taguchi process, the main appropriate situations for the highest drop time of Total petroleum hydrocarbon (TPH) in petro oily sludge are achieved. The outcome verified that the greater part of actual constraints on the results of the Fenton method are as follows: pH, the mass fraction of H_2O_2 to trial, the reaction time, and the molar relation of H_2O_2 to Fe (II). The most favorable state of pH, the mass fraction of H_2O_2 to trial, and the molar portion of H_2O_2 to Fe (II) are 5, 15, and 10, respectively. The lessening speed of TPH was 36.47% at most favorable ailment. Increasing the dampness content by weakening with water had a role in enhancing the decrease proportion up to 73.07%. Eventually, the seepage of petro hydrocarbon is 35 g/kg. This process can be an appropriate pretreatment for giving OS and addition an opposite treatment phase is essential for accomplishment the most constant numbers (Farzadkia et al. 2014).

12.3.2 Composting for Remediation of Oil Contamination

Even after decades of research, successful biological remediation remains a challenge for OS in the environment. There are so many are already reported like physical, chemical, and biological approach of which the landfilling, landfarming, and bioreactor treatments can hinder the degradation of OS, making them incompletely efficient and sometimes too costly. It is essential to investigate new, cheaper, and more nature-friendly options that can accelerate bioremediation of OS. These options should be able to support the restrictions of the earlier processes while improving OS bioremediation. As a result, composting process are involved for the control and addition of nutrients, tilling, watering, accumulation of suitable potential microbial flora with bulking agents (hay or wood chips) were measured another alternative way to boost the bioremediation process of OSW (De-qing et al. 2007). The procedure leads to the generation of carbon dioxide, minerals, water, and neutralized organic matter (Ubani et al. 2013).

Beaudin et al. (1996) studied total hydrocarbon sludge composting with maple and alfalfa vegetation and found it useful for the mitigation of hydrocarbon OS pollution. The soils polluted with grease and oils (40% aliphatic, 28% aromatic hydrocarbons, and 32% polar) at the concentration of 17,000 mg, the sum hydrocarbons per kilogram (54% aromatic, 60% aliphatic, and 83% polar substances) had despoiled subsequent to 180 days for the period of the composting. The result revealed that no more than 50% mineral oil and grease in the soil had degraded. After 287 days, no more than 73% of mineral oil and grease had broken down. Milne et al. (1998) also found that different bulking substances are chiefly helpful for the acceleration of OS composting. The bulking agents (sliced barley warmth up treated mulch moss, straw and Solv-II, which is mulch moss supplemented with many nutrients along with oil-degrading potential microbes) were used equally and are excellent for activating the biodegradation process during the entire composting period. On the other hand, there is a decrease (25%) in the concentration of total petroleum carbons in the decomposing supplemented with barley and the mulch moss if the composting phase does not go beyond 800 hours. The composting process is containing the Solv-II, which is reduced the entirety petroleum hydrocarbon (55%) along with high CO_2 generation due to the high microbial action. The study recommended that composting be combined with bioaugmentation for the thriving bioremediation of OS residues.

Van Gestel et al. (2003) studied soil mixed with diesel oil and other biowaste (fruit/vegetables and garden waste) at the proportion of 1:10 and composted for 12 weeks. The biowaste is composted in equal amounts. In sequence to differentiate, the effect of temperature from the additional effect of biowaste on diesel breakdown, one beneficiary with polluted soil particles was hold at ambient temperature, while an additional was worked at the optimal temperature of composting. The identifications and enumerations of compost microbes demonstrated the additive effect of the contaminated soil showed different effects on the process of composting. Zytner et al. (2006) also examined the various factors widely affecting the effectiveness of composting; these factors include microbial population, environmental conditions and composition of the hydrocarbon, impact of composition mixture on the individual compounds, and the whole degradation process.

12.3.3 Energy Recapture and Conversion to Natural Intermediates

The oil industry provides a huge amount of global energy, in addition to the preponderance of petrochemical mediators wanted for new products such as polymers, different dyes, diluters, and pharmaceutical chemicals (Schlepp et al. 2001; Brown et al. 2017). Crude/raw petroleum oil cannot be used in its raw state and must be refined before it can be used in the manufacturing of other products (Fumoto et al. 2004). Crude/raw oil is purified in the petroleum factory and converted into many transitional end products like gasoline, diesel fuel, jet fuel, kerosene, heating oil, petroleum naphtha, petrochemical feedstocks, lubricating oil, asphalt, liquefied and wax petroleum products, etc. (Lucia et al. 2006; Varjani 2017). The purification procedure includes (1) chemical decontamination, (2) catalysis, and (3) products purification (Abed and Koster 2005; Anyika et al. 2015).

Catalysis plays an important function in nearly all parts of the oil petroleum trade (Abed and Koster 2005). Widely held of these pollutants posture an overweight on the surroundings ecosystem throughout their exploit and at the conclusion of their life succession (Waigi et al. 2015). Petroleum refinery activities discharge most of the hydrocarbon pollutants into the natural environment (Schlepp et al. 2001; Lucia et al. 2006; Meckenstock et al. 2016). Problems regarding the disposal of hydrocarbon impurities must be given priority when investigating their harmful effects (Tong et al. 2013; Varjani and Srivastava 2015). There is an urgent need to mitigate the pollution resulting from the petrochemical industry. In the attempts to understand the green technological application of biotic forms for removal of petroleum impurities, microbes are currently at the center of researchers' attention (Waigi et al. 2015; Varjani and Upasani 2016c). In addition, the utilization of different types of biomass as a feedstock must be use as a substitute fossil fuel bases for energy purposes (Lucia et al. 2006).

Several conventional physical and chemical refinement methods are more exclusive by reason of price associated among the mine and moving of huge amounts of hydrocarbon dirty matters for ex-situ management, that is, soil washing, natural stabilization (usage of hydrogen peroxide either or both potassium permanganate as an elemental oxidant to transform nonaqueous pollutants for petroleum case in point), and waste burning (Farhadian et al. 2008). Additional physical and chemical processes used for the identical principle are dilution, dispersal, volatilization, sorption, and environmental transformations (Varjani and Upasani 2012). The increasing cost and limited effectiveness of these conventional physical and chemical processes have led to the search for substitute technologies for in-situ wide application, in particular, the biological removal capabilities of microorganisms and plants (Singh and Jain 2003; Farhadian et al. 2008). Sustainable green technologies for management of contaminants by means of microorganisms are applied for the biological remediation of hydrocarbon waste–polluted area(s) (Varjani and Srivastava 2015; Harmsen and Rietra 2018). Bioremediation is the ability of living organisms to detoxify/degrade contaminants (Anyika et al. 2015; Varjani and Upasani 2016b). This green-efficient process is a financially, well-organized, multipurpose, and ecologically sound course of action (Varjani et al. 2014a, 2014b; Ron and Rosenberg 2014).

Bioremediation is a state-of-the-art technique in which potential microbes mitigate, degrade, or reduce hazardous carbon-based pollutants to more harmless mixtures, for example, CO_2, H_2O, CH_4, and many other biomasses with hazardous environmental effects. The biodegradation process, using the combined action of mixed microbes, is the primary complex mechanism used for the improvement of hydrocarbon contaminant removal. Petroleum hydrocarbons are energy-loaded materials degraded by the organisms present in the environment. The use of indigenous microorganisms as removal tools takes advantage of the microorganisms' catalytic abilities to boost the pollutant degradation commonly used for oil recapture (Varjani and Upasani 2016c; Brown et al. 2017).

12.4 Microbial Degradation of Hydrocarbon Contamination

Microbial bioremediation is used extensively for the management of petroleum hydrocarbon pollution from different segments in the petroleum industry (Al-Hawash et al. 2018). Various studies related to hydrocarbon impurity biodegradation have been conducted (Al-Hawash et al. 2018). Many researchers have described general pathways and mechanisms for the biodegradation of oily hydrocarbon (Zabbey et al. 2017). Microorganisms play a vital function in cleaning the environment and building a sustainable ecosystem. They are also the indicators of pollution and thus are useful for biomonitoring as well as bioremediation (Chandra et al. 2013; Varjani and Upasani 2016c). The microorganisms release energy during the degradation of the pollutants and utilize them for their metabolism (Figure 12.3) (Abbasian et al. 2015). Very few microbes can fully break down the aromatic and alkane compounds from the environment (Zabbey et al. 2017). Microorganisms living in contaminated areas acclimate to the conditions as a result of mutations or genetic changes, which may mean that successive generations can play a prominent role in biodegrading hydrocarbon. Hydrocarbon degradation by microorganisms in uncontaminated environments usually means fewer microbes; however, this number may vary or increase in the areas surrounding oil and sludge contamination (Atlas 1981; Brown et al. 2017). In contrast, polluted ecosystems do not support, in general, microbial growth. Native microbes play a vital function in breaking down petroleum hydrocarbon

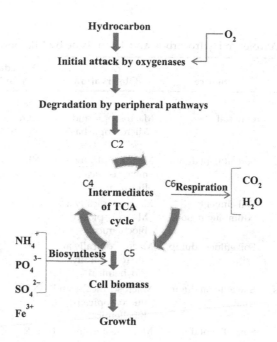

FIGURE 12.3
Microbial hydrocarbon breakdown process.

and their products (Varjani and Upasani 2016c). Microorganisms also have the ability to degrade hydrocarbon contaminants (Table 12.2) (Atlas 1981; Boonchan et al. 2000; Wilkes et al. 2016).

12.4.1 Bacteria

Many microorganisms can break down aliphatics. Some organisms can break down the polyaromatics or monoaromatics, while few can degrade resins. The breakdown of petro hydrocarbons and their varieties by microbes is shown in Table 12.1. Bacterial species like *Arthrobacter* sp., *Bacillus cereus*, *Citrobacter* species, *Citrobacter amalonaticus*, *Cronobacter muytjensii*, *Enterobacter aerogenes*, *Enterobacter helveticus*, *Klebsiella pneumoniae*, *Klebsiella* species, *Proteus* species, *Pseudomonas aeruginosa*, *Staphylococcus aureus*, and *Streptococcus* species can degrade hydrocarbon (Anyika et al. 2015).

Many bacteria have been evaluated for their ability to degrade hydrocarbon. Some examples are *Marinobacter*, *Oleispira*, *Thalassolituus*, *Cycloclasticus*, and *Alcanivorax*, which were obtained from different polluted areas with hydrocarbon contaminants. These native bacteria occur in undetectable numbers prior to pollution but are found to dominate in polluted regions (Varjani and Upasani 2013). *Alcanivorax* strains develop in the presence of branched alkanes and n-alkanes, although they cannot develop in the presence of

TABLE 12.2

Degradation of Petroleum Hydrocarbon and Their Type by Microorganisms

Name of Microbe	Source	Observation	Degradation Activity	Reference
Fungi				
Penicillium sp.	Oily soil	Macroscopic and Microscopic basis	Petro hydrocarbon	Vanishree et al. (2014)
Fusarium species	Soil effluent dump site	Macroscopically and microscopically, Biochemical	NR	Eze et al. (2013)
Aspergillus species	Soil effluent dumping regions	Macroscopically and Microscopically, Biochemical	NR	Eze et al. (2013)
Rhizopus species	Soil effluent dump site	Macroscopically and microscopically, Biochemical	NR	Eze et al. (2013)
Trichoderma species	Soil effluent dump site	Macroscopically and microscopically, Biochemical	NR	Eze et al. (2013)
Penicillium species	Soil effluent dump site	Macroscopically and microscopically, Biochemical	NR	Eze et al. (2013)
Mucor species	Soil effluent dump site	Macroscopically and microscopically, Biochemical	NR	Eze et al. (2013)
Geotricum species	Soil effluent dump site	macroscopically and microscopically, Biochemical	NR	Eze et al. (2013)
Bacteria				
Arthrobacter sp.	Oil contaminated soil	Molecular level	Diesel	AlDisi et al. (2016)
Bacillus cereus	Oil contaminated soil	Molecular level	Diesel	AlDisi et al. (2016)
Bacillus species	Soil effluent dump site	Macroscopically and Microscopically, Biochemical	NR	Eze et al. (2013)
Citrobacter amalonaticus	Oil contaminated soil	Molecular level	Diesel	AlDisi et al. (2016)
Citrobacter sp.	Oil contaminated soil	Molecular level	diesel	AlDisi et al. (2016)
Citrobacter species	Soil effluent dump site	Macroscopically and Microscopically, Biochemical	NR	Eze et al. (2013)

(Continued)

TABLE 12.2 (*Continued*)

Degradation of Petroleum Hydrocarbon and Their Type by Microorganisms

Name of Microbe	Source	Observation	Degradation Activity	Reference
Cronobacter muytjensii	Oil contaminated soil	Molecular level	Diesel	AlDisi et al. (2016)
Enterobacter aerogenes	Soil effluent dump site	Macroscopically and microscopically, Biochemical	NR	Eze et al. (2013)
Enterobacter helveticus	Oil contaminated soil	Molecular level	Diesel	AlDisi et al. (2016)
Klebsiella pneumoniae	Oil contaminated soil	Molecular level	Diesel	AlDisi et al. (2016)
Klebsiella species	Soil effluent dump site	macroscopically and microscopically, Biochemical	NR	Eze et al. (2013)
Proteus species	Soil effluent dumping regions	Macroscopically and Microscopically, Biochemical	NR	Eze et al. (2013)
Pseudomonas aeruginosa	Oily soil, Soil effluent dumping place	Molecular level, Macroscopically and Microscopically, Biochemical	Diesel	Eze et al. (2013); AlDisi et al. (2016)
Staphylococcus aureus	Top soil waste matter dump site	Macroscopically and Microscopically, Biochemical	NR	Eze et al. (2013)
Streptococcus species	Soil effluent dumping place	Macroscopically and Microscopically, Biochemical	NR	Eze et al. (2013)

NR: Not reported.

amino acids or sugars. *Cycloclasticus* strains develop in the presence of naphthalene, phenanthrene, aromatic hydrocarbons, and anthracene. Oleispira strains develop in the presence of alkanoles, and aliphatic and alkanoate hydrocarbons. It is also important to study the degradation process in the presence of light in order to recognize the entire native community metabolic potentials (Brown et al. 2017).

12.4.2 Fungi

In the environment, fungi play a crucial function by removing harmful hydrocarbon chemicals. Sediment particles polluted with crude OS are preferred natural niches in which fungi live; they use the polluted particles as sources of carbon. Numerous studies have been published on the biodegradation function of fungi on petroleum intermediates and the majority of biodegrader fungi like *Aspergillus, Alternaria, Candida, Cladosporium, Cephalosporium, Fusarium, Gliocladium, Geotrichum, Mucor, Paecilomyces, Pleurotus, Penicillium, Polyporus, Rhodotolura, Rhizopus, Saccharomyces, Torulopsis,* and *Talaromyces* (Vanishree et al. 2014).

12.5 Microbial Physiology and Metabolism

Biodegradation of pollutants involves a sequence using a cocktail of enzyme catalysis (Abbasian et al. 2015). Hydrocarbons are processed by the enzyme that is secreted by a single microbial strain or a collection of microbial strains (Al-Hawash et al. 2018). A mixture of potential enzyme-secreting strains has been found to be more effective than a single strain for processing or degrading a wide-range of hydrocarbon concentrations (Deziel et al. 1996; Deppe et al. 2005). The biocatalysts for biodegradation are fixed on specific plasmids. Whyte et al. (1998) emphasized the utility and involvement of Q15 plasmid for the breakdown of petro hydrocarbon. These plasmids are OCT, Q15, TOL, pND140, pND160, and NAH7, with the genes like alkM, alkA, alkB, theA, assA1, assA2, nahA-M, and LadA (Anyika et al. 2015). The *Acinetobacter* sp. are incomparable with affections to existence of biocatalysts for the biodegradation. In *Acinetobacter* sp., the plasmids are situated on the chromosomes (Wilkes et al. 2016).

The n-alkanes are the normal degrading mechanism in a hydro-petroleum combination. The branching of methyl usually enhances the development of resistance mechanism toward the contagious attack (Abbasian et al. 2015; Varjani and Srivastava 2015). Cycloalkanes are principally resistant to interruption by the microbes. The majority of persistent parts of petroleum contamination in the ecosystem are hopanes, pristine, and phytane (Ron and Rosenberg 2014; Louvado et al. 2015; Abdel-Shafy and Mansour 2015). In biodegradation, the microbes acquire dynamism or digest the petro hydrocarbons into the cell materials. The microbial degradation of oily hydrocarbon contaminants is involved in the metabolic actions accelerated by the enzymes consortium (Varjani 2017). The enzymes that accelerate the process of petroleum hydrocarbon biodegradation include peroxidases, oxygenases, hydroxylases, reductases, and dehydrogenases. They are principally apprehensive with the anaerobic and aerobic route of the elimination of petro hydrocarbon (Peixoto et al. 2011; Mbadinga et al. 2011).

12.5.1 Biocatalysts Involved in Degradation of Hydrocarbons

The ability of microbes to break down organic compounds is a complex process resulting from the organism's genetic materials. The many complex reactions are intricate in metabolism facilitated by enzymes. The bacterial

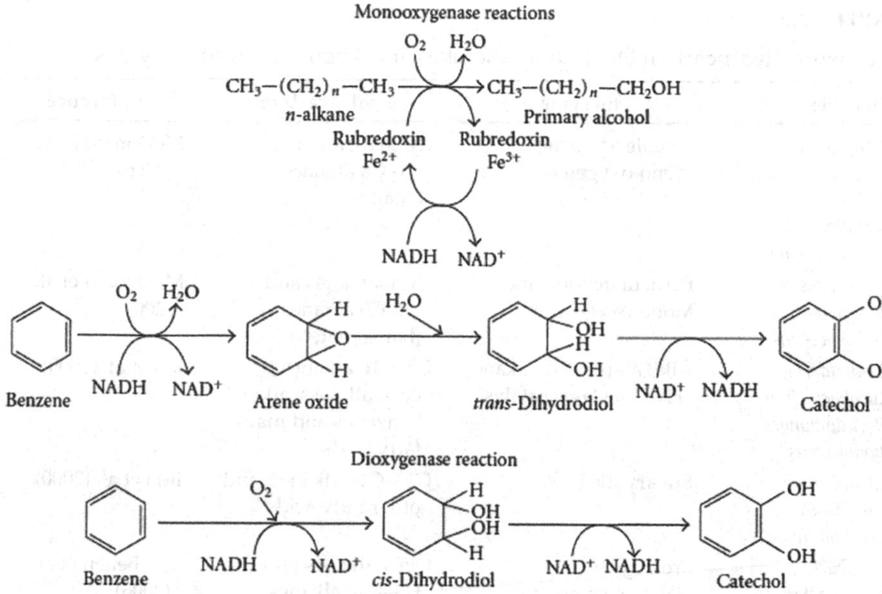

Monooxygenase reactions

O_2 H_2O

$CH_3-(CH_2)_n-CH_3 \longrightarrow CH_3-(CH_2)_n-CH_2OH$

n-alkane   Primary alcohol

Rubredoxin   Rubredoxin
Fe^{2+} Fe^{3+}

NADH NAD^+

Benzene — O_2 H_2O / NADH NAD^+ → Arene oxide — H_2O → trans-Dihydrodiol — NAD^+ NADH → Catechol

Dioxygenase reaction

Benzene — O_2 / NADH NAD^+ → cis-Dihydrodiol — NAD^+ NADH → Catechol

FIGURE 12.4

Enzymatic petro-hydrocarbon degradation course of action. (From Das, N., and Chandran, P., *Biotechnol. Res. Int.*, Article ID 941810, 13, 2011.)

enzymes as alkane hydroxylases (cytochrome P450) contain Heme-thiolate Mono-oxygenases, dynamically mixed up in the microbial metabolism of OS (Van Beilen and Funhoff 2007). The enzymes release the oxygen into the substrate to initiate biodegradation (Figure 12.4 and Tables 12.3 and 12.4). Higher eukaryotes microorganisms contain numerous different P450 types that may subsidize as an assembly of isoforms to the metabolic breakdown of specified substrate. The enzyme P450 can solitary be occurred in very less strain (Zimmer et al. 1996). Cytochrome P450 is also involved in the breakdown of OS waste. The ability of different yeast types to utilize n-alkanes and supplementary aliphatic hydrocarbons as a chief basis of carbon and dynamism is regulated by the presence of various P450 microsomal. The cytochrome P450 enzyme has been purified from many diverse yeast isolates like *Candida tropicalis*, *Candidamaltosa*, and *Candida apicola* (Scheuer et al. 1998). The multiplicity of alkaneoxygenase complex in microorganisms that are dynamically sharing in the breakdown of alkanes under the oxygenic conditions like essential membrane di-iron alkane hydroxylases, Cytochrome P450 complex, membrane attached copper with methane monooxygenases and dissolved di-iron methane monooxygenases (Van Beilen and Funhoff 2007).

TABLE 12.3

Petroleum Hydrocarbon Degrading Microbes and Their Different Enzymes

Microbes	Enzyme	Hydrocarbon	Reference
Methylocella, *Methylococcus,* *Methylosinus,* *Methylocystis,* *Methylomonas*	Soluble Methane Mono-oxygenases	Cycloalkanes and C1–C8 alkenes alkanes	McDonald et al. (2006)
Methylocystis, *Methylobacter,* *Methylococcus,*	Particulate Methane Mono-oxygenases	Cycloalkanes and C1–C5 alkanes (halogenated)	McDonald et al. (2006)
Burkholderia, *Mycobacterium,* *Pseudomonas,* *Rhodococcus*	AlkB associated Alkane Hydroxylases catalyst	C5–C16 alkanes,, cycloalkanes, alkyl benzenes and many fatty acids	Jan et al. (2003)
C. tropicalis, *C. maltosa,* *Yarrowia lipolytica*	Eukaryotic P450	C10–C16 alkanes, and other fatty acids	Iida et al. (2000)
Caulobacter, *Mycobacterium,* *Acinetobacter*	Prokaryotic P450 oxygenase	Cycloalkanes and C5–C16 alkanes	Van Beilen et al. (2006)
Acinetobacter sp.	Dioxygenases catalyst	C10–C30 alkanes	Maeng et al. (1996)

TABLE 12.4

Genetically Improved Microbes for the Biodegradation of Different Hydrocarbons

Pollutants	Microbial Strains	Application	References
PCB	*Alcaligenes eutrophus*	Development monitoring	Van Dyke et al. (1996)
BTEX, TCE	*Pseudomonas putida*	Development monitoring	Applegate et al. (1998)
Naphthalene and Anthracene	*Pseudomonas fluorescens*	Monitoring of development	Sayler and Ripp (2000)
2,4-D	*Burkholderia cepacia*	Strain monitor	Masson et al. (2002)
BTEX	*Pseudomonas fluorescens*	Pressure response	Sousa et al. (1997)
2, 4-dinitrophenol hydroquinone	*Pseudomonas* sp.	Toxicity investigation	Kelly et al. (1999)
Narcotics	*Alcaligenes eutrophus*	Product investigation	Layton et al. (1999)

Source: Das, N. and Chandran, P., *Biotechnol. Res. Int.*, Article ID 941810, 13, 2011.

12.5.2 Molecular Methods

Molecular methods are important tools for learning the molecular potential of microbes (Liu et al. 2016). Current information about microbial groups is the result of the use of molecular methods (Liu et al. 2010; Yang et al. 2016; Li et al. 2016). The 16S rRNA sequence, reverse genome probing (RGP), restriction fragment length polymorphism (RFLP), oligo-nucleotide matrix array hybridization (OMAH), and denaturing gradient gel electrophoresis (DGGE) are universally accepted techniques for the molecular characterization of microbes' diversity isolated from the hydrocarbon contamination sites (Al-Hawash et al. 2018). Polymerase chain reaction–based fingerprinting method is less effort-concentrated method that has been widely used to examine microbial diversity. The examination of 16S rDNA has improved from the samples discovered beforehand unidentified diversity in a range of territories (Li et al. 2010; Peixoto et al. 2011; Feng et al. 2011; Zhou et al. 2013; Yang et al. 2016; Liu et al. 2016). The study of diversity has been examined for the explanation of petroleum hydrocarbon microbial diversity (Tardy-Jacquenod et al. 1996; Sette et al. 2007).

Orphan et al. (2000) have established the contagious diversity of thermophilic collections in an important elevated temperature, oil-bearing development in California, via 16S rDNA phylogenetic exploration. They examined the sequence of the worldwide genomic collection and showed that clones were extremely similar to known microbes recovered from related places. The rDNA libraries contain *Thermoanaerobacter, Desulfothiovibrio, Acidaminococcus, Aminobacterium, Halomonas, Sphingomonas, Acinetobacter, Desulfomicrobium, Methylobacterium, Pseudomonas,* and *Thermococcus*. For that reason, it is of enormous attention to read the diversity of microbes for the utilization of indigenous microbes through their optimistic property (Kaster et al. 2009).

12.5.3 Genetically Improved Microbes

Petroleum hydrocarbon contamination is a serious environmental problem (Peixoto et al. 2011; Varjani 2017). Numerous microbes and plants have the capacity to use hydrocarbon petroleum contaminants as growth nutrients and thus clean the surrounding environment. Genetically improved microbes (GIMs) with bioremediation capabilities can be used for the application of complex techniques. An enormous microbe has the ability to utilize alkanes as well as well as low- and high-molecular-weight materials like acenaphthene, naphthalene, anthracene, chrysene, fluoranthene, and pyrene used as carbon energy sources (Wilkes et al. 2016). Exploration efforts have been conducted to find GIMs for petroleum hydrocarbon contaminant degradation (Urgun-Demirtas et al. 2006).

GIMs can be formed in test site environment with the help of advanced molecular practices by means of plasmids transfer comprising essential genetic substance as of exogenous in the direction of indigenous microbes. GIMs thus formed can be used to bioremediate petroleum hydrocarbon contaminated sites (Sana 2015). Research areas include regulation and production of innovative pathways, exploitation of substrate, pathway expansion not including detrimental metabolites, degradative enzyme affinity and specificity alteration, and development of genetic immovability and enhance pollutant bioavailability. Problems occur through persistence, existence, and antagonism of living cells in the normal situation of bioremediation is extremely essential. Possible biosafety risks and conservation issues, i.e., free of GIMs in surroundings wants to be mentioned (Varjani 2014).

12.5.4 Genes for Biodegradation

Various key genes to understand the anaerobic and aerobic biodegradation of hydrocarbon contaminants are alkM, alkB, alkA, LadA, nahA-M, assA1, assA2, napA, dsrAB, mcrA, amoA etc. (Li et al. 2011; Feng et al. 2011; Guan et al. 2013; Liu et al. 2016; Varjani and Upasani 2017). The primary breakdown of organic contaminants is an oxidative course of action. The peroxidases and oxygenases catalyze establishment as well as absorption of oxygen molecule. Tangential pathways for degradation convert organic contaminants systematically into metabolic intermediates like Tricarboxylic Acid (TCA) (Abbasian et al. 2015). The construction of cell biomass establishes from essential microbe's products precursor like pyruvate, acetyl-CoA, and succinate (Das and Chandran 2011). The different useful genes used for the breakdown of petro hydrocarbon are well recognized previously (Foght 2008). The most important genes participating in microbial degradation of hydrocarbon contaminants are XylE gene: catechol dioxygenase, AlkB gene: alkane monooxygenase, and naphthalene dioxygenase similar to the gene NahAc (Meckenstock et al. 2016).

12.5.5 Plasmid-Borne Gene's Involvement

Genetic material in all microbes is in the form of deoxyribose nucleic acid (DNA). The evidence is substantially present in contagious cells in two dissimilar forms: the plasmids and the chromosome. The microbial genetic material is a double-strand, highly folded, single circular DNA. In addition to DNA, bacteria also have some extrachromosomal deoxyribose nucleic acid as a plasmid (Zylstra and Gibson 1991). Many different plasmids contain different genes, which code many different enzymes necessary for the pathways essential for bioremediation. Enzymes participating in the breakdown of toluene, salicylate, octane, naphthalene, etc., are encoded in plasmid (Barbly and Barbour 1984; Nelson 1990).

Plasmid genes also play an important function in the development of strains with greater biodegradation capability. With the help of molecular tools, DNA can be cut into many different pieces that possess genes targeting specific biodegradative pathways. These plasmid genes can be transferred into a recipient host, and the recombinant or genetically modified host possesses improved biodegradative capacities (Brand et al. 1992). These organisms can mitigate hydrocarbon contamination in polluted sites through processes such as bioaugumentation by microbes (McClune et al. 1989; Phillips et al. 1989; Focht 1998). Favorable environmental conditions are crucial for the activity of these microorganisms. The diverse form has extra chromosomal DNA found in the plasmids of prokaryotic cells. Thus, identifying the bacterial plasmids is essential. The many agents like physical and chemical (DNA intercalating substances, surfactants, antibiotics, and other metals) may enhance the rate of elimination of plasmid genes in bacterial species (Caro et al. 1984; Stanisich 1984).

12.5.6 Biodegradation of Hydrocarbon by Immobilized Cells

Isolation of petroleum-degrading microbes is required to immobilize them on an appropriate carrier. The application of immobilized microbial cells in contaminated areas is an essential bioremediation process (Cunningham et al. 2004; Lu et al. 2009). It is important to discover new substances to be used as carriers to increase the oily hydrocarbon pollutant degradation efficiency of microbes (Gentili et al. 2006). Different carriers, such as chitosan, sodium alginate, chitin, wheat straw and wood chips, mollusk shells, and biochar, can be applied for microbial cell immobilization (Zhang et al. 2016). Suitable carrier parts must have elevated stability, inflexibility, and no harmful impact on the ecosystem (Cunningham et al. 2004). Shen et al. (2015) found that immobilized microbial cells could give a higher biodegradation rate (47%) of crude oil compared to free microbes. They also noticed that immobilized microbial cells became more suitable to work in the presence of alkaline, acidic, low temperatures, and NaCl.

Microbial cell immobilization for the treatment of petroleum sector wastewater has received more research attention (Zhao et al. 2006; Lu et al. 2009; Tong et al. 2013). The biological aerated filter (BAF) is a form of immobilized bioreactor used to maintain higher hydraulic loading rates and give a high biomass percentage to mitigate the environmental impact; it results in less sludge generated and promotes microbial growth (Lu et al. 2001; Tong et al. 2013). Zhao et al. (2006, 2016) applied the immobilized B350 and B350M microbe consortia on the carriers in the BAF reactors to pretreat oil field wastewater before desalination. The efficiency rate of PAH degradation in BAF immobilized B350 and B350M group microbes was 84% and 90%, respectively. The filamentous fungus has played a crucial function in the immobilization action of BAFs.

12.6 Factors That Affect Oil and Sludge Hydrocarbon Degradation

Many conditions influence petroleum hydrocarbon biodegradation (Brusseau 1998). The natural biodegradability of the oil and sludge hydrocarbon contaminants is the initial and principal significant consideration when appropriate remediation methods are investigated. Temperature plays a vital role in hydrocarbon biodegradation because of its direct influence on the chemistry of the pollutants as well as on microbial diversity and their physiology. Low temperature also affects oil viscosity, while low-molecular-weight toxic hydrocarbon volatility was abridged to reduce the biodegradation (Atlas 1975). The biodegradation process is widely affected by numerous environmental factors (Atlas 1981; Leahy and Colwell 1990; Boopathy 2000; Baldwin et al. 2003; Rojo 2009; Meckenstock et al. 2016), including:

- The properties of the contaminants, for example, hydrocarbon type, length and availability, volatilization, and diffusion into liquid stage
- Microbial cell metabolic pathway and structural variations
- Various environmental conditions like temperature, pH, salinity, water content, oxygen availability, and other nutritional factors like the nitrogen and carbon source
- Soil physiochemical characteristics such as pH, moisture, and water-holding ability

It is essential to assess all the different factors prior to choosing the options for bioremediation (Atlas 1981; Okoh 2006). The hydrocarbon contaminants' bioavailability plays an essential function in the successful mitigation of pollutants from the soil (Sugiura et al. 1997; Chaudhry et al. 2005). Many diverse petro products like biopolymers, solvents, biosurfactants, and other acids are also formed by hydrocarbon-degrading strains (Mulligan 2005; Varjani et al. 2014a). Mixed bacterial consortium (heterogeneous) confirmed the suitable consequences with the maximum degradation ability (Cerqueira et al. 2011). Okoh (2006) and Aislabie et al. (2006) have found that high temperatures enhance the hydrocarbon contaminant solubility and reduce the thickness (Varjani and Upasani 2016a, 2016b, 2016c). Thamer et al. (2013) found that elevated temperatures as well as salinity reduce the microbial action and the yield. Leahy and Colwell (1990) noticed that petroleum hydrocarbons are a rich supply of energy and carbon matter. They do not include major concentrations of additional nutrients (such as phosphorous and nitrogen) necessary for microbial growth. Carbon (C), nitrogen (N), phosphorous (Ph), and potassium (K) ratios can adjusted by adding urea, NKP fertilizers, phosphate, phosphate salts, and ammonium, all of which enhance the breakdown course of action (Boopathy 2000; Ron and Rosenberg 2014).

12.7 Future Opportunities and Challenges

The oil and gas businesses face important challenges in meeting the rising global demand for hydrocarbons in an environmentally sound and socially suitable way whereas limitation its own power utilization. Oil contamination is caused by human error or carelessness and sometimes by unexpected disasters such as earthquakes or hurricanes. Intentional activity during war, acts of terrorism, acts of sabotage, or law breaking has proved, however, that oil contamination is not always an accident. The petro hydrocarbon oil radioactivity is a significant environmental issue for oil contamination. Sediments bind with the bulk of oil radionuclides, which then migrate with the sediments and with water in small part. Studies of the movement of radioactive substances in an aqueous medium carried out for other situations may be applied to the analysis of the migration of oil spill radionuclides. The monitoring scheme before revealing of oil contaminations and for monitoring the contamination in the post-emergency period are most effective if the remote (fluorescence lidar) and contact methods are used together.

With the increase in efficient green technologies comes more jobs for the labor forcey. Government-funded research will remain vital, especially for promising technologies that are not yet ready to be commercialized. Public budgets for oil and gas green technology research remains well below the levels reached after the oil shocks of the 1970s, and they have been cut in many cases. There is an urgent need for the public and private sectors to work together to develop green technologies. The oil and gas industries are responsible for ensuring efficient energy use and conservation in their own industrial activities. But they also have an interest—and, in some cases, a legal obligation—to promote energy-efficient use of their products once they are used by consumers too, particularly since the possibility for redeemable energy is noticeably higher.

Some European nations have already introduced the scheme of the white certificate, which outlines the obligations and commitments of producers, suppliers, and distributors of oil, gas, and electricity to undertake energy-efficiency measures that ensure that their final users save an amount of energy equal to a predefined percentage of their once-a-year power delivery. White certificates are papers certifying a definite mitigation of power utilization. European nations like Great Britain introduced the scheme for bulk suppliers to conserve energy sources with the possibility of trading certificates. Italy started the white certificate scheme in January 2005 and France a year later, while Denmark and the Netherlands are considering introducing it in the near future. Certainly, this is an era of great promise for the prudent progress of sustainable technologies. The integration of human ingenuity, innovative ideas, and novel technologies continues to develop novel, cleaner, sustainable technologies for handling hydrocarbon contamination. Biotechnology has developed into a significant tool for novel approaches in petroleum business throughout the oil production, decontamination, and processing units in addition to managing contaminants out of harm's way and clearance practices.

12.8 Conclusion

All promising technologies, including bioremediation, must be not only environmentally friendly process but also cost effective. Up-and-coming metagenomic, nano-technological, and transgenic approaches may transform the field, but their function is too limited because of high cost, greater maintenance needs, and harsh regulations. The multi-principle remediation technological process can transform the industry because it will bring environmental, social, and economic profit to all stakeholders; on the other hand, additional technological improvement is essential to make the remediation process less labor intensive, more cost effective, and more efficient. Still, potential economic rewards may be forthcoming for the clean-up of certain wastes if success is achieved. Many research areas must be pursued to discover and refine effective bioremediation methods for the clean-up of oil and sludge hydrocarbon contaminants.

Acknowledgments

The authors are grateful for the financial support from the Distinguished High-Level Talents Research Grant from the Guizhou Science and Technology Corporation Platform Talents Fund (Grant No.: [2017]5733-001 & CK-1130-002). They are also grateful to Zunyi Medical University, Zunyi, China, for its advanced research facilities. The authors also thank to their key laboratory colleagues and research staff members for their constructive advice and help.

References

Abbasian, F., Lockington, R., Mallavarapu, M., and Naidu, R. 2015. A comprehensive review of aliphatic hydrocarbon biodegradation by bacteria. *Applied Biochemistry and Biotechnology*, 1–30, doi:10.1007/s12010-015-1603-5.
Abdel-Shafy, H.I., and Mansour, M.S.M. 2015. A review on polycyclic aromatic hydrocarbons: Source, environmental impact, effect on human health and remediation. *Egyptian Journal of Petroleum*. doi:10.1016/j.ejpe.2015.03.011.

Abed, R.M.M., and Koster, J. 2005. The direct role of aerobic heterotrophic bacteria associated with cyanobacteria in the degradation of oil compounds. *International Biodeterioration & Biodegradation,* 55, 29–37.

Ahad, J.M.E., and Slater, G.F. 2008. Carbon isotope effects associated with Fenton-like degradation of toluene: Potential for differentiation of abiotic and biotic degradation. *Science of the Total Environment,* 401, 194–98. doi:10.1016/j.scitotenv.2008.02.048.

Aislabie, J., Saul, D., and Foght, J. 2006. Bioremediation of hydrocarbon-contaminated polar soils. *Extremophiles,* 10, 171–179.

AlDisi, Z., Jaoua, S., Al-Thani, D., AlMeer, S., and Zouari, N. 2016. Isolation, screening and activity of hydrocarbon-degrading bacteria from harsh soils. Proceedings of the World Congress on Civil, Structural, and Environmental Engineering (CSEE'16). Prague, Czech Republic, March 30–31, 2016, Paper No. ICESDP 104, doi:10.11159/icesdp16.104.

Al-Hawash, A.B., Dragh, M.A., Li, S., Alhujaily, A., Abbood, H.A., Zhang, X., and Ma, F. 2018. Principles of microbial degradation of petroleum hydrocarbons in the environment. *The Egyptian Journal of Aquatic Research,* 2018. doi:10.1016/j.ejar.2018.06.001.

Alloway, B.J., and Ayres, D.E. 1997. *Chemical Principles of Environmental Pollution,* 2nd ed. London, UK: Blackie Academic and Professional.

Alloway, J.B., and Ayres, D.C. 1993. *Chemical Principles of Environmental Pollution.* London, UK: Chapman & Hall.

Anyika, C., Abdul Majid, Z., Ibrahim, Z., Zakaria, M.P., and Yahya, A. 2015. The impact of biochars on sorption and biodegradation of polycyclic aromatic hydrocarbons in soils—A review. *Environment Science and Pollution Research,* 22, 3314–3341. doi:10.1007/s11356-014-3719-5.

Applegate, B.M., Kehrmeyer, S.R., and Sayler, G.S. 1998. A chromosomally based tod-luxCDABE whole-cell reporter for benzene, toluene, ethybenzene, and xylene (BTEX) sensing. *Applied and Environmental Microbiology,* 64(7), 2730–2735.

Atlas, R.M. 1975. Effects of temperature and crude oil composition on petroleum biodegradation. *Journal of Applied Microbiology,* 30(3), 396–403.

Atlas, R.M. 1981. Microbial degradation of petroleum hydrocarbons: An environmental perspective. *Microbiology Review,* 45(1), 180–209.

Atlas, R.M., and Bartha, R. 1972. Degradation and mineralization of petroleum by two bacteria isolated from coastal waters. *Biotechnology and Bioengineering,* 14, 297–308.

Baldwin, B.R., Nakatsu, C.H., and Nies, L. 2003. Detection and enumeration of aromatic oxygenase genes by multiplex and real-time PCR. *Applied and Environmental Microbiology,* 69(6), 3350–3358.

Barbly, R.C., and Barbour, M.G. 1984. The degradation of aromatic compounds by the meta and gentisate pathways. In: *Microbial Degradation of Organic Compounds.* D.T. Gibson (Ed.) Marcel Dekker, pp. 253–294.

Beaudin, N., Caron, R.F., Legros, R., Ramsay, J., Lawlor, L., and Ramsay, B. 1996. Cocomposting of weathered hydrocarbon-contaminated soil. *Compost Science and Utilization,* 4, 37–45.

Boonchan, S., Britz, M.L., and Stanley, G.A. 2000. Degradation and mineralization of high-molecular-weight polycyclic aromatic hydrocarbons by defined fungal bacterial cocultures. *Applied Environment Microbiology,* 66(3), 1007–1019.

Boopathy, R. 2000. Factors limiting bioremediation technologies. *Bioresource Technology,* 74, 63–67.

Brand, D.R., Lee, S.G., and Yoon, B.D. 1992. Isolation of a novel pentachloro-phenol-degrading bacterium, Pseudomonas sp. Bu 34. *Journal of Applied Microbiology,* 85, 1–8.

Brooijmans, R.J.W., Pastink, M.I., and Siezen, R.J. 2009. Hydrocarbon-degrading bacteria: The oil-spill clean-up crew. *Microbial Biotechnology,* 2(6), 587–594.

Brown, L.D., Cologgi, D.L., Gee, K.F., and Ulrich, A.C. 2017. Chapter 12—bioremediation of oil spills on land. In Fingas, M. (Ed.), *Oil Spill Science and Technology,* 2nd ed. Boston, MA: Gulf Professional Publishing, pp. 699–729.

Brusseau, M.L. 1998. The impact of physical, chemical and biological factors on biodegradation. In Serra, R. (Ed.), *Proceedings of the International Conference on Biotechnology for Soil Remediation: Scientific Bases and Practical Applications.* Milan, Italy: C.I.P.A. S.R.L., pp. 81–89.

Canadian Council of Ministers of the Environment (CCME). 2008. Canada-wide Standard for Petroleum Hydrocarbon (PHC) in Soil. User Guidance. PN 1398. Canada. http://www.ccme.ca/files/Resources/csm/phc_cws/pn_1398_phc_user_guide_1.1_e.pdf. Accessed December 10, 2017.

Canadian Council of Ministers of the Environment (CCME). 2010. Canadian Soil Quality Guidelines for the Protection of Environmental and Human Health (polycyclic aromatic hydrocarbons). http://ceqg-rcqe.ccme.ca/download/en/320. Accessed: December 12, 2017.

Caro, L., Chruchwar, G.D., and Chandler, M. 1984. Study of plasmid replication in vivo. *Methods in Microbiology,* 17, 72–122.

Cerqueira, V.S., Hollenbach, E.B., Maboni, F., Vainstein, M., Camargo, F., Do-Carmo, R.P.M., and Bento, F.M. 2011. Biodegradation potential of oily sludge by pure and mixed bacterial cultures. *Bioresource Technology,* 102(23), 11003–11010.

Chamarro, E., Marco, A., and Esplugas, S. 2001. Use of Fenton reagent to improve organic chemical biodegradability. *Water Research,* 35, 1047–1051. doi:10.1016/S0043-1354(00)00342-0.

Chandra, S., Sharma, R., Singh, K., and Sharma, A., 2013. Application of bioremediation technology in the environment contaminated with petroleum hydrocarbon. *Annals of Microbiology,* 63, 417–431.

Chaudhry, Q., Blom-Zandstra, M., Gupta, S., and Joner, E.J. 2005. Utilizing the synergy between plants and rhizosphere microorganisms to enhance breakdown of organic pollutants in the environment. *Environmental Science and Pollution Research,* 12(1), 34–48.

Costa, A.S., Romao, L.P., Araujo, B.R., Lucas, S.C., Maciel, S.T., Wisniewski Jr., A., and Alexandre, M.R. 2012. Environmental strategies to remove volatile aromatic fractions (BTEX) from petroleum industry wastewater using biomass. *Bioresource Technology,* 105, 31–39.

Cunningham, C.J., Ivshina, I.B., Lozinsky, V.I., Kuyukina, M.S., and Philp, J.C. 2004. Bioremediation of diesel-contaminated soil by microorganisms immobilized in polyvinyl alcohol. *International Biodeterioration Biodegradation,* 54, 167–174.

Das, N., and Chandran, P. 2011. Microbial degradation of petroleum hydrocarbon contaminants: An overview. *Biotechnology Research International,* Article ID 941810, 13.

Deppe, U., Richnow, H.H., Michaelis, W., and Antranikian, G. 2005. Degradation of crude oil by an arctic microbial consortium. *Extremophiles,* 9(6), 461–470.

De-qing, S., Jain, Z., Zhao-long, G., Jian, D., Tian-Li, W., Murygina, V., and Kalyuzhnyi, S. 2007. Bioremediation of oil sludge in shengli oilfield. *Water, Air, & Soil Pollution,* 185, 177–184.

Desforges, J.W., Sonne, C., Levin, M., Siebert, U., Guise, S.D., and Dietz, R. 2016. Immunotoxic effects of environmental pollutants in marine mammals. *Environment International*, 86, 126–139.

Deziel, E., Paquette, G., Villemur, R., Lepine, F., and Bisallon, J. 1996. Biosurfactant production by soil Pseudomonas strain growing on polycyclic aromatic hydrocarbons. *Applied Environmental Microbiology*, 62(6), 1908–1912.

Eze, V.C., Owunna, N.D., and Avoaja, D.A. 2013. Microbiological and physicochemical characteristics of soil receiving palm oil mill effluent in Umuahia, Abia State, Nigeria. *Journal of Natural Sciences Research*, 3(7), 163–169.

Farhadian, M., Vachelard, C., Duchez, D., and Larroche, C. 2008. In situ bioremediation of monoaromatic pollutants in groundwater: A review. *Bioresource Technology*, 9, 5296–5308.

Farzadkia, M., Dehghani, M., and Moafian, M. 2014. The effects of Fenton process on the removal of petroleum hydrocarbons from oily sludge in Shiraz oil refinery Iran. *Journal of Environmental Health Science & Engineering*, 12, 31

Feng, W.W., Liu, J.F., Gu, J.D., and Mu, B.Z. 2011. Nitrate-reducing community in production water of three oil reservoirs and their responses to different carbon sources revealed by nitrate-reductase encoding gene (napA). *International Biodeterioration & Biodegradation*, 65, 1081–1086.

Ferrari, M.D., Neirotti, E., Albornoz, C., Mostazo, M.R., and Cozzo, M. 1996. Biotreatment of hydrocarbons from petroleum tank bottom sludges in soil slurries. *Journal of Biotechnology Letters*, 18, 1241–1246.

Focht, D.D. 1998. Performance of biodegradative microorganisms in soil; xenobiotic chemicals as unexploited metabolic niches, In: *Environmental Biotechnology* (ed). Plenum Press, pp. 15–30.

Foght, J.M. 2008. Anaerobic biodegradation of aromatic hydrocarbons: Pathways and prospects. *Journal of Molecular Microbiology and Biotechnology*, 15, 93–120.

Fumoto, E., Tago, T., Tsuji, T., and Masuda, T. 2004. Recovery of useful hydrocarbons from petroleum residual oil by catalytic cracking with steam over Zirconia supporting iron oxide catalyst. *Energy Fuels*, 18, 1770–1774.

Gentili, A.R., Cubitto, M.A., Ferrero, M., and Rodriguez, M.S. 2006. Bioremediation of crude oil polluted seawater by a hydrocarbon-degrading bacterial strain immobilized on chitin and chitosan flakes. *International Biodeterioration Biodegradation*, 57, 222–228.

Guan, J., Xia, L.P., Wang, L.Y., Liu, J.F., Gu, J.D., and Mu, B.Z. 2013. Diversity and distribution of sulfate-reducing bacteria in four petroleum reservoirs detected by using 16S rRNA and dsrAB genes. *International Biodeterioration & Biodegradation*, 76, 58–66.

Harayama, S., Kasai, Y., and Hara, A. 2004. Microbial communities in oil-contaminated seawater. *Current Opinion in Biotechnology*, 15(3), 205–214.

Harmsen, J., and Rietra, R.P.J.J. 2018. 25 years monitoring of PAHs and petroleum hydrocarbons biodegradation in soil. *Chemosphere*, 207, 229–238, doi:10.1016/j.chemosphere.2018.05.043.

Helmy, Q., Kardena, E., Nurachman, Z., and Wisjnuprapto. 2010. Application of biosurfactant produced by Azotobacter Vinelandii AV01 for enhanced oil recovery and biodegradation of oil sludge. *International Journal of Civil & Environmental Engineering IJCEE-IJENS*, 1(10), 6–12.

Igunnu, E.T., and Chen, G.Z. 2012. Produced water treatment technologies. *International Journal of Low-Carbon Technologies*, 1–21.

Iida, T., Sumita, T., Ohta, A., and Takagi, M. 2000. The cytochrome P450ALK multigene family of an n-alkane-assimilating yeast, Yarrowia lipolytica: Cloning and characterization of genes coding for new CYP52 family members. *Yeast,* 16(12), 1077–1087.

Jan, B., Beilen, V., Neuenschwunder, M., Suits, T.H.M., Roth, C., Balada, S.B., and Witholt, B. 2003. Rubredoxins involved in alkane degradation. *The Journal of Bacteriology,* 184(6), 1722–1732.

Kadali, K.K., Simons, K.L., Skuza, P.P., Moore, R.B., and Ball, A.S. 2012. A complementary approach to identifying and assessing the remediation potential of hydrocarbonoclastic bacteria. *Journal of Microbiology Methods,* 88, 348–355.

Kaster, K.M., Bonaunet, K., Berland, H., Kjeilen-Eilertsen, G., and Brakstad, O.G. 2009. Characterisation of culture-independent and -dependent microbial communities in a high-temperature offshore chalk petroleum reservoir. *Antonie Van Leeuwenhoek,* 96, 423–439.

Kelly, C.J., Lajoie, C.A., Layton, A.C., and Sayler, G.S. 1999. Bioluminescent reporter bacterium for toxicity monitoring in biological wastewater treatment systems. *Water Environment Research,* 71(1), 31–35.

Lal, B., and Khanna S. 1996. Mineralization of [14C] octacosane by *Acinetobacter calcoaceticus* S30. *Can Journal of Microbiology,* 42, 1225–1231.

Layton, A.C., Gregory, B., Schultz, T.W., and Sayler, G.S. 1999. Validation of genetically engineered bioluminescent surfactant resistant bacteria as toxicity assessment tools. *Ecotoxicology and Environmental Safety,* 43(2), 222–228.

Leahy, J.H., and Colwell, R. 1990. Microbial degradation of hydrocarbons in the environment. *Microbiological Review,* 54(3), 305–315.

Li, C.Y., Zhang, D., Li, X.X., Mbadinga, S.M., Yang, S.Z., Liu, J.F., Gu, J.D., and Mu, B.Z. 2016. A biofilm properties and its correlationship with high-molecular-weight polyacrylamide degradation in a water injection pipeline of Daqing oilfield. *Journal of Hazardous Materials,* 304, 388–399.

Li, H., Chen, S., Mu, B.Z., and Gu, J.D. 2010. Molecular detection of anaerobic ammonium-oxidizing (anammox) bacteria in high-temperature petroleum reservoirs. *Microbial Ecology,* 60(4), 771–783.

Li, H., Mu, B.Z., Jiang, Y., and Gu, J.D. 2011. Production processes affected prokaryotic amoA gene abundance and distribution in high-temperature petroleum reservoirs. *Geomicrobiology Journal,* 28, 692–704.

Lin, G., Nie, J.Y., Zhu, N.W., Wang, L., Yuan, H.P., and Shou, Z. 2012. Enhanced Fenton's degradation of real naphthalene dye intermediate wastewater containing 6-nitro-1-diazo-2-naphthol-4-sulfonic acid: A pilot scale study. *Journal of Chemical Engineering,* 189–190, 108–116.

Liu, J.F., Mbadinga, S.M., Sun, X.B., Yang, G.C., Yang, S.Z., Gu, J.D., and Mu, B.Z. 2016. Microbial communities responsible for fixation of CO_2 reveled by using mcrA, cbbM, cbbL, fthfs, fefe-hydrogenase genes as molecular biomarkers in petroleum reservoirs of different temperatures. *International Biodeterioration & Biodegradation,* 114, 164–175.

Liu, W., Luo, Y., Teng, Y., Li, Z., and Ma, L.Q. 2010. Bioremediation of oily sludge-contaminated soil by stimulating indigenous microbes. *Environmental Geochemistry and Health,* 32, 23–29.

Louvado, A., Gomes, N.C.M., Simões, M.M.Q., Almeida, A., Cleary, D.F.R., and Cunha, A. 2015. Polycyclic aromatic hydrocarbons in deep sea sediments: Microbe-pollutant interactions in a remote environment. *Science of the Total Environment,* 526, 312–328. doi:10.1016/j.scitotenv.2015.04.048.

Lu, M., Zhang, Z.Z., Yu, W.Y., and Zhu, W. 2009. Biological treatment of oilfield-produced water: A field pilot study. *International Biodeterioration Biodegradation*, 63, 316–321.

Lu, M.C., Lin, C.J., Liao, C.H., Ting, W.P., and Huang, R.Y. 2001. Influence of pH on the dewatering of activated sludge by Fenton's reagent. *Water Science Technology*, 44(10), 327–332.

Lucia, L.A., Argyropoulos, D.S., Adamopoulos, L., and Gaspar, A.R. 2006. Chemicals and energy from biomass. *Canadian Journal of Chemistry*, 84, 960–970.

Macaulay, B.M., and Rees, D. 2014. Bioremediation of oil spills: A review of challenges for research advancement. *Annals of Environmental Science*, 8, 9–37.

Maeng, J.H.O., Sakai, Y., Tani, Y., and Kato, N. 1996. Isolation and characterization of a novel oxygenase that catalyzes the first step of n-alkane oxidation in *Acinetobacter* sp. strain M-1. *Journal of Bacteriology*, 178(13), 3695–3700.

Mandal, A.K., Sarma, P.M., Singh, B., Jeyaseelan, C.P., Channashettar, V.A., Lal, B., and Datta, J. 2011. Bioremediation: A sustainable eco-friendly biotechnological solution for environmental pollution in oil industries. *Journal of Sustainable Development and Environmental Protection*, 1, 5–23.

Mang, L., Zhongzhi, Z., Wei, Q., Xiaofang, W., Yueming, G., Qingxia, M., and Yingchun, G. 2010. Remediation of petroleum-contaminated soil after composting by sequential. Treatment with Fenton-like oxidation and biodegradation. *Bioresource Technology*, 2010, 101: 2106–2113. doi:10.1016/j.biortech.2009.11.002.

Masson, L., Tabashnik, B.E., Mazza, A., Préfontaine, G., Potvin, L., Brousseau, R., and Schwartz, J.L. 2002. Mutagenic analysis of a conserved region of domain III in the Cry1ac toxin of Bacillus thuringiensis. *Applied and Environmental Microbiology*, 68(1), 194–200.

Mbadinga, S.M., Wang, L.Y., Zhou, L., Liu, J.F., Gu, J.D., and Mu, B.Z. 2011. Microbial communities involved in anaerobic degradation of alkanes. *International Biodeterioration & Biodegradation*, 65(1), 1–13.

McClune, N.C., Weightman A.J., and Fry, J.C. 1989. Survival of Pseudomonas putida UWCI containing cloned catabolic genes in a model activated–sludge unit. *Applied and Environment Microbiology*, 55, 2627–2634.

McDonald, I.R., Miguez, C.B., Rogge, G., Bourque, D., Wendlandt, K.D., Groleau, D., and Murrell, J.C. 2006. Diversity of solublemethane monooxygenase-containing methanotrophs isolated from polluted environments. *FEMS Microbiology Letters*, 255(2), 225–232.

Meckenstock, R.U., Boll, M., Mouttaki, H., Koelschbach, J.S., Tarouco, P.C., Weyrauch, P., Dong, X., and Himmelberg, A.M. 2016. Anaerobic degradation of benzene and polycyclic aromatic hydrocarbons. *Journal of Molecular Microbiology and Biotechnology*, 26, 92–118.

Megharaj, M., Ramakrishnan, B., Venkateswarlu, K., Sethunathan, N., and Naidu, R. 2011. Bioremediation approaches for organic pollutants: A critical perspective. *Environment International*, 37, 1362–1375.

Milne, B.J., Baheri, H.R., and Hill, G.A. 1998. Composting of a heavy oil refinery sludge. *Environmental Progress*, 17, 24–27.

Mulligan, C.N. 2005. Environmental applications for biosurfactants. *Environmental Pollution*, 133, 183–198.

Nam, K., Rodrigeuz, W., and Kukor, J.J. 2001. Enhanced degradation of polycyclic aromatic hydrocarbons by biodegradation combined with a modified Fenton reaction. *Chemosphere*, 45, 11–20. doi:10.1016/S0045-6535(01)00051-0.

Nelson, A.H. 1990. The biodegradation of halogenated organic compounds. *Journal of Applied Bacteriology*, 69, 445–470.

Neyens, E., and Baeyens, J. 2003. A review of classic Fenton's peroxidation as an advanced oxidation technique. *Journal of Hazardous Materials*, 98, 33–50. doi:10.1016/S0304-3894(02)00282-0.

Okoh, A.I. 2006. Biodegradation alternative in the cleanup of petroleum hydrocarbon pollutants. *Biotechnology and Molecular Biology Reviews*, 1(2), 38–50.

Orphan, V.J., Taylor, L.T., Hafenbradl, D., and Delong, E.F. 2000. Culture-dependent and culture-independent characterization of microbial assemblages associated with high-temperature petroleum reservoirs. *Applied and Environmental Microbiology*, 66(2), 700–711.

Overton, E.B., Sharp, W.D., and Roberts, P. 1994. Toxicity of petroleum. In Basic Environ. Toxicol. CRC Press: Boca Raton, FL., 133-156.

Paulauskiene, T., Zabukas, V., and Vaitiekunas, P. 2009. Investigation of volatile organic compound (VOC) emission in oil terminal storage Tank Park. *Journal of Environmental Engineering and Landscape Management*, 17(2), 81–89.

Peixoto, R.S., Vermelho, A.B., and Rosado, A.S. 2011. Petroleum-degrading enzymes: Bioremediation and new prospects. *Enzyme Research*, 2011, 1–7. doi:10.4061/2011/475193. Article ID 475193.

Perez-Cadahia, B., Lafuente, A., Cabaleiro, T., Pásaro, E., Méndez, J., and Laffon, B. 2007. Initial study on the effects of Prestige oil on human health. *Environment International*, 33(2), 176–185.

Perez-Sanz, A., Millan, R., and Sierra, M.J. 2012. Mercury uptake by Silene vulgaris grown on contaminated spiked soils. *Journal of Environmental Management*, 95(3), S233–S237.

Phillips, S.J., Dalgarn, D.S., and Young, S.K. 1989. Recombinant DNA in wastewater. pB 322 degradation kinetics. *Journal of Water Pollution Control Federation*, 61, 1588–1595.

Phillips, T.M., Liu, D., Seech, A.G., Lee, H., and Trevors, J.T. 2000. Monitoring bioremediation in creosote-contaminated soils using chemical analysis and toxicology tests. *Journal of Industrial Microbiology and Biotechnology*, 65, 627.

Rojo, F. 2009. Degradation of alkanes by bacteria. *Environmental Microbiology*, 11, 2477–2490.

Ron, E.Z., and Rosenberg, E. 2014. Enhanced bioremediation of oil spills in the sea. *Current Opinion Biotechnology*, 27, 191–194.

Sana, B. 2015. Bioresources for control of environmental pollution. *Advances in Biochemical Engineering/Biotechnology*, 147, 137–183.

Sayler, G.S., and Ripp, S. 2000. Field applications of genetically engineered microorganisms for bioremediation processes. *Current Opinion in Biotechnology*, 11(3), 286–289.

Scheuer, U., Zimmer, T., Becher, D., Schauer, F., and Schunck, W.-H. 1998. Oxygenation cascade in conversion of n-alkanes to α,ω-dioic acids catalyzed by cytochrome P450 52A3. *Journal of Biological Chemistry*, 273(49), 32528–32534.

Schlepp, L., Elie, M., Landais, P., and Romero, M.A. 2001. Pyrolysis of asphalt in the presence and absence of water. *Fuel Process Technology*, 74, 107–123.

Sette, L.D., Simioni, K.C., Vasconcellos, S.P., Dussan, L.J., Neto, E.V., and Oliveira, V.M. 2007. Analysis of the composition of bacterial communities in oil reservoirs from a southern offshore Brazilian basin. *Anton Van Leeuwenhoek*, 91, 253–266.

Shen, T., Pi, Y., Bao, M., Xu, N., Li, Y., and Lu, J. 2015. Biodegradation of different petroleum hydrocarbons by free and immobilized microbial consortia. *Environmental Science: Processes & Impacts*, 17, 2022–2033.

Singh, K., and Chandra, S. 2014. Treatment of petroleum hydrocarbon polluted environment through bioremediation: A review. *Pakistan Journal of Biological Sciences*, 17(1), 1–8.

Singh, O.V., and Jain, R.K., 2003. Phytoremediation of toxic aromatic pollutants from soil. *Applied Microbiology and Biotechnology*, 63, 128–135.

Sousa, C., De Lorenzo, V., and Cebolla, A. 1997. Modulation of gene expression through chromosomal positioning in Escherichia coli. *Microbiology*, 143(6), 2071–2078.

Stanisich, A. 1984. Identification and analysis of plasmids at the genetic level. *Methods in Microbiology*, 17, 5–32.

Sugiura, K., Ishihara, M., Shimauchi, T., and Harayama, S. 1997. Physicochemical properties and biodegradability of crude oil. *Environmental Science and Technology*, 31(1), 45–51.

Sung-Ho, K., Watts, R.J., and Choi, J.H. 1998. Treatment of petroleum-contaminated soils using iron mineral catalyzed hydrogen peroxide. *Chemosphere*, 37, 1473–1482. doi:10.1016/S0045-6535(98)00137-4.

Tardy-Jacquenod, C., Caumette, P., Matheron, R., Lanau, C., Arnauld, O., and Magot, M. 1996. Characterization of sulfate-reducing bacteria isolated from oil-field waters. *Canadian Journal of Microbiology*, 42, 259–266.

Testa, S.M., and Winegardner, D.L. 1991. *Restoration of Petroleum-Contaminated Aquifers*. Chelsea, MI: Lewis Publishers.

Thamer, M., Al-Kubaisi, A.R., Zahraw, Z., Abdullah, H.A., Hindy, I., and Khadium, A.A. 2013. Biodegradation of Kirkuk light crude oil by Bacillus thuringiensis North. Iraqian. *National Science*, 5(7), 865–873.

Tong, K., Zhang, Y., Liu, G., Ye, Z., and Chu, P.K. 2013. Treatment of heavy oil wastewater by a conventional activated sludge process coupled with an immobilized biological filter. *International Biodeterioration & Biodegradation*, 84, 65–71.

Ubani, O., Atagana, H., and Thantsha, M.S. 2013. Biological degradation of oil sludge: A review of the current state of development. *African Journal of Biotechnology*, 12(47), 6544–6567.

Udotong, I.R., Udotong, I.J., Inam, E., and Kim, K. 2011. Bioconversion of crude oil production into soil conditioner using sawdust as organic amendment. *Geosystem Engineering*, 14(2), 51–58.

Urgun-Demirtas, M., Stark, B., and Pagilla, K. 2006. Use of genetically engineered microorganisms (GEMs) for the bioremediation of contaminants. *Critical Review in Biotechnology*, 26, 145–164.

Van Beilen, J.B., and Funhoff, E.G. 2007. Alkane hydroxylases involved in microbial alkane degradation. *Applied Microbiology and Biotechnology*, 74(1), 13–21.

Van Beilen, J.B., Funhoff, E.G., and Funhoff, E.G. 2006. Cytochrome P450 alkane hydroxylases of the CYP153 family are common in alkane-degrading eubacteria lacking integral membrane alkane hydroxylases. *Applied and Environmental Microbiology*, 72(1), 59–65.

Van Dyke, M.I., Lee, H., and Trevors, J.T. 1996. Survival of luxAB-marked Alcaligenes eutrophus H850 in PCB-contaminated soil and sediment. *Journal of Chemical Technology and Biotechnology*, 65(2), 115–122.

van Gestel, K., Mergaert, J., Swings, J., Coosemans, J., and Ryckeboer, J. 2003. Bioremediation of diesel oil contaminated soil by composting with biowaste. *Environmental Pollution*, 125, 361–368.

Vanishree, M., Thatheyus, A.J., and Ramya, D. 2014. Biodegradation of petrol using the fungus *Penicillium* sp. *Science International*, 2(1), 26–31. doi:10.17311/sciintl.2014.26.31.

Van Meter, R.J., Spotila, J.R., and Avery, H.W. 2006. Polycyclic aromatic hydrocarbons affect survival and development of common snapping turtle (*Chelydra serpentina*) embryos and hatchlings. *Environmental Pollution*, 142(3), 466–475.

Varjani, S.J. 2014. Hydrocarbon degrading and biosurfactants (bioemulsifiers) producing bacteria from petroleum oil wells. PhD Thesis. Kadi SarvaVishwavidyalaya, Gujarat, India.

Varjani, S.J. 2017. Microbial degradation of petroleum hydrocarbons. *Bioresource Technology*, 223, 277–286.

Varjani, S.J., Rana, D.P., Bateja, S., Sharma, M.C., and Upasani, V.N. 2014a. Screening and identification of biosurfactant (bioemulsifier) producing bacteria from crude oil contaminated sites of Gujarat, India. *International Journal of Innovative Research in Science Engineering and Technology*, 3(2), 9205–9213.

Varjani, S.J., and Srivastava, V.K. 2015. Green technology and sustainable development of environment. *Renewable Research Journal*, 3(1), 244–249.

Varjani, S.J., Thaker, M.B., and Upasani, V.N. 2014b. Optimization of growth conditions of native hydrocarbon utilizing bacterial consortium "HUBC" obtained from petroleum pollutant contaminated sites. *Indian Journal of Applied Research*, 4(10), 474–476.

Varjani, S.J., and Upasani, V.N. 2012. Characterization of hydrocarbon utilizing Pseudomonas strains from crude oil contaminated samples. *International Journal of Science and Computer*, 6(2), 120–127.

Varjani, S.J., and Upasani, V.N. 2013. Comparative studies on bacterial consortia for hydrocarbon degradation. *International Journal of Innovative Research in Science Engineering and Technology*, 2(10), 5377–5383.

Varjani, S.J., and Upasani, V.N. 2016a. Core flood study for enhanced oil recovery through ex-situ bioaugmentation with thermo- and halo-tolerant rhamnolipid produced by Pseudomonas aeruginosa NCIM 5514. *Bioresource Technology*, 220, 175–182.

Varjani, S.J., and Upasani, V.N. 2016b. Carbon spectrum utilization by an indigenous strain of Pseudomonas aeruginosa NCIM 5514: Production, characterization and surface active properties of biosurfactant. *Bioresource Technology*, 221, 510–516.

Varjani, S.J., and Upasani, V.N. 2016c. Biodegradation of petroleum hydrocarbons by oleophilic strain of Pseudomonas aeruginosa NCIM 5514. *Bioresource Technology*, 222, 195–201.

Varjani, S.J., and Upasani, V.N. 2017. A new look on factors affecting microbial degradation of petroleum hydrocarbon pollutants. *International Biodeterioration & Biodegradation*, 120, 71–83.

Vasudevan, N., and Rajaram, P. 2001. Bioremediation of oil sludge-contaminated soil. *Environment International*, 26, 409–411.

Vaziri, A., Panahpour, E., and Beni, M.H.M. 2013. Phytoremediation, a method for treatment of petroleum hydrocarbon contaminated soils. *International Journal of Farming and Allied Sciences*, 2(21), 909–913.

Waigi, M.G., Fuxing, K., Carspar, G., Wanting, L., and Yanzheng, G. 2015. Phenanthrene biodegradation by sphingomonads and its application in the contaminated soils and sediments: A review. *International Biodeterioration & Biodegradation*, 104, 333–349.

Walker, C.H. 2006. *Principles of Ecotoxicology*. Boca Raton, FL: CRC, Taylor & Francis Group.

Watts, R.J., and Dilly, S.E. 1996. Evaluation of iron catalysis for the Fenton-like remediation of diesel-contaminated soils. *Journal of Hazardous Materials*, 51, 209–224. doi:10.1016/S0304-3894(96)01827-4.

Whyte, L.G., Hawari, J., Zhou, E., Bourbonnière, L., Inniss, W.E., and Greer, C.W. 1998. Biodegradation of variable-chain-length alkanes at low temperatures by a psychrotrophic *Rhodococcus* sp. *Applied Environmental Microbiology*, 64, 2578–2584.

Wilkes, H., Buckel, W., Golding, B.T., and Rabus, R. 2016. Metabolism of hydrocarbons in n-Alkane utilizing anaerobic bacteria. *Journal of Molecular Microbiology and Biotechnology*, 26, 138–151.

Yang, G.C., Zhou, L., Mbadinga, S.M., You, J., Yang, H.Z., Liu, J.F., Yang, S.Z., Gu, J.D., and Mu, B.Z. 2016. Activation of CO2-reducing methanogenesis in oil reservoir after addition of nutrient. *Journal of Bioscience and Bioengineering*, 122(6), 740–747.

Yeh, C.K., Wu, H.M., and Chen, T.C. 2003. Chemical oxidation of chlorinated nonaqueous phase liquids by hydrogen peroxide in natural sand systems. *Journal of Hazardous Materials*, 96, 29–51. doi:10.1016/S0304-3894(02)00147-4.

Yoon, J., Lee, Y., and Kim, S. 2001. Investigation of the reaction pathway of OH radicals produced by Fenton oxidation in the conditions of wastewater treatment. *Water Science Technology*, 44(5), 15–21.

Yun, W.K., and Kyung, Y.H. 2000. Effects of reaction conditions on the oxidation efficiency in the Fenton process. *Water Research*, 34, 2786–2790. doi:10.1016/S0043-1354(99)00388-7.

Zabbey, N., Sam, K., and Onyebuchi, A.T., 2017. Remediation of contaminated lands in the Niger Delta, Nigeria: Prospects and challenges. *Science of the Total Environment*, 586, 952–965. doi:10.1016/j.scitotenv.2017.02.075.

Zhang, H., Tang, J., Wang, L., Liu, J., Gurav, R.G., and Sun, K. 2016. A novel bioremediation strategy for petroleum hydrocarbon pollutants using salt tolerant Corynebacterium variable HRJ4 and biochar. *Journal of Environment Science*, 47, 7–13.

Zhao, F., Zhou, J.-D., Ma, F., Shi, R.-J., Han, S.-Q., Zhang, J., and Zhang, Y. 2016. Simultaneous inhibition of sulfate-reducing bacteria, removal of H2S and production of rhamnolipid by recombinant Pseudomonas stutzeri Rhl: Applications for microbial enhanced oil recovery. *Bioresource Technology*, 207, 24–30.

Zhao, X., Wang, Y., Ye, Z., Borthwick, A.G.L., and Ni, J. 2006. Oil field wastewater treatment in biological aerated filter by immobilized microorganisms. *Process Biochemistry*, 41, 1475–1483.

Zhou, F., Mbadinga, S.M., Liu, J.F., Gu, J.D., and Mu, B.Z. 2013. Evaluation of microbial community composition in thermophilic methane-producing incubation of production water from a high-temperature oil reservoir. *Environmental Technology*, 34(18), 2681–2689.

Zimmer, T., Ohkuma, M., Ohta, A., Takagi, V., and Schunck, W.H. 1996. The CYP52 multigene family of *Candida maltosa* encodes functionally diverse n-alkane-inducible cytochromes p450. *Biochemical and Biophysical Research Communications*, 224(3), 784–789.

Zylstra, G.J., and Gibson D.T. 1991. Aromatic hydrocarbon degradation: A molecular approach. In Setlow, J.K. (Ed.), *Genetic Engineering Principles and Methods*. New York: Plenum Press, pp. 183–203.

Zytner, R.G., Salb, A.C., and Stiver, W.H. 2006. Bioremediation of diesel fuel contaminated soil: Comparison of individual compounds to complex mixtures, *Soil and Sediment Contamination*, 15, 277–297.

Index

Note: Page numbers in italic and bold refer to figures and tables respectively.

Printed in the United States
by Baker & Taylor Publisher Services